Plant demography in vegetation succession

Tasks for vegetation science 26

Series Editors

HELMUT LIETH

University of Osnabrück, F.R.G.

HAROLD A. MOONEY

Stanford University, Stanford, Calif., U.S.A.

Plant demography in vegetation succession

by

KRYSTYNA FALIŃSKA

SPRINGER-SCIENCE+BUSINESS MEDIA, B.V.

Library of Congress Cataloging-in-Publication Data

Falińska, Krystyna.
Plant demography in vegetation succession / Krystyna Falińska.
p. cm. -- (Tasks for vegetation science ; 26)
Includes index.
ISBN 978-94-010-5441-6 ISBN 978-94-011-3266-4 (eBook)
DOI 10.1007/978-94-011-3266-4
1. Plant succession. 2. Plant populations. 3. Vegetation dynamics. I. Title. II. Series.
QK910.F34 1991
581.5'24--dc20 90-22407

ISBN 978-94-010-5441-6

Printed on acid-free paper

TABLE OF CONTENTS

PREFACE

This book is the synthesis of research on the dynamics of vegetation conducted in the Białowieża Primeval Forest (North-Eastern Poland), following the first book on this subject published in 1986 by Dr W. Junk Publishers as volume 8 in the Geobotany series [Faliński J. B. (with the assistance of K. Falińska), *Vegatation Dynamics in Temperate Lowland Primeval Forests. Ecological Studies in Białowieża Forest*]. Like the first book, this one resulted from the work of the Białowieża Geobotanical Station of Warsaw University and the Workshop for Plant Demography of the Botanical Institute of the Polish Academy of Sciences in Kraków.

This book summarizes the results of 15 years of research on the behaviour of plant populations in the course of secondary succession. Preliminary results of these investigations have already been published in the first monography and in several papers (Falińska 1985, 1986, 1989, 1990).

The results of long-term studies on the demography of plants in the course of vegetation changes in abandoned meadows made it possible to verify many hypotheses and theories on the causes, processes, course, and mechanisms of succession. In the book much attention was paid to the relationships between the dynamics of communities and the dynamics of plant populations.

I am grateful to my colleagues and friends; especially to Professors J. B. Faliński, K. Zarzycki, R. Olaczek, H. Piotrowska, and J. L. Olszewski, and to participants of the seminars conducted in Białowieża in 1980—1990 for their cooperation and discussions about the research and preparation of the book. Professor J. B. Faliński and his co-workers substantially helped in organizing the study area and providing financial support in the course of field work.

I gratefully thank Professor K. Zarzycki for his unfailing support and help in these long-term studies and for financing the studies with the grant, CPBP 04.04.B. The final stages of preparation of this book (graphics and translations) were financed by the grant CPBP 04.10.07. I am grateful to Professors A. Szujecki and R. Andrzejewski for the administering of the grant.

I would especially like to thank my friend, Prof. Franco Pedrotti from Camerino University (Italy), for arranging the pleasant atmosphere during my several visits to the Institute of Botany and Ecology in 1983–1990, when I worked on this book.

I thank Professor Helmuth Lieth (Osnabrück) for his initiative to publish this book.

My special thanks go to the former and present technicians of the Białowieża Geobotanical Station of Warsaw University: Mrs. Bożena Buszko, Walentyna Maciejewska, Alicja Wiktoruk and Mr. Roman Wołkowycki, who helped me to collect and elaborate the material during many years of their field and laboratory work. I am deeply grateful to Mrs. Halina Kościelecka for her thorough and creative typing and preparation of the reference list, and to Mrs. Irena Wołkowycka for her drawings. Mr. Mieczysław Kwieciński did a great job of translating the book into English.

Białowieża,
13 March, 1991

Krystyna Falińska

PART 1

THEORETICAL AND METHODICAL FUNDAMENTALS

CHAPTER 1

INTRODUCTION AND AIM

Ecological succession has been the object of theoretical and methodological considerations as well as of stationary and long-term studies for 80 years. Succession processes are analysed from two viewpoints: system ecology and population ecology. In the former case it is assumed that there exists a system-development trend (Margalef 1968, Odum 1969, Whittaker 1975), whereas a succession is a sequence of communities, that is, specific floristic compositions under specific ecological conditions. With such an approach population dynamics is understood as a phytocoenosis dynamics. Succession studies therefore try to explain the fate of a particular phytocoenoses from its beginning to the dynamic-equilibrium state attained by it, or the fate of the phyotocoenoses following one another (Figure 1.1).

Clements' (1916) theory concerning community sequence in a succession, that is "relay floristics" is the basis of many studies and theoretical considerations. The theory has recently been criticized by many authors (e.g. Egler 1954, Colinvaux 1973, Drury and Nisbet 1973, Horn 1974, Connell and Slatyer 1977).

Following the success of population genetics and plant demography, succession has more and more often been treated as a species-turnover process, whereas the role of species in vegetation changes is related to their life strategies, and par-

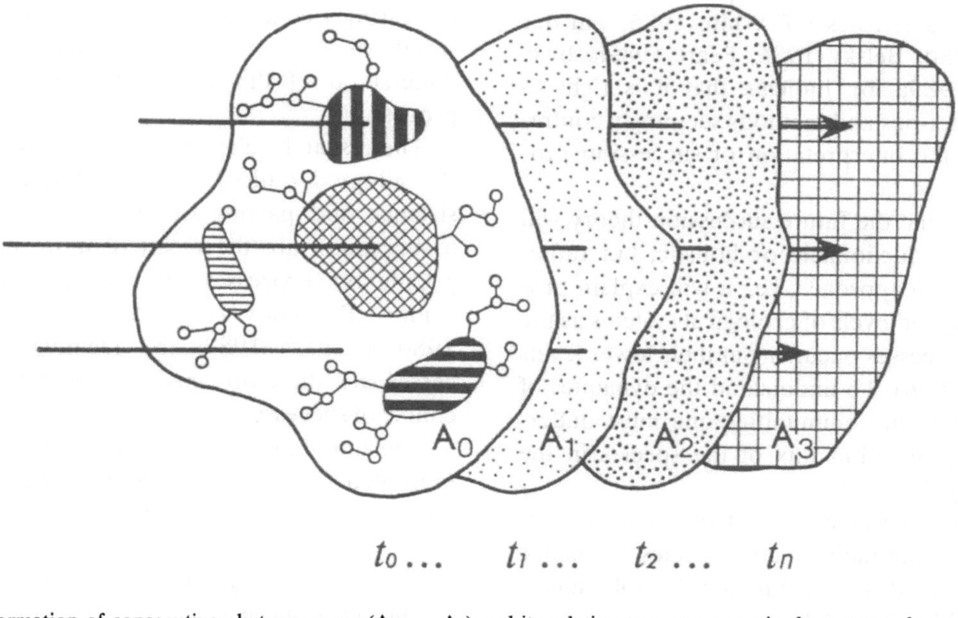

Fig. 1.1 Formation of consecutive phytocoenoses (A_0 ... A_3) and its relation to some space in the course of succession. After V. N. Sočava (1972), changed by Faliński (1989).

ticularly to: life history, growth type and reproduction biology (Harper 1977, Grime 1979, White 1985, van der Maarel 1980).

In the theory of succession a clear approach-dichotomy can be noticed (Shugart 1984). Some authors assume that succession occurs at the ecosystem level, while others think it is the result of inter-species and species-environment interactions (MacMahon 1980). The former view follows from the reproducibility of the ecosystem structure and functions. In this case succession is defined as the effect of an intra-biocoenotic regulation.

In their interpretation of succession, authors (Harper 1977, Grime 1979, White 1985, Rabotnov 1985) holding the view different from that expressed by the holists, emphasize the importance of population processes in the dynamics of biocoenoses. They also ascribe a significant role to environment disturbances (man-caused and natural) and the causes of vegetation instability (Grime 1979).

Until not long ago the above two theories were considered alternative, but recently it has been stressed that they are complementary rather than alternative (Shugart 1984, Prentice 1986). Long-term studies of individual, population and community characteristics (cf. White 1985, van der Maarel 1980, Belsky 1986, Peart 1989a, b, c, Falińska 1989a, b), indicate that it is right to combine theories and methods seemingly contradictory in the interpretation of succession processes.

It seems that the future succession theory will not be reproductionistic or holistic, but rather hierarchical in approach, i.e., it will explain processes at higher levels of ecological systems as the result of processes occurring at their lower levels (Prentice 1986). To understand the dynamics of changes in plant communities one must understand the population flux of the species in the community (Figure 1.2).

Connell and Slatyer (1977) suggest that there exist three alternative seral species sequence models defined as "facilitation-", "tolerance-" and "inhibition-" models:

1. The "facilitation" model describes situations where earlier occupants facilitate the successive encroachment of a number of species (e.g. by preparing the soil). This model is closest to the classical model of succession (Clements 1916, 1936).
2. The "tolerance" model defines situations where an expected species sequence follows from the existence of various food-source finding strategies. Species that appear at a later stage of succession are able to tolerate lower food supply levels than are those which appear earlier. They can therefore colonize an area and grow in the presence of the earlier settlers.
3. The "inhibition" model represents situations in which late settlers cannot grow to reach the mature stage in the presence of early settlers. This model sugests that all species "resist" new colonizers and competitors. Early colonizers occupy space, prevail and thereby exclude or impede the growth of late colonizers. Only after the death or damage of the former are food supplies released, and only then can late colonizers attain maturity.

In studies of the mechanisms of the species turnover during a succession the opinion becomes increasingly common that it will only be understood when the population dynamics has been understood of the species taking part in the process.

The research relayed in this book stems from the adopted outlook that the formation, persistence and passing of phytocoenoses during a succession are the result of intra- and inter-population processes, as indicated by some studies (Rabotnov 1985, Peet and Christensen 1980, Falińska 1989a, b). Demographic processes are therefore made responsible for the course of succession.

The aim of the studies, of which the results are presented in this book, was to understand the following:

1. The nature and rate of vegetation changes during a succession in a situation where a forest returns to fertile and floristically rich meadows. Attention was particularly focused

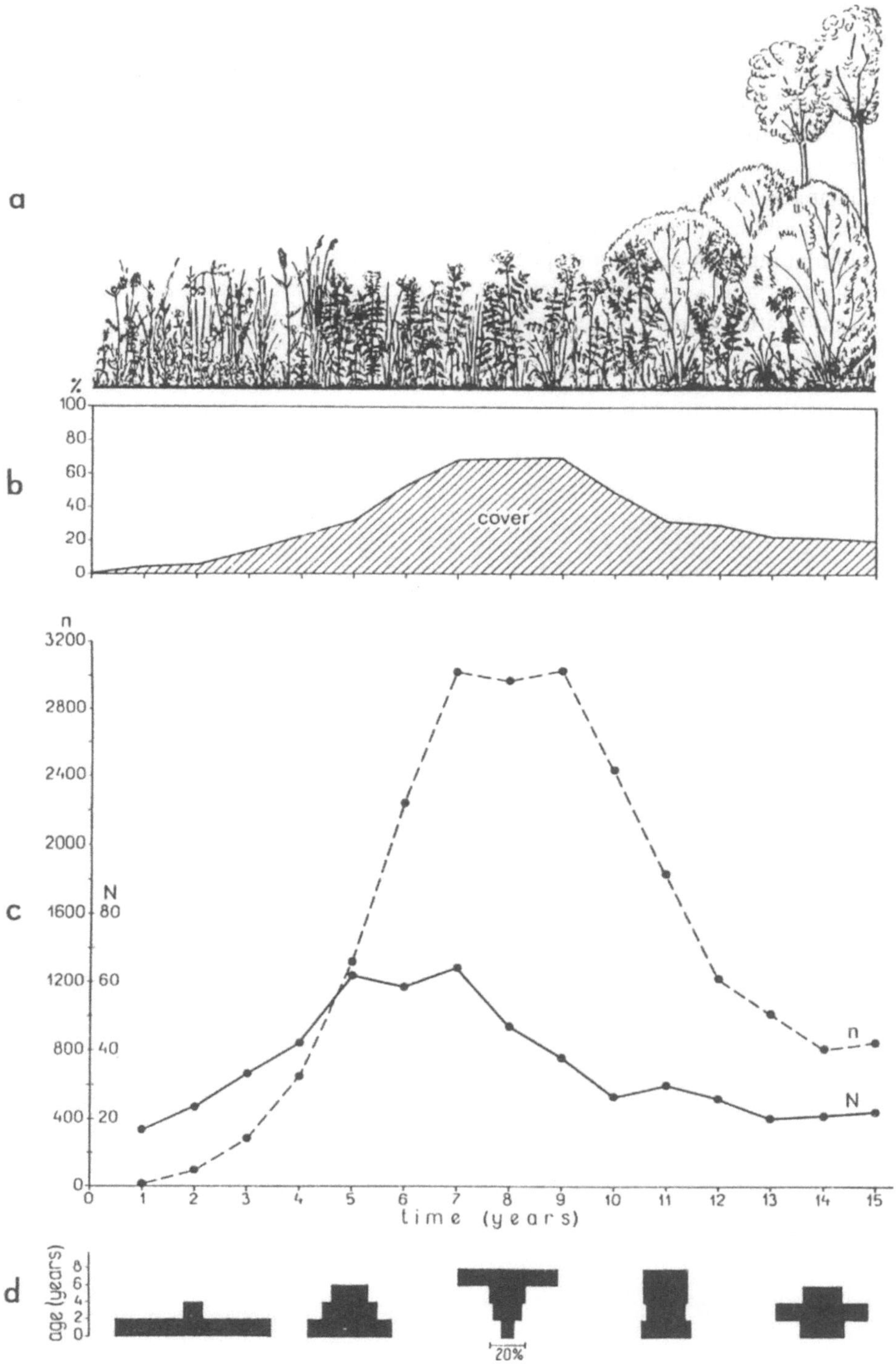

Fig. 1.2. Behaviour of *Filipendula ulmaria* population, the main promoter of succession. Change in some traits of population in consecutive phases of succession: a) meadow, macroforb, brushwood and forest communities, b) change of coverage, c) abundance dynamics of a population: *N* — number of individuals; *n* — number of shoots, d) age structure (Falińska 1985).

on the fate of the phytocoenoses that appear during a succession, causes of their formation, presence and passing.

2. The biological mechanisms of species turnover; an attempt to answer the question: Which of the species contribute to meadow vegetation changes towards forest communities, and which inhibit this process?
3. The fate populations, especially of species present throughout the succession process, that is: (a) what relations exist between population dynamics phases and succession stages; (b) whether the population growth of species present throughout a succession lasts as long as does the return of a forest to abandoned meadows, or whether it is a series of successive populations; (c) whether there is any relationship or co-occurrence between population dynamics phases and succession stages.
4. Characters of the species deciding whether a species will become a succession promoter or inhibitor.

CHAPTER 2

ASSUMPTIONS AND TERMINOLOGY

Population structure of the vegetation

The concept of population structure of phytocoenoses is based on the finding that in the hierarchic structure of the plant kingdom populations hold a particular position since they lie "at the intersection" of the genetic and spatial levels of the biological organization. This concept is supported by the following findings: (a) populations of the same species in different phytocoenoses exhibit different dynamics and structure (Zarzycki 1964, 1968, Wilkoń-Michalska 1976, 1985, White 1985, Symonides 1985, Falińska 1979, 1985, 1986 b, c, 1989, a, b, Zhukova and Ermakova 1985, Vorontzova and Zaugolnova 1985); (b) individual variation within populations may be the result of modification by the environment, or it may be also of ecotype nature (McMillan 1964, Antonovics and Levin 1980, Solbrig 1980); (c) the density and distribution of individuals of various species in the biochore of a phytocoenosis are not a mirror image of the soil conditions, and do not always indicate the influence of competition; (d) no species occupies the entire territory which it could take into possession on account of the soil and climate, its tendencies to propagation clash with those of other species capable of existing in the same ecological conditions (Whittaker 1975); (e) the final species composition under specific conditions is determined by the interaction of populations of various species known as "phytocoenotic selection" (Rabotnov 1985).

Individual and population

The commonness and morphological diversity of modular plant organisms make investigators seek theoretical and methodological solutions to plant demography (Harper 1977, 1981, Harper and White 1974, Harper and Bell 1979, White 1979, 1980, 1985, Falińska 1986a).

Within the several or a dozen or so years of their life, individuals of perennials considerably build up their underground organs and multiply the number of shoots, forming assemblages varying in size and permanence. In many plant-growth forms, parts of a plant body die, while other parts continue to grow (Figure 1.3). Due to this specific plant growth and life-history, as well as to the commonness of vegetative propagation, in natural populations individuals are difficult to distinguish. Consequently, population abundance assessment, and understanding the demographic processes (reproductiveness, mortality) are difficult. For this reason, in perennials, various demographic entities are established according to their morphological features. The units distinguished are usually parts of a perennial organism, notably shoots, rosettes, or even single leaves.

The populations of many plant species, of herbaceous species in particular, have in fact, a two-level structure (Harper 1980). In the populations of those species two processes coincide: exchange of individuals, and exchange of their structure elements. In such a case a full population-dynamics description should take into account the exchange rate of individuals and of rooted units. For it has been found that within

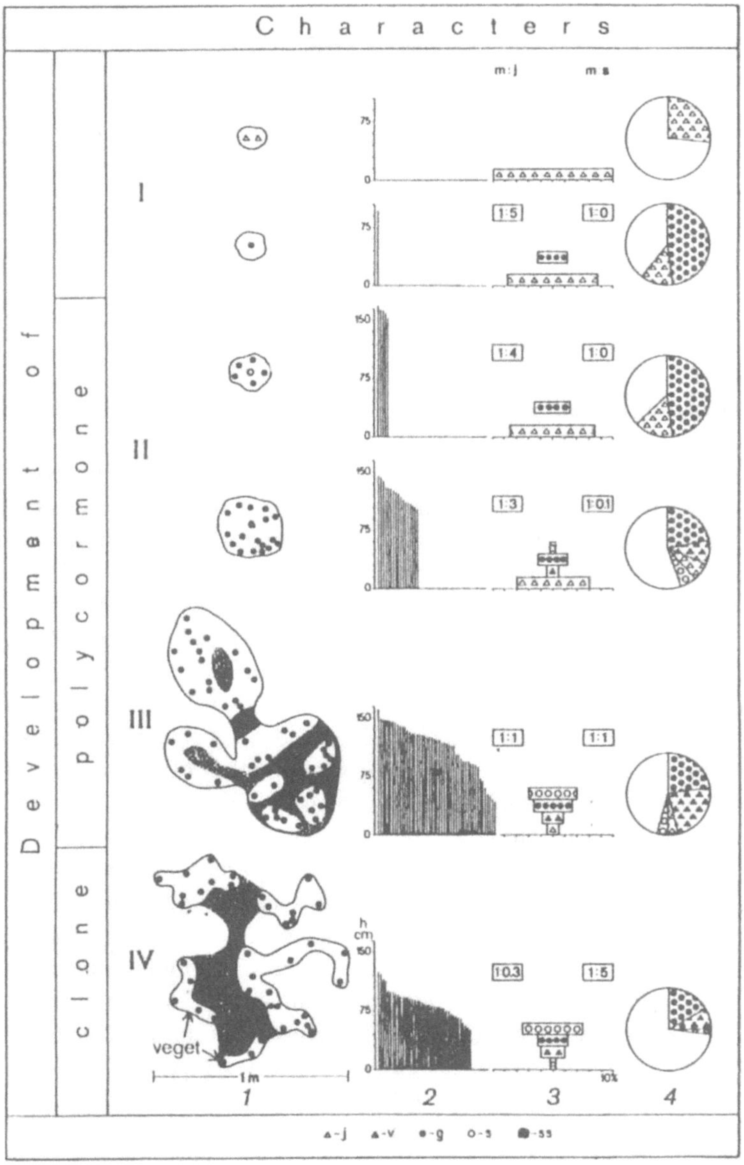

Fig. 1.3. The diagram of development and growth of polycormone of *Filipendula ulmaria L.* in 10-year life history (Falińska 1986a). Development phases: I — juvenile, II — mature, III — senile, IV — disintegration of polycormone and clone formation. As the individual ages, the following change: size of the area (1), number and distribution of above-ground shoots (dots), their height (2), contribution of shoots to different development phases (3), ratio of the above-ground part to the underground one (empty fields) in total biomass of polycormone (4), the ratio of mature shoots to juvenile ones (m:j), mature to senile ones (m:s); development phases of shoots; j — juvenile, m — mature, g — generative, s — senile, ss — dying parts in polycormone.

the same population and during the same time periods variations in the number of individuals and shoots may follow different growth models (Figure 1.4).

This means that for species whose individuals' growth is iterative the concepts and demographic study methods should be analysed anew. Harper (1986) stresses that it seems a strange fact that

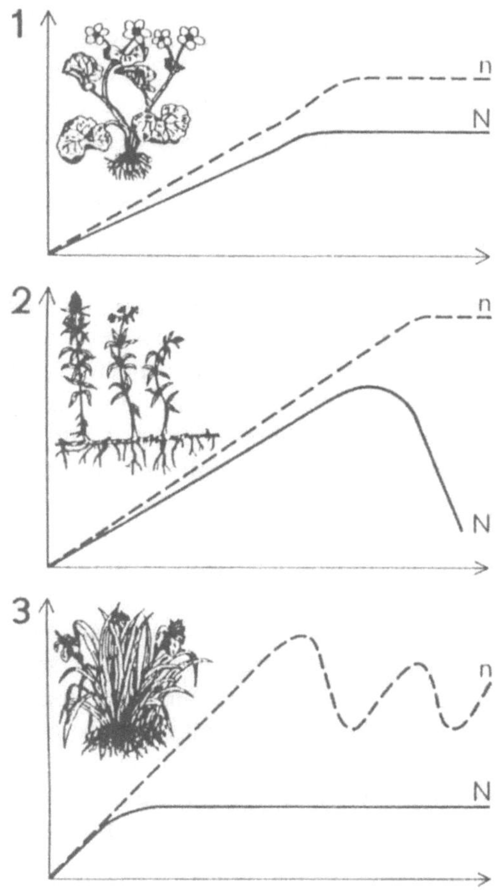

Fig. 1.4. Model of the relationship between the growth in number of individuals and the shoots in populations of species having different morphological and biological properties (Falińska 1986a). (1) Type one is the population of species with individuals producing a small number of shoots growing, as a rule, in the nearest vicinity; (2) type two is the population of species with individuals having a linear type of growth; (3) type three is the population of species with individuals having a fast growth rate and forming, in a short time, aggregations of shoots varying in size and durability.

ecological and evolutional studies have been based to such an extent on the behaviour of plant organisms with a unitary type of growth when the iterative growth type is so common among plants.

Though it has often been emphasized that the possibility to apply the demographic theory to phytoecology depends on the knowledge of the morphological characters of plants (Harper and White 1974, Harper 1977, 1980, 1986, White 1979, 1980, Falińska 1986a, b), adequate methods have not yet been developed for identifying individuals and describing their characters (age, fecundity, life-history) under natural conditions.

The diversity of criteria used for identifying the basic demographic entities in a population is the result of both the morphological diversity of plants and the adoption of various concepts in the studies of population ecology. For demographic problems are solved on the basis of a genetical (derivation) or morphological (distinctness, structure) definition of individuals.

From a genetic-evolutional viewpoint, an entity or entities that take its/their origin from a single zygote are considered to be one organism, for which Harper (1977) has suggested the term of "genet". According to this approach, a genet can be a seedling, a monocormonal (one-shoot), or a polycormonal (multi-shoot) individual, as also an assemblage of plants of the same genotype, that is, a clone.

If the life-history is defined as a model of the zygote-to-zygote passage, then modular organisms pass it by iterative growth. This means that there is no germ-plasm continuity, and that some tissues, in their entirety or in part remain omnipotent, in contrast to non-modular (unitary) organisms in which cell differentiation leads on to the formation of somatic parts while omnipotence disappears during the growth of the zygote.

Though the genetical definition of the individual is unique, i.e., it is an organism of a unique genotype, its use in demographical analyses of species is not always possible. For this reason, investigators often distinguish a great life-history (from zygote to zygote) and a small life-history, the latter being understood as the development of the particular parts of which a plant organism is made (e.g. single shoots, shoot assemblages, clumps etc.).

New information on the biology and evolution of modular organisms (White 1979, Harper 1977, 1986) throws some new light on the demographic methods and population dynamics definitions of species whose individuals are of such a body type. In such cases population dynamics is often assessed on the basis of variations in the number of modules, their structure, permanence and reproduction (e.g. Bishop *et al.* 1978, Noble *et al.* 1979, Callaghan 1980, Fenner 1987, Callaghan *et al.* 1986).

Such an approach is fully justifiable in the case of a stolonate growth type, that is, where individuals are very distinct, while the distinctness of the boundaries of an organism is poor. The approach arouses, however, doubts in situations where perennial organisms form compact structures, that is, are made up of many rooted shoots interconnected with strong, durable common underground organs.

From the ecological viewpoint, an individual is often considered to be a separate, morphologically and physiologically, uninterrupted organism. It may consist of one rooted shoot (monocormonal) or of many shoots (polycormonal). It is an ontogenetic entity of generative or vegetative origin. Then the criterion used for identifying individuals is their distinctness. Ecological assignment of the status of individual to single shoots, shoot clumps and shoot assemblages etc. is substantiated by the fact that each such unit: (a) occupies a specified area, (b) is connected by various relationships with similar units of the same and other species, (c) takes up water and mineral salts, (d) produces generative and/or vegetative propagules, hence participates in the reproduction process. Thus each shoot, shoot clump and shoot assemblage in a specific way determines the population structure and dynamics.

If we assume that in their evolution, plant organisms increased the number of shoots and roots to prolong their life-span and to intensify their colonizing capabilities, then the recognition of the short-lived parts of a perennial as demographic entities seems to be an artificial, though not infrequently necessary, approach. The more so as in similar cases the growth, more precisely growth in width, of plants is considered tantamount to vegetative propagation, such individuals being called clonal organisms (Harper 1986).

If we assume, however, that for the understanding of population characteristics the understanding of the properties (architecture, fecundity, age) of population members is needed, then population distinctness, besides population origin, must be considered to be an equally important character. For from the viewpoint of population formation and growth it is not unimportant whether a population is made up of independently living individuals, or of ones living in a co-existence determined by perennial underground organs. The use of the distinctness criterion for the identification of individuals is justified by the fact that shoots connected with common rhizomes represent a community not only in the physical, but also in the physiological sense, because assimilates are known to be transmitted between shoots, usually from young to old shoots. This regularity has been found, e.g. in *Carex arenaria* (Noble and Marshall 1983). Apart from this, individuals consisting of several dozen shoots, each show certain group features, that is, have a specific spatial structure and an inner dynamics, that is, seasonal and long-term shoot abundance variations (Figure 1.3).

In this book the following assumptions have been made (Figure 1.5):

1. An individual is a morphologically and physiologically uninterrupted organism, regardless of how many basic units (modules) it is made of. It represents an ontogenetic entity developed from a generative or a vegetative propagule.
2. For one-shoot individuals the term "monocormonal organisms" has been adopted, and for those made up of many shoots the term "polycormonal organisms".
3. Other terms for "adopted demographic units" (e.g. single rooted shoots, shoot clumps or shoot assemblages) have been used where polycormones did not represent sufficiently distinct integral wholes, but shoots or shoot assemblages of which they consist grew at certain distances from one another and stood out well (e.g. *Lysimachia vulgaris*).
4. The term "clone" has been reserved for a

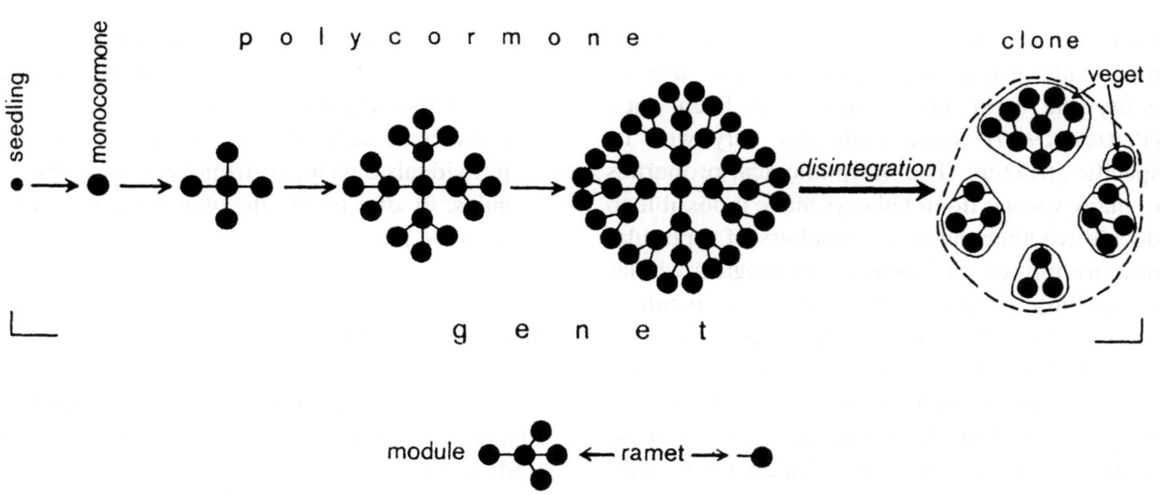

Fig. 1.5. Diagram of growth of polycormone and clone formation (Falińska 1986a). This process is described by means of currently applied terms in plant demography: after Harper (1977) — genet, module, ramet; after Penzes (1960) — polycormone; after Falińska (1986a) — veget.

collection of individuals developed from propagules (e.g. *Caltha palustris*), through rhizome fragmentation (e.g. *Myosotis scorpioides*), or by polycormone disintegration (e.g. *Filipendula ulmaria*). Interrupted morphological links between daughter units and the parent organism was the criterion used for the identification of vegetative individuals (Figures 1.3; 1.5).

Problems of terminology

It will be well to consider to what extent the concepts and terms worked out in phytoecology can be used for describing demographic phenomena. For instance for the description of units of different origin, Harper (1977) proposed the terms *genet* and *ramet*. As regards to units arising due to generative reproduction the term genet fully describes their nature and origin, whereas the term ramet is derived from a morphological term (ramification), and is used for describing units that form on various types of above-ground and underground stolones which may, but need not, detach from the parent organism (e.g. in *Ranunculus repens*). It is known, however, that vegetative-derivation individuals can also develop from bulbs, vegetative propagules, or through plant fragmentation. On account of its etymology and biological connotation, ramet does not seem to be the right term for individuals developed, for example, the special vegetative propagules, or as a result of disintegration (Figure 1.3).

If a unique term is used to describe individuals developed by way of generative reproduction (from one zygote), then a specific term should also be used for vegetative-origin individuals. As such individuals develop in a variety of ways, the term cannot refer to one particular specialized type, but must, as in the case of generative offspring, also be of an all-embracing nature. It seems that the right term to name individuals that have come into being due to asexual propagation would be *veget* (Falińska 1986a). *Genet-veget* would then constitute a complementary term pair, uniquely indicating the different origins of the individuals of a population. The adoption of this principle requires, however, a verification of the range of existing terms used in plant demography (Figure 1.5). For instance, the division of a genet (individual), or detachment from it of vegetative propagules (or its parts) leads on to the formation of a number of entities, that is, development of a clone. For this reason, the term *clone* cannot refer to an individual consisting of many rooted shoots interconnected with common underground organs, because for such an entity the term *polycormone* has been used (Penzes 1960) which highlights very well its specific structure. The morphological properties of many species do not always make it possible to decide whether the basic members of a population are unified and form morphological wholes (organism), or are separate and independent. Therefore the various biological processes, like the growth and life-histories of organisms, formation and growth of new shoots, and vegetative propagation through the detachment of these elements and their becoming independent individuals, are treated as one process. Though from the genetical point of view such an approach arouses no doubts, because the processes lead on to the propagation of a particular genotype, it does not seem right from the viewpoint of population ecology. For such an approach leads on to an indiscriminate treatment of processes occurring at the level of the individual (growth, necrosis, senescence and renewal of senescent structures), and their effects noticeable at the population level. The difficulty of a separate treatment of these phenomena lies in the fact that many of the processes take place in the underground organs of plants.

The rationalization, suggested in this book, of the concepts and terms hitherto used in plant demography has provided a basis for a fuller assessment of the phenomena occurring at the level of the individual, and at the level of the population, especially by differentiating between the following, and uniquely defining them:

- origin of an individual (generative: genet; vegetative: veget) from its specific structure, e.g. modular;
- growth and life-history of a perennial plant made up of many units interconnected with common underground organs (polycormone) from the formation of a collection of independent entities (vegets) of the same genotype, i.e., a clone;
- abundance dynamics of modules (shoots, rosettes, tussocks), that is, body structure elements (n) from the abundance dynamics of individuals (N), separate morphologically and physiologically (Figure 1.6),
- individuals developed from seeds (N_G) from individuals developed from vegetative propagules, or due to the disintegration of organisms (N_V).

Methodological implications

The fate of individuals in a population depends to a large extent on how many rooted shoots they are made of, and on the nature of their connections — permanent, or temporary. Individuals consisting of many rooted shoots are known to be longaeval, and possess a high reproductive potential (more generative shoots) and competitive power. It is difficult to correctly trace the boundaries of such an individual because of the necrosis and renewal processes that go on in its underground organs (Figure 1.3). One of the more straightforward methods for studying the development of perennial individuals is the mapping method in which

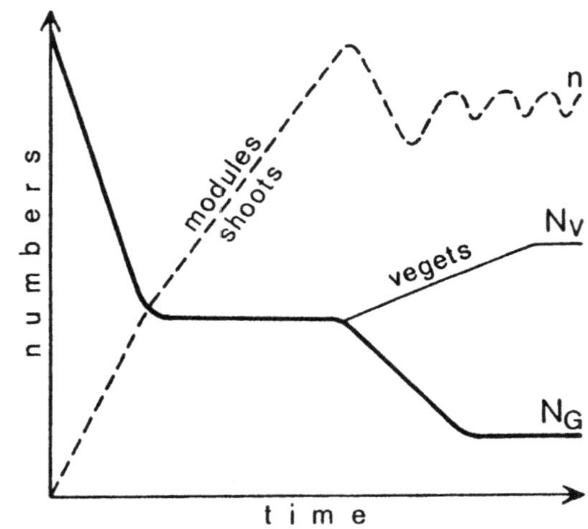

Fig. 1.6. Model of population size dynamics of perennial species, the individuals of which grow into multishoot polycormones and have two kinds of reproduction (Falińska 1985). N_G — changes in abundance of individuals of generative origin; N_V — changes in abundance of individuals of vegetative origin; n — changes in abundance of rooted shoots of which perennial individuals are constructed.

the distribution of individuals is marked, at an appropriate scale, on specially prepared charting sheets, with accompanying information on the number and kind of units they are made of. Distribution maps of this kind (large-scale maps) made for many years are the source of information on the origin, structure, growth rate, permanency and disintegration (Figures 1.7; 1.8).

Studies carried out so far of the population abundance dynamics in which the fate was followed of labelled individuals, indicate that the course of variations in the number of individuals is not always similar to variation in the number of shoots. In many populations the growth in number of individuals (N) was found to follow a logistic model and that of shoots, an exponential model (Figure 1.4). A progressing population abundance stabilization is accompanied by a rapid growth in numbers of the shoots. So if abundance curves are used, e.g., for the description of population growth phases, then for the same population a different estimate is arrived at if the estimation is based on N than when it is based on n. The shoot abundance curve may, for instance, indicate a population growth phase and the numbers curve of individuals — an equilibrium phase, or even a population decline phase (Figure 1.2). It must be stressed that perennial-shoot abundance curves are usually similar in shape to those for annual or biennial populations. This is natural because the above-ground parts of many perennials live only for one or two growing seasons.

A full description of the population dynamics of species with a polycormonal growth of individuals and two reproduction modes should, therefore, contain descriptions of: (a) abundance variations of generative and vegetative individuals; (b) variations in numbers of the shoots or other basic units of which the individuals of a particular population are made; (c) number of clones and vegets of which they consist, that is, into how many parts a plant organism has fragmented (Figure 1.8).

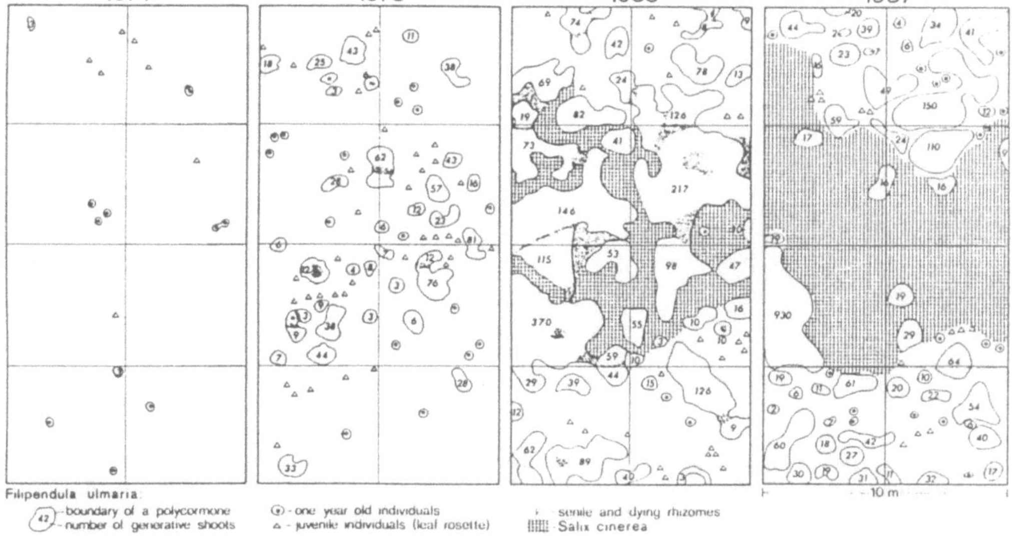

Fig. 1.7 Observations in 200 m² study areas — the method enabling to follow the fate of individuals and population, the case of one selected species (Falińska 1985, completed).

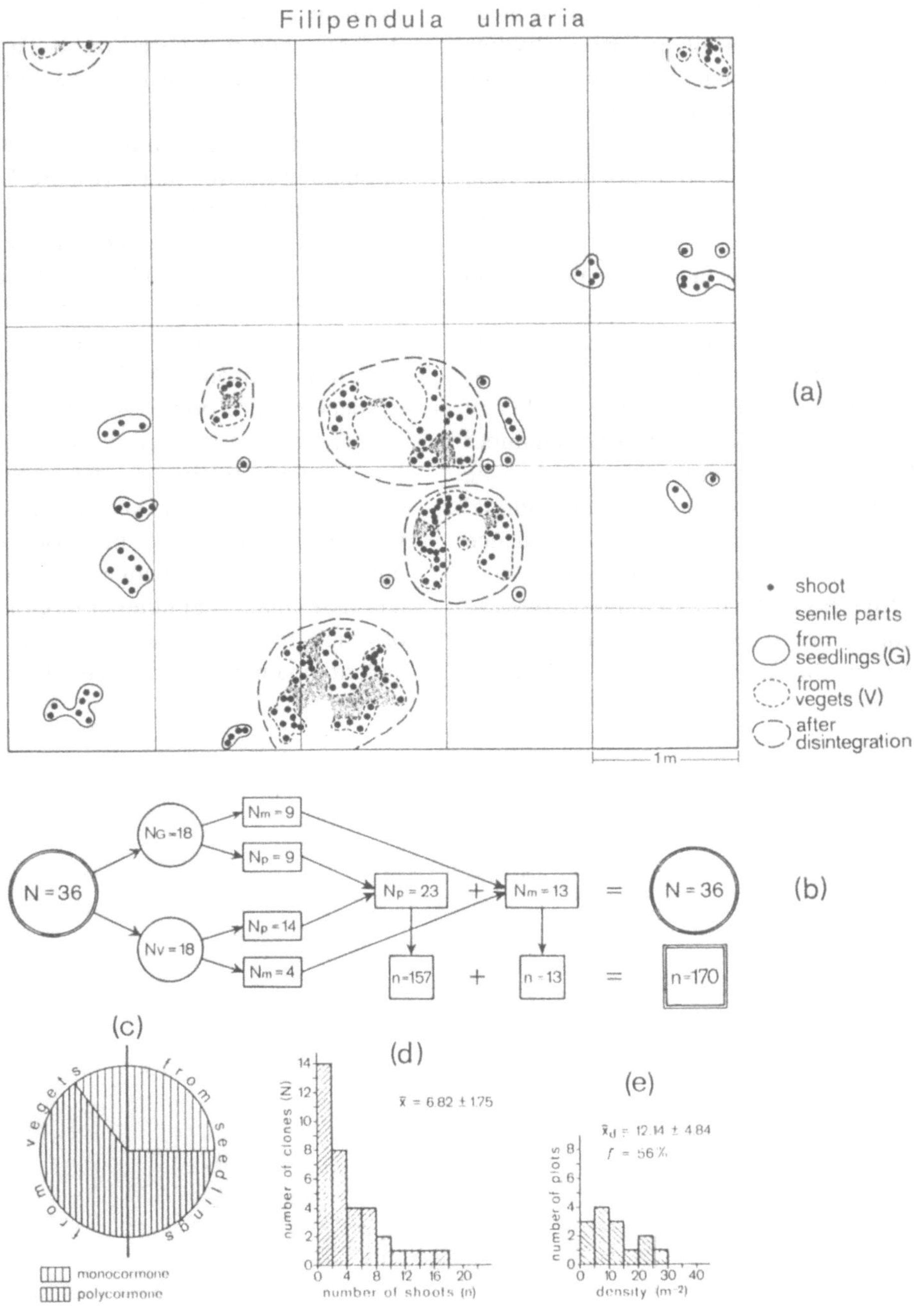

Fig. 1.8 Basic plots applied to observe the fates of individuals of different origin in population and methods for the study of population: a) mapping — spatial organization of population; b) numbers of individuals (N) and shoot (n) in the population; N_p — number of polycormones, N_G — number of individuals derived from seedlings, N_V — number of vegets (independent parts of a clone), N_m — number of monocormone; c) percentage of individuals of different derivation; d) the distribution of number of shoots (n) per polycormone (N), mean (± SD) of shoots per polycormone is indicated; e) frequency of 1 m x 1 m² plots with different shoot density mean (± SD) density (m^{-2}) is indicated.

CHAPTER 3

STUDY AREA AND METHODOLOGICAL BASIS

Study area and choice criteria

To understand the relationship between the fate of populations and seral vegetation changes, many methodological conditions must be satisfied. Population growth must, in such cases, be followed for many years, i.e., from the initial phase to the phase of a dynamics-equilibrium and/or a complete, or at least partial retreat of species. The fulfillment of methodological conditions is possible when a particular species colonizes new areas, or returns to an area previously abandoned, or considerably alters its abundance due to the effect of an agency, e.g. cessation of man's management in meadows and farmlands (Figure 1.2).

As regards to perennials, and particularly those with a polycormonal growth type and two reproduction modes, gathering data for demographic analyses requires long-term studies, special study-plot systems, and various methods (Figure 1.9).

The fate of populations during a long-term succession in abandoned meadows was studied over the period 1974—1989 in the Reski Range located in the Narewka River valley (forest-free of 200 years) in the centre of the Białowieża Primeval Forest complex. The studies covered 15 ha of meadows abandoned at different, but known times during the period 1960—1975.

Succession processes, that is, species-turnover mechanisms and rates, were followed at different vegetation-organization levels (individual, population, phytocoenosis), using a hierarchical plot system (Figure 1.9). The plots were set up according to the cluster method, i.e., in a descending order of sizes. The Reski Range was divided into the following 4 sectors, according to the meadow abandonment times and phytocoenosis types: A — meadows abandoned in the year when the studies were started (1974); B —meadows unmown for 5 years, i.e., since 1970 (herbaceous communities); C — meadows unmown for 10 years, i.e., since 1965 (willow scrub); D —meadows abandoned in 1960, that is, unmown for 15 years (forest initiation phase). In each sector a 1 ha area was marked out for the phytocoenoses typical of the successive phases. In each of them six 200 m^2 plots were set up, and each of these was subdivided into eight 25 m^2 (5 m by 5 m) subplots. In each area one 25 m^2 subplot was divided into 25 squares, 1 m^2 in size.

Since in the cluster method, plots are nested, i.e., smaller plots are inside bigger ones, data from a smaller plot (e.g. 1 m^2) could be used for the interpretation of the phenomena occurring in larger areas.

The hierarchical plots system made it possible to carry out observations and gather data on vegetation changes at different levels of vegetation organization, viz.: (a) changes in the plant landscape in a 15 ha area; (b) photocoenosis dynamics in four 1 ha areas; (c) population dynamics in twenty-four 200 m^2 plots; (d) architecture growth and life-histories of individuals in six 25 m^2 subplots; the plots were also used for studying the seed bank and seedling survivorship.

Observation and material

1. *Community dynamics.* To follow the vegetation changes after the cessation of meadow mowing, every 5 years a plant community map of the

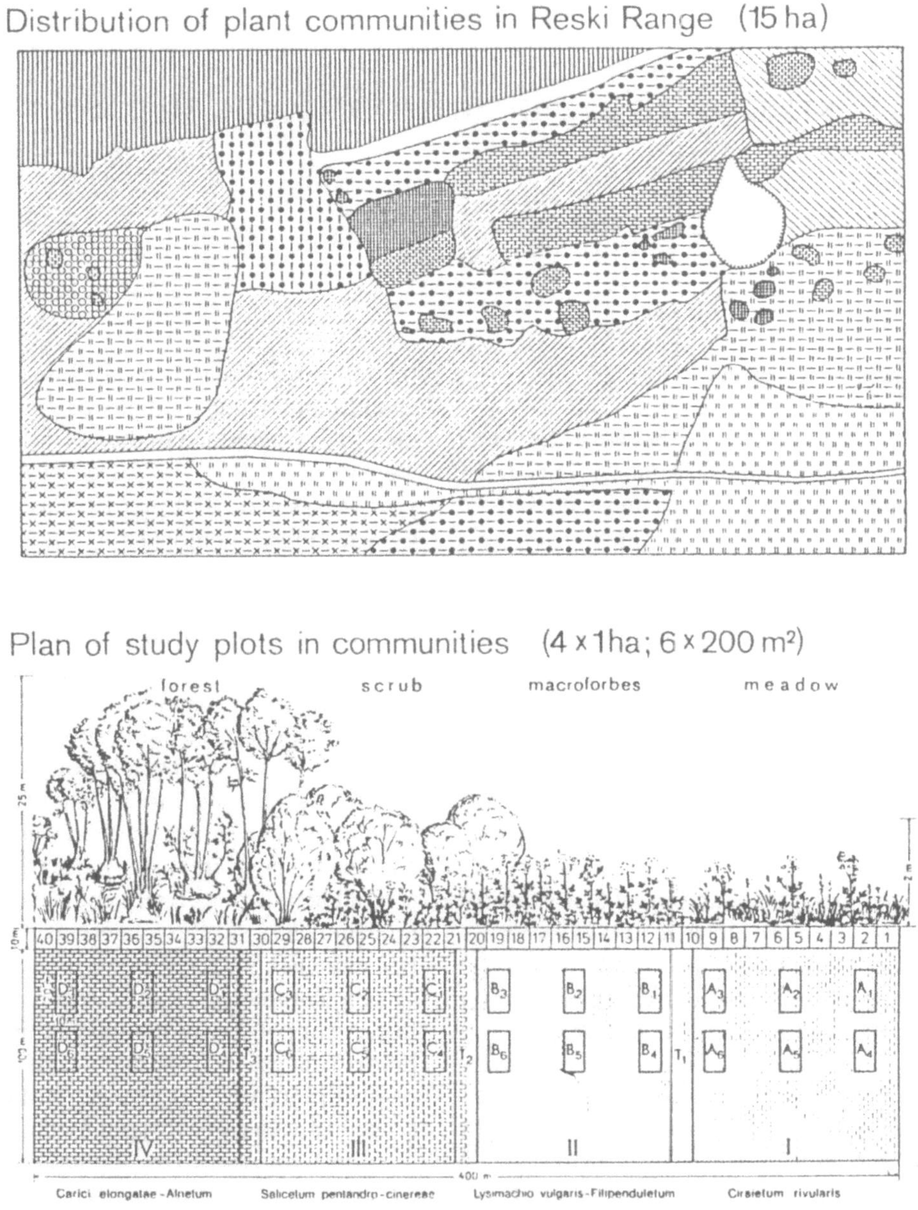

Fig. 1.9

whole study area (15 ha) was prepared. To do this, a system of squares, 10 × 10 m each, was used.

2. *Population spatial structure.* Every growing season (1975–1985) the distribution was mapped at the 1:100 scale of the populations of all the species in the 200 m² plots (Figure 1.7), and flowering and vegetative shoots of all the species in the particular subplots (25 m²) were counted (Photo 1.1).

3. *Numbers of individuals and shoots.* Individual architecture, growth and life-history were studied

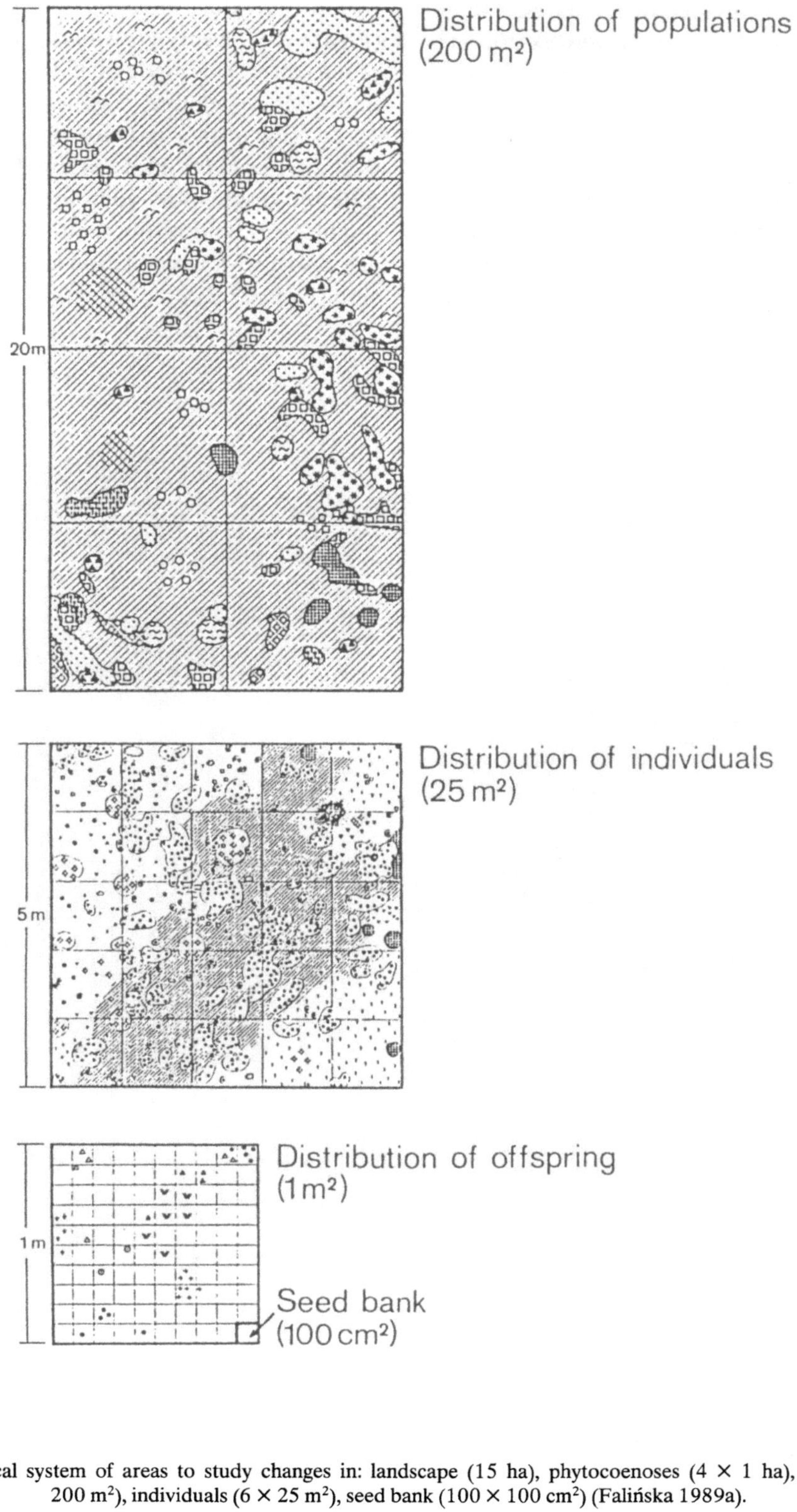

Fig. 1.9 Hierarchical system of areas to study changes in: landscape (15 ha), phytocoenoses (4 × 1 ha), populations (24 × 200 m^2), individuals (6 × 25 m^2), seed bank (100 × 100 cm^2) (Falińska 1989a).

in six 25 m^2 subplots divided into squares, 1 m in side length. The subplots were inside six 200 m^2 plots, selected from 24 such plots, found to be the most representative of the course of succession in abandoned meadows (Figure 1.8).

Every growing season a 1:10 scale map was prepared with the following marked on it: (a) distribution of the individuals of the main phytocoenosis components; (b) shape and size of the areas occupied by them; (c) shoot distribution within individual areas of polycormonal species (Figure 1.8). Every growing season in every subplot a census was made of all the generative (seedlings) and vegetative (fragments of disintegrated parent organisms) individuals (Figure 1.8). The data were used for the determination of the role of different-origin individuals in the dynamics of the populations, and of the relationship between the age structure of a population and its role in succession (Figure 1.2).

Photo 1.1. A basic study plot for observing the fate of individuals over their life cycle. The plot, 25 m^2 in size (5 × 5 m), has been subdivided into squares 1 m in side length each.

Photo: J. B. Faliński

PART 2

VEGETATION DYNAMICS

CHAPTER 4

OLD AND PRESENT DIVERSITY OF THE RESKI RANGE VEGETATION

The Reski Range as a study area

The Reski Range is situated in the Narewka River valley (unforested for 200 years) in the vicinity of the Białowieża National Park, in the middle of the Białowieża Primeval Forest complex.

The mosaic vegetation pattern over the 15 ha area of the Reski Range reflects the heterogeneity of a carr habitat replaced with meadows, and is also the result of the abandonment of the latter by farmers in different years over the 1960—1974 period. Some of the meadow allotments were still sporadically mown in 1978, due to which there

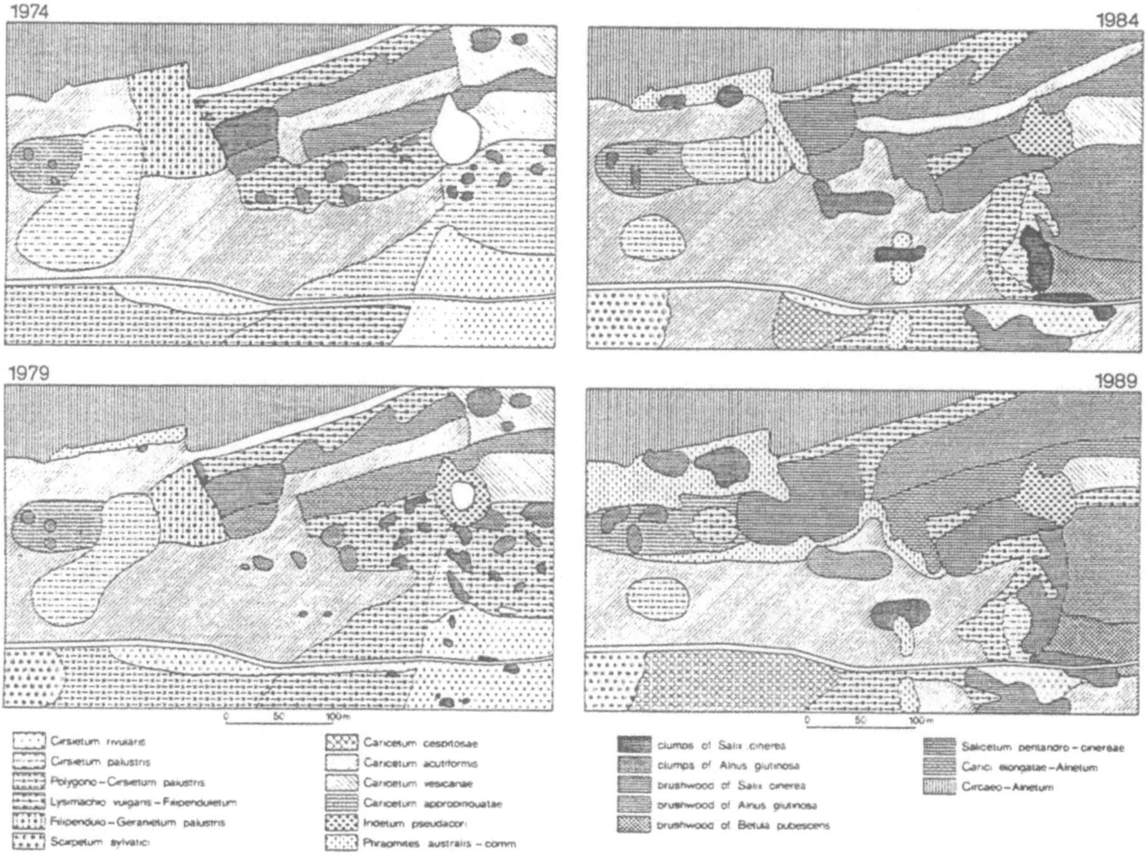

Fig. 2.1. Changes in the vegetation of the Reski Range on 15 ha of abandoned meadows.

Table 2.1. Cirsietum rivularis (Ralski 1931)

Consecutive record number	1	2	3	4	5	6	7	8	9	10	11	12	13	14	15	16	17	18	19	Frequency		
Record number	89	90	91	92	93	94	95	96	97	98	99	100	101	102	103	104	105	106	107			
Sector	A_2	A_2	A_2	A_2	A_2	A_3	A_3	A_3	A_3	A_3	A_3	B_3	B_3	B_3	B_3	B_3	B_3	B_3	B_3			
Coverage of herb layer c%	90	90	90	90	90	90	90	90	90	90	90	90	90	90	90	90	90	90	80	number of records	%	constancy class
Record surface m^2	25	25	25	25	25	25	25	25	25	25	25	25	25	25	25	25	25	25	25			
Number of species	34	39	42	35	27	30	31	26	29	29	27	25	35	31	35	32	37	36	38			
Ch.Ass: *Cirsietum rivularis*																						
Cirsium rivulare	1.2	+.2	+.2	+.2	.	+.2	+.1	.	+.2	r.1	+.1	.	1.2	+.2	+.2	3.3	+.2	+.2	+.2	16	84	V
Ch.All: *Calthion*																						
Caltha palustris	.	.	.	.	1.2	r.1	+.1	1.2	+.2	.	+.2	+.2	1.2	+.2	2.3	+.2	+.2	1.2	1.2	14	74	IV
Geum rivale	2.3	2.3	.	2.3	+.2	+.2	2.4	2.3	2.4	2.3	2.3	.	2.3	+.2	1.2	2.2	2.3	+.2	+.2	17	89	V
Polygonum bistorta	+.1	+.2	.	.	.	.	+.1	+.1	+.1	+.1	+.1	1.1	+.1	+.1	.	+.1	+.1	+.1	r.1	14	74	IV
Scirpus sylvaticus	.	.	.	.	.	.	.	.	.	.	+.1	+.1	.	.	+.1	+.1	.	.	2.5	5	26	II
Myosotis scorpioides	1.2	+.2	+.2	.	.	r.1	.	+.1	+.2	+.1	.	.	.	.	+.1	.	.	.	.	8	42	III
Ch.O: *Molinietalia*																						
Filipendula ulmaria	+.2	2.2	+.2	+.2	+.2	2.2	+.2	+.2	.	.	+.2	+.2	r.1	+.1	2.2	1.2	+.2	r.1	+.2	16	89	V
Lysimachia vulgaris	+.1	+.1	+.1	.	+.1	+.2	+.2	+.2	+.1	+.2	1.3	+.1	+.1	1.1	+.2	+.2	.	+.1	+.2	17	89	V
Lythrum salicaria	+.1	+.1	+.2	+.1	r.1	+.1	+.1	+.1	1.2	+.2	+.1	+.1	+.1	+.1	+.2	+.2	r.1	+.1	+.2	19	100	V
Cirsium palustre	.	+.1	+.1	.	.	r.1	.	+.1	r.1	+.1	+.1	+.1	+.1	+.1	r.1	.	.	.	.	11	58	III
Deschampsia cespitosa	+.2	.	.	+.1	.	.	+.1	+.1	+.2	+.1	.	.	+.2	.	+.1	1.1	.	+.2	.	10	52	III
Galium uliginosum	1.1	1.1	1.2	1.2	1.1	.	1.2	+.1	+.1	+.1	+.1	+.1	+.1	+.1	+.1	1.2	+.1	+.1	+.1	18	95	IV
Juncus effusus	1.2	2.2	2.3	2.2	+.2	+.2	+.2	+.2	+.2	+.2	.	.	+.2	.	+.1	.	.	.	+.3	13	68	IV
Lychnis flos cuculi	+.1	+.1	+.2	+.2	+.1	.	+.2	+.1	+.1	+.1	.	+.1	+.1	+.1	+.1	1.2	+.1	+.1	+.1	17	89	V
Ch.U: *Molinio-Arrhenatheretea*																						
Rumex acetosa	+.2	1.2	+.1	1.2	1.1	+.1	+.2	+.1	.	.	+.1	+.2	+.1	+.1	+.1	+.1	+.1	+.1	+.1	17	89	V
Poa pratensis	+.1	+.2	+.2	+.1	+.1	+.2	2.2	+.2	+.1	+.1	.	1.1	+.1	+.1	+.1	+.1	+.1	+.1	.	17	89	V
Carex cespitosa	2.2	+.2	.	.	+.2	2.2	3.2	2.2	+.2	+.2	2.2	1.2	4.2	+.2	2.2	.	+.2	2.2	3.2	16	84	V
Ranunculus acris	+.1	+.1	1.2	+.1	+.2	.	1.1	.	+.1	+.1	+.1	+.2	+.1	+.1	+.1	+.1	+.1	+.1	+.1	17	89	V
Lathyrus pratensis	1.1	3.3	4.4	3.4	1.2	.	.	.	.	.	.	.	.	2.3	+.1	2.3	2.4	2.2	+.1	11	58	III
Lotus corniculatus	.	.	.	.	.	3.4	3.4	4.4	2.3	.	.	.	.	.	.	.	.	.	+.1	5	26	II
Veronica chamaedrys	.	.	.	+.1	.	.	1.3	.	.	+.1	+.1	.	.	+.1	+.1	.	.	r.1	+.1	8	42	III
Dactylis glomerata	+.1	+.2	+.2	+.2	.	.	+.2	+.1	.	.	.	.	+.2	.	.	1.1	+.1	+.1	+.1	11	58	III
Holcus lanatus	+.2	+.2	+.2	+.2	.	.	.	.	.	.	.	.	.	.	.	.	+.1	.	.	5	26	II
Carex hirta	2.3	+.1	.	+.2	+.1	.	.	.	.	.	.	.	.	+.2	.	.	+.1	+.1	.	7	37	II
Phleum pratense	.	.	+.2	+.1	.	.	+.2	.	.	.	.	.	.	.	.	1.1	+.1	+.1	+.1	7	37	II
Plantago lanceolata	+.1	+.1	+.1	.	.	.	.	.	.	.	.	.	r.1	.	.	.	r.1	.	r.1	6	32	II
Anthoxanthum odoratum	+.1	+.1	.	.	.	.	1.2	.	.	.	.	.	+.1	.	.	.	+.1	+.1	+.1	7	37	II
Taraxacum officinale	.	+.1	+.1	.	.	.	.	.	.	.	.	+.1	+.1	+.2	r.1	.	+.1	.	r.1	8	42	III
Campanula patula	.	+.2	r.1	.	.	.	.	.	.	.	.	.	.	.	.	.	.	.	.	2	10	I
Alopecurus pratensis	+.1	+.2	+.2	+.2	+.2	+.1	+.2	.	.	.	.	1.1	+.1	+.1	r.1	1.2	+.1	+.1	+.1	15	79	IV
Cerastium holosteoides	.	.	+.1	+.1	+.2	.	.	.	.	.	.	.	.	.	.	.	.	.	.	3	16	I
Galium palustre	.	.	.	.	.	+.1	.	.	.	.	.	.	.	.	1.1	.	.	.	.	2	10	I
Festuca rubra	.	.	.	.	.	.	+.2	.	.	.	.	.	+.2	+.1	.	.	.	+.1	.	4	21	II
Achillea millefolium	.	.	.	.	.	.	.	.	.	.	+.1	.	.	.	.	.	+.1	r.1	.	3	16	I
Ch: *Scheuchzerio-Caricetea fuscae*																						
Carex nigra	4.5	2.5	2.3	2.5	4.5	1.2	.	3.5	2.3	3.4	4.4	4.4	2.3	5.5	2.3	2.3	4.4	+.2	3.5	18	95	V
Viola palustris	1.4	+.1	.	2.3	1.1	1.4	1.4	1.4	2.4	3.4	3.4	2.3	2.3	+.2	1.3	1.4	1.4	1.4	1.4	18	95	V
Potentilla palustris	2.4	2.3	2.2	.	.	+.2	.	+.2	2.2	3.4	.	.	.	+.1	+.1	r.1	.	.	+.2	11	58	III
Epilobium palustre	+.1	+.2	+.2	+.1	.	.	.	.	.	.	+.1	.	.	.	.	r.1	+.1	+.2	.	8	42	III
Carex flava	.	+.1	.	.	.	.	.	.	.	+.1	.	.	.	.	.	.	.	.	.	2	10	I
Carex canescens	.	.	r.1	.	.	.	.	.	.	.	.	.	+.2	+.2	.	.	1.1	+.2	1.2	6	32	II
Carex leporina	.	.	+.1	.	.	.	.	.	.	.	.	.	.	.	.	.	.	.	+.2	2	10	I
Carex echinata	.	.	.	.	.	.	.	.	+.1	.	.	+.1	.	+.1	+.1	.	+.2	.	+.2	6	32	II
Ch: *Phragmitetea*																						
Carex rostrata	2.2	+.2	+.2	+.1	2.2	2.3	1.3	2.2	+.1	2.2	3.5	2.2	3.3	2.2	1.3	1.2	3.2	3.3	1.2	19	100	V
Equisetum fluviatile	+.1	1.2	1.1	1.1	2.4	2.1	1.2	1.2	1.1	+.2	1.1	.	+.1	+.1	+.1	1.1	1.2	+.1	1.2	18	95	V
Carex acutiformis	.	+.2	+.2	2.5	2.5	3.5	3.5	.	3.5	3.3	1.2	.	+.1	.	3.3	1.2	+.2	2.2	.	15	79	IV
Scutellaria galericulata	+.1	.	+.1	+.1	1.1	1.2	3.2	+.1	+.1	+.1	+.1	+.1	+.1	.	+.1	+.1	+.1	+.2	r.1	17	89	V
Carex gracilis	.	.	+.1	+.2	+.2	2.5	.	.	1.2	.	1.3	.	.	.	.	.	.	2.2	.	7	37	II
Carex vesicaria	.	.	+.2	2.3	+.1	.	1.3	.	.	+.1	.	.	+.2	.	2.2	1.3	+.2	+.2	.	10	53	III
Carex panicea	+.1	+.1	1.1	+.2	.	.	.	.	.	+.1	.	.	.	.	.	.	.	.	1.2	6	32	II
Peucedanum palustre	.	.	.	+.1	.	+.2	.	.	.	.	r.1	+.1	.	.	.	.	.	.	.	4	21	II
Carex appropinquata	.	.	.	.	.	.	.	.	+.2	+.2	.	+.2	+.2	.	+.2	+.2	.	.	+.2	7	37	II
Accompanying species:																						
Equisetum arvense	.	3.4	1.1	1.1	2.4	2.2	1.2	2.3	2.3	2.4	1.4	+.1	2.3	+.1	+.1	r.1	1.3	1.2	1.2	18	95	V
Ranunculus repens	2.2	2.3	2.3	2.3	+.2	+.2	4.4	3.4	2.4	.	+.1	2.3	+.2	+.2	2.2	2.4	2.3	1.2	+.2	18	95	V
Salix cinerea	.	.	.	+.2	.	.	.	.	3.4	3	2	.	.	.	.	.	.	.	r	5	26	II
Potentilla anserina	+.1	2.3	2.3	2.2	.	.	.	.	.	.	.	.	.	.	.	.	.	.	.	4	21	II
Potentilla erecta	+.1	1.2	1.1	1.2	+.2	.	.	.	.	.	.	.	.	.	.	.	.	.	.	5	26	II
Urtica dioica	.	.	r.1	.	.	r.1	.	.	.	.	.	.	.	.	.	.	.	.	+.2	3	16	I
Lycopus europaeus	.	+.1	+.2	.	.	.	.	.	.	.	.	.	.	.	r.1	.	.	.	.	3	16	I
Agrostis vulgaris	.	.	.	.	.	+.2	1.2	.	.	.	.	.	.	+.1	.	.	.	.	.	3	16	I
Luzula campestris	+.1	+.1	+.1	.	.	.	.	.	.	.	.	.	.	.	.	.	+.1	.	.	4	21	II
Carex elongata	.	.	.	.	.	+.2	.	.	.	.	.	+.2	.	.	.	.	.	+.2	.	3	16	I

Sporadic species: *Crepis paludosa* 108/r.1; *Avenochloa pubescens* 106/r.1; *Vicia cracca* 97/r.1; *Briza media* 99/+.1; *Carex otrubae* 107/+.2; *Stellaria palustris* 91/+.1; *Orchis* sp. 110/r.1.

Table 2.2. Lysimachio vulgaris-Filipenduletum (Balatova-Tulackova 1979)

Consecutive record number	1	2	3	4	5	6	7	8	9	10	11	12	13	14	15	16	17	18	Frequency		
Record number	113	114	115	116	117	118	119	120	121	122	123	124	125	126	127	128	129	130			
Sector	B_2	B_2	B_2	B_2	B_2	B_2	B_2	B_2	B_1	B_1	B_1	B_1	B_1	B_1	B_1	B_1	B_1	B_1			
Coverage of herb layer c%	90	90	90	90	90	90	90	90	90	90	90	90	90	90	80	80	90	80	number of records	%	constancy class
Record surface m^2	25	25	25	25	25	25	25	25	4	4	4	4	4	4	4	4	4	4			
Number of species	18	14	14	22	18	13	12	14	13	12	15	15	9	10	10	12	11	9			
Ch.Ass: *Lysimachio vulgaris-Filipenduletum*																					
Filipendula ulmaria	3.4	4.4	3.4	2.3	3.3	5.4	4.4	3.2	3.5	4.5	3.5	4.5	5.5	5.5	3.3	3.5	3.5	5.5	18	100	V
Lysimachia vulgaris	+.1	+.1	+.1	+.1	.	+.1	+.1	+.1	+.1	r.1	.	.	r.1	.	.	+.1	.	.	11	61	IV
Lythrum salicaria	2.2	+.2	+.1	+.2	3.2	+.2	+.2	3.2	r.1	r.1	r.1	+.2	+.1	+.1	+.1	+.1	+.1	.	17	94	V
Ch.All.O.Ch: *Filipendulo-Petasition, Molinietalia, Molinio-Arrhenatheretea*																					
Geum rivale	2.4	2.3	2.3	1.2	2.3	1.2	1.3	1.3	+.1	.	+.2	2.2	.	+.2	2,2	+.2	2.2	2.2	16	89	V
Lathyrus pratensis	1.2	1.2	+.1	+.1	.	.	.	1.2	+.1	r.1	1.1	1.1	r.1	1.1	1.1	1.1	1.1	1.1	15	83	V
Cirsium rivulare	.	.	+.2	.	1.2	+.1	+.2	+.2	r.1	+.2	+.2	+.2	.	.	r.1	+.2	.	.	11	61	IV
Polygonum bistorta	+.1	+.1	+.2	1.2	+.1	+.1	.	+.1	+.1	r.1	+.2	+.2	+.1	1.1	1.1	1.2	+.1	+.2	17	94	V
Caltha palustris	+.1	.	.	+.1	+.1	.	.	.	.	.	.	.	.	.	.	.	.	.	3	17	I
Rumex acetosa	.	.	.	+.1	.	+.1	.	.	+.1	.	.	.	1.1	.	.	.	1.1	.	5	28	II
Galium uliginosum	.	.	.	+.1	+.1	.	.	.	.	.	.	1.1	.	.	r.1	.	r.1	.	5	28	II
Dactylis glomerata	+.1	.	.	.	.	.	.	.	.	.	.	.	.	.	.	.	.	.	1	5	I
Phleum pratense	+.1	.	.	.	.	.	.	.	.	.	.	.	.	.	.	.	.	.	1	5	I
Alopecurus pratensis	+.2	.	.	.	.	.	.	+.1	.	1.1	+.2	+.1	.	.	.	.	.	.	5	25	II
Lychnis flos-cuculi	.	.	.	.	r.1	.	.	.	r.1	.	.	.	.	.	.	.	.	.	2	11	I
Equisetum palustre	+.1	+.1	.	.	.	.	.	.	.	.	+.1	.	.	.	.	.	.	.	3	17	I
Veronica longifolia	.	.	.	+.2	r.1	.	.	.	.	.	.	.	.	.	.	.	.	.	2	11	I
Scirpus sylvaticus	2.3	.	.	.	.	.	.	2.2	.	1.1	.	.	.	.	.	.	.	.	5	28	II
Myosotis scorpioides	.	.	.	r.1	.	.	.	.	.	.	.	.	.	.	.	.	.	.	1	5	I
Achillea millefolium	.	.	.	+.1	.	.	.	.	.	.	.	.	.	.	.	.	.	.	1	5	I
Poa pratensis	.	.	.	+.1	.	.	.	.	.	.	.	.	.	.	.	.	.	.	1	5	I
Crepis paludosa	.	.	.	.	.	.	.	.	.	.	r.1	.	.	.	.	.	.	.	1	5	I
Cirsium palustre	.	.	.	.	.	.	.	.	.	.	.	.	.	.	.	.	.	r.1	1	5	I
Ch. *Scheuchzerio Caricetea-fuscae*																					
Viola palustris	1.4	1.3	2.4	3.4	2.4	+.2	+.3	+.1	+.1	r.1	+.1	1.3	1.1	+.1	+.2	1.1	1.1	.	17	94	V
Carex nigra	.	.	.	+.1	+.2	.	.	.	.	.	.	.	.	.	.	.	.	.	2	11	I
Potentilla palustris	.	.	.	+.2	r.1	.	.	.	.	.	.	.	.	.	.	.	.	.	2	11	I
Epilobium palustre	.	.	.	+.1	.	.	.	.	.	.	.	.	.	.	.	.	.	.	1	5	I
Ch. *Phragmitetea*																					
Equisetum fluviatile	+.1	+.1	+.1	+.1	+.1	+.1	+.1	+.1	.	.	.	.	.	.	.	.	.	.	8	44	III
Scutellaria galericulata	+.1	+.1	.	+.1	1.1	.	.	.	.	.	.	.	.	.	.	.	.	.	4	22	II
Peucedanum palustre	.	.	r.1	+.1	+.1	.	.	.	.	.	.	.	.	.	.	.	.	.	3	17	I
Carex gracilis	.	.	.	.	.	.	+.2	.	.	.	.	.	.	.	.	.	.	.	1	7	I
Carex acutiformis	.	.	.	.	+.1	.	.	.	.	.	+.2	+.2	+.1	.	.	2.2	1.1	+.1	7	39	II
Accompanying species:																					
Salix cinerea	.	+	2	.	.	2	3	.	.	.	.	.	.	.	.	.	.	.	4	22	II
Equisetum arvense	+.1	+.1	+.1	+.1	+.1	+.1	+.1	+.1	.	r.1	+.1	+.1	+.1	+.1	.	1.1	.	+.1	15	83	V
Carex cespitosa	2.2	+.2	2.2	.	4.2	2.2	+.2	2.2	+.2	+.2	2.2	2.2	.	+.2	3.2	+.2	+.2	.	15	83	V
Urtica dioica	.	+.2	+.1	+.1	.	+.1	+.2	+.1	+.1	+.2	.	+.2	.	.	.	.	.	+.1	10	55	III
Prunella vulgaris	r.1	.	.	.	.	.	.	.	.	.	.	.	.	.	.	.	.	.	1	5	I
Ranunculus repens	.	.	.	.	.	.	.	.	+.1	.	+.1	+.2	.	+.2	+.2	2.2	2.2	+.2	8	44	III
Galeopsis speciosa	.	.	.	.	.	.	.	.	.	.	.	.	.	+.2	.	.	.	.	1	5	I

have survived small parts of floristically rich meadows within the range area, with an advanced stage of succession (Figure 2.1).

When the studies were started, 47% of the Reski Range surface area was grown up with sedges (*Caricetum acutiformis*), 23% of it represented meadows (*Cirsietum rivularis*), 25% herb communities (*Filipendulo-Geranietum* and *Lysimachio vulgaris-Filipenduletum*), and 5% willow scrub (Tables 2.1; 2.2; 2.3).

Cirsietum rivularis. This community represents a floristically rich meadow type where numerous grass species prevail, e.g. *Dactylis glomerata,*

Table 2.3. Caricetum acutiformis (Sauer 1937)

Consecutive record number	1	2	3	4	5	6	7	8	9	10	11	12	13	14	15	16	17	18	19	20	21	22	23	24	25	26	27	28	29	30	31	32	Frequency		
Record number	1	2	3	4	5	6	7	8	9	10	15	16	17	18	19	20	21	22	23	33	34	40	63	64	68	69	94	95	96	97	103	104			
Sector	C_1	C_1	C_1	C_1	C_1	C_1	C_1	C_1	C_2	C_2	C_2	C_2	C_3	C_3	C_3	C_3	C_3	C_3	C_3	D_1	D_1	D_1	D_4	D_4	D_5	D_5	A_2	A_2	A_2	A_3	A_3	A_3			
Coverage of herb layer c%	90	90	90	90	90	90	90	90	90	90	90	90	90	90	90	90	90	90	90	90	90	90	90	90	80	80	90	90	80	90	90	90	number of records	%	constancy class
Record surface m²	25	25	25	25	25	25	25	25	25	25	25	25	25	25	25	25	25	25	25	25	25	25	25	25	25	25	25	25	25	25	25	25			
Number of species	9	10	14	11	13	11	11	12	16	17	16	14	11	10	8	9	11	9	9	13	14	12	8	7	9	7	29	22	19	27	27	17			
Ch.Ass: *Caricetum acutiformis*																																			
Carex acutiformis	5.5	5.5	5.5	5.5	5.5	5.5	5.5	5.5	5.5	5.5	5.5	5.5	3.5	5.5	5.5	5.5	5.5	5.5	5.5	5.5	4.5	5.5	4.5	5.5	4.5	5.5	4.5	3.5	5.5	3.5	4.5	5.5	32	100	V
Ch: *Magnocaricion, Phragmitetalia, Phragmitetea*																																			
Equisetum fuviatile	+.1	1.1	+.1	+.1	+.1	+.1	+.1	+.1	1.1	+.1	+.1	+.1	+.1	1.1	1.1	+.2	+.1	+.1	+.1	+.1	+.1	+.1	r.1	.	.	.	1.1	1.1	1.1	2.1	1.1	.	28	87	V
Carex appropinquata	.	+.2	.	.	.	.	.	.	.	.	+.2	1.2	2.2	2.2	.	+.2	.	+.2	+.2	.	+.2	.	+.2	.	+.2	.	.	.	.	.	.	.	11	34	II
Phragmites australis	.	.	.	.	.	.	.	.	+.1	2.2	2.4	1.1	.	.	.	.	.	.	.	+.1	3.5	3.5	5.5	.	.	.	.	.	.	.	.	.	8	25	II
Scutellaria galericulata	.	.	.	.	.	.	.	.	+.1	+.1	.	.	.	.	.	.	.	.	.	.	+.1	.	.	.	.	.	+.1	+.1	.	1.2	+.1	.	7	22	II
Carex vesicaria	.	.	.	.	.	.	.	.	+.1	.	+.1	+.1	.	.	.	.	.	.	.	.	.	+.1	.	.	.	.	2.2	.	.	+.2	.	.	3	9	I
Poa palustris	.	.	.	.	.	.	.	.	.	.	.	.	.	.	.	.	.	.	.	.	+.1	.	.	.	+.2	.	.	.	.	+.2	.	.	3	9	I
Iris pseudacorus	.	.	.	.	.	.	.	.	.	.	.	.	.	.	.	.	.	.	.	.	.	.	.	.	2.3	+.2	.	.	.	.	.	.	2	6	I
Carex gracilis	.	.	.	.	.	.	.	.	.	.	.	.	.	.	.	.	.	.	.	.	.	.	.	.	.	.	.	.	+.2	2.5	+.2	.	3	9	I
Carex rostrata	.	.	.	.	.	.	.	.	.	.	.	.	.	.	.	.	.	.	.	.	.	.	.	.	.	.	+.	+.1	+.1	2.3	2.2	+.1	6	19	I
Peucedanum palustre	.	.	.	.	.	.	.	.	.	.	.	.	.	.	.	.	.	.	.	.	.	.	.	.	.	.	.	.	.	+.2	r.1	r.1	4	9	I
Galium palustre	.	.	.	.	.	.	.	.	.	.	.	.	.	.	.	.	.	.	.	.	.	.	.	.	.	.	.	.	.	+.1	.	.	1	3	I
ChP: *Molinio-Arrhenatheretea*																																			
Lythrum salicaria	+.1	1.1	+.1	+.1	1.1	+.1	+.1	+.1	+.2	+.2	+.1	+.1	+.2	+.1	1.2	2.2	+.1	+.2	+.1	+.1	r.1	.	.	+.2	+.2	+.2	+.1	+.1	+.1	+.1	2.2	+.2	30	94	V
Filipendula ulmaria	2.2	1.2	3.2	2.2	2.2	2.2	2.2	2.1	2.2	r.1	+.2	.	.	+.2	1.2	2.2	2.2	2.1	1.2	+.1	.	+.1	+.1	+.1	2.4	+.2	2.2	+.2	.	2.2	+.2	+.1	28	87	V
Scirpus sylvaticus	2.2	+.1	+.1	.	3.6	3.4	3.5	3.4	3.3	3.2	.	+.2	4.5	2.2	2.2	2.2	2.2	2.2	3.5	2.2	.	1.2	.	3.2	+.1	+.1	.	.	+.1	.	.	2.2	24	75	IV
Lysimachia vulgaris	+.1	1.1	+.1	.	1.1	+.1	.	.	+.1	+.2	+.1	1.1	+.1	.	2.2	2.2	+.1	+.1	.	+.1	+.1	+.1	+.2	+.2	.	.	+.2	+.1	+.1	+.2	1.1	+.2	25	78	IV
Caltha palustris	.	.	.	.	.	2	r.1	.	r.1	.	+.1	+.1	+.2	+.2	+.1	.	r.1	+.1	1.2	1.2	+.1	1.2	.	.	.	.	.	.	.	r.1	.	.	15	47	III
Cirsium rivulare	r.	.	r.1	2.2	2.2	2.2	2.2	+.1	r.1	+.2	+.1	r.1	.	.	.	r.1	+.1	.	.	+.1	+.1	1.2	.	.	.	.	+.1	.	+.1	+.2	.	+.2	20	63	IV
Geum rivale	.	.	+.1	r.1	2.2	1.4	r.1	+.1	+.2	r.1	1.2	.	.	.	.	.	.	.	.	1.1	+.1	1.2	.	.	.	.	+.1	2.2	2.3	+.2	2.3	1.2	18	56	III
Galium uliginosum	.	.	+.1	+.1	+.1	+.1	+.1	+.2	.	r.1	+.1	r.1	.	+.1	.	.	.	.	.	.	.	+.1	.	.	.	.	1.2	+.1	+.1		+.1		15	47	III

Polygonum bistorta	+.1	r.1	+.1	1.1	+.1	r.1	2.1	+.1	.	.	.	.	.	.	.	.	.	.	.	r.1	.	.	.	.	.	.	+.1	+.1	.	.	+.1	+.1	13	41	III
Lychnis flos-cuculi	.	.	.	.	.	.	.	.	.	r.1	.	.	.	.	.	.	+.2	.	.	.	.	.	.	.	.	.	+.2	+.1	.	.	.	+.1	5	16	I
Veronica longifolia	.	.	.	r.1	.	.	.	.	.	.	.	.	.	.	.	.	.	.	.	.	.	.	.	.	.	.	.	.	.	.	.	.	1	3	I
Symphytum officinale	.	.	.	.	.	.	r.1	.	.	.	.	.	.	.	.	.	.	.	.	.	.	.	.	.	.	.	.	.	.	.	.	.	1	3	I
Juncus effusus	.	.	.	.	.	.	.	.	+.1	.	.	+.1	.	.	.	.	.	.	.	.	.	.	.	.	.	.	2.2	.	+.1	+.2	.	.	5	16	I
Cirsium palustre	.	.	r.1	.	.	.	.	.	.	.	.	.	.	.	.	.	.	.	.	.	.	.	.	.	.	.	.	.	+.2	r.1	r.1	.	4	13	I
Carex cespitosa	.	.	.	.	.	.	.	.	.	.	.	.	.	.	.	.	.	.	.	.	.	.	.	.	.	.	+.2	+.2	.	2.2	2.2	+.2	5	16	I
Vicia cracca	.	.	.	.	.	.	.	.	.	.	.	.	.	.	.	.	.	.	.	.	.	.	.	.	.	.	.	.	.	r.1	.	.	1	3	I
Rumex acetosa	.	.	.	.	.	.	.	.	.	.	.	.	.	.	.	.	.	.	.	.	.	.	.	.	.	.	+.1	+.1	.	r.1	+.2	.	4	13	I
Achillea millefolium	.	.	.	.	.	.	.	.	.	.	.	.	.	.	.	.	.	.	.	.	.	.	.	.	.	.	.	.	.	.	.	+.1	1	3	I
Veronica chamaedrys	.	.	.	.	.	.	.	.	.	.	.	.	.	.	.	.	.	.	.	.	.	.	.	.	.	.	.	.	.	.	+.1	+.2	2	6	I
Myosotis scorpioides	.	.	.	.	.	.	.	.	.	.	.	.	.	.	.	.	.	.	.	.	.	.	.	.	.	.	+.1	.	.	.	+.1	.	2	6	I
Lotus corniculatus	.	.	.	.	.	.	.	.	.	.	.	.	.	.	.	.	.	.	.	.	.	.	.	.	.	.	.	.	.	3.4	1.3	+.1	3	9	I
Angelica sylvestris	.	.	.	.	.	.	.	.	.	.	.	.	.	.	.	.	.	.	.	.	.	.	.	.	.	.	.	.	.	.	+.1	.	1	3	I
Lathyrus pratensis	.	.	.	.	.	.	.	.	.	.	.	.	.	.	.	.	.	.	.	.	.	.	.	.	.	.	3.4	1.3	+.1	.	.	.	3	9	I
Carex hirta	.	.	.	.	.	.	.	.	.	.	.	.	.	.	.	.	.	.	.	.	.	.	.	.	.	.	+.2	+.2	.	.	.	.	2	6	I
Anthoxanthum odoratum	.	.	.	.	.	.	.	.	.	.	.	.	.	.	.	.	.	.	.	.	.	.	.	.	.	.	+.2	.	.	.	.	.	1	3	I
Accompanying species:																																			
Equisetum arvense	.	.	+.1	+.1	.	.	.	1.1	1.1	1.1	+.1	+.1	+.1	1.1	+.1	+.1	+.1	.	+.1	.	.	.	.	.	.	.	1.1	1.1	1.1	.	1.1	1.1	18	56	III
Viola palustris	1.2	+.2	+.1	1.1	2.4	2.4	1.4	2.4	+.2	.	.	.	.	.	.	.	.	.	.	1.2	.	.	.	.	.	.	1.4	.	1.4	1.4	4.4	1.4	15	47	III
Impatiens noli-tangere	.	.	.	.	.	.	.	.	.	r.1	+.1	r.1	.	.	.	.	r.1	r.1	.	.	2.3	.	+.2	+.1	r.1	+.2	.	.	.	.	.	.	10	31	II
Humulus lupulus	.	.	.	.	.	.	.	.	.	.	.	.	r.1	.	.	.	.	.	.	.	.	.	.	+.1	r.1	.	.	.	.	.	.	.	3	9	I
Alnus glutinosa	.	.	.	.	.	.	.	.	r	.	.	.	.	.	.	.	.	.	.	.	.	.	.	r	.	.	.	.	.	.	.	.	2	6	I
Lycopus europaeus	.	.	.	.	.	.	.	.	.	r.1	r.1	.	.	.	.	.	.	.	.	.	.	.	.	.	.	.	+.1	.	.	.	.	.	3	9	I
Urtica dioica	.	.	.	.	.	.	.	.	.	+.1	.	r.1	.	.	.	.	.	.	.	.	2.4	.	.	.	.	.	+.2	.	.	r.1	+.3	.	5	16	I
Salix cinerea	.	.	.	.	.	.	.	.	.	.	.	.	.	.	.	.	.	.	.	.	.	.	.	.	.	.	.	.	.	.	+	.	1	3	I
Thelypteris palustris	r	r.1	+.2	+.2	+.1	.	.	.	.	.	.	.	.	.	.	.	.	.	.	.	.	.	.	.	.	.	.	.	.	.	.	.	4	13	I
Eupatorium cannabinum	.	.	.	.	.	.	.	.	.	.	.	.	.	.	.	.	.	.	.	.	.	.	.	.	.	.	.	.	.	.	.	.	1	3	I
Galium spurium	.	.	.	.	.	.	.	+.1	+.1	.	.	.	r.1	r.1	.	.	.	.	.	.	.	.	.	.	.	.	.	.	.	.	.	.	4	13	I
Dryopteris spinulosa	.	.	.	.	.	.	.	.	.	.	.	.	.	.	.	.	.	.	.	.	.	+.2	.	.	.	.	.	.	.	.	.	.	1	3	I
Solanum dulcamara	.	.	.	.	.	.	.	.	.	.	.	.	.	.	.	.	.	.	.	+.1	.	.	+.1	+.1	.	.	.	.	.	.	.	.	3	9	I
Carex nigra	.	.	.	.	.	.	.	.	.	.	.	.	.	.	.	.	.	.	.	.	.	.	.	.	.	.	+.2	2.3	+.1	1.2	2.5	.	5	16	I
Potentilla palustris	.	.	.	.	.	.	.	.	.	.	.	.	.	.	.	.	.	.	.	.	.	.	.	.	.	.	2.3	2.2	2.3	+.2	2.3	+.1	6	19	I
Epilobium palustre	.	.	.	.	.	.	.	.	.	.	.	.	.	.	.	.	.	.	.	.	.	.	.	.	.	.	+.1	+.1	1.2	.	+.1	.	4	13	I
Ranunculus repens	.	.	.	.	.	.	.	.	.	.	.	.	.	.	.	.	.	.	.	.	.	.	.	.	.	.	1.2	.	.	+.2	1.4	.	3	9	I
Ranunculus acris	.	.	.	.	.	.	.	.	.	.	.	.	.	.	.	.	.	.	.	.	.	.	.	.	.	.	+.1	1.1	+.1	.	.	.	3	9	I
Carex canescens	.	.	.	.	.	.	.	.	.	.	.	.	.	.	.	.	.	.	.	.	.	.	.	.	.	.	.	+.2	.	.	.	.	1	3	I
Potentilla erecta	.	.	.	.	.	.	.	.	.	.	.	.	.	.	.	.	.	.	.	.	.	.	.	.	.	.	.	+.2	.	.	.	.	1	3	I
Agrostis tenuis	.	.	.	.	.	.	.	.	.	.	.	.	.	.	.	.	.	.	.	.	.	.	.	.	.	.	.	+.2	.	+.2	.	.	1	3	I

Alopecurus pratensis, Festuca pratensis, Lolium perenne, Anthoxanthum odoratum, Holcus lanatus, Poa pratensis and *Deschampsia cespitosa.* Among the grasses the following species are represented by single plants or aggregations: *Lychnis flos-cuculi, Ranunculus acris, Cirsium palustre, Cirsium rivulare, Rumex acetosa, Lathyrus pratensis* and *Lotus corniculatus.* In depressions and damper sites, the following species form larger aggregations: *Carex acutiformis, Filipendula ulmaria, Geum rivale, Potentilla palustris, Carex cespitosa* (Table 2.1).

Over the periods during which the meadows were used the above fragments were not mown every growing season, and after the use of the range had ceased completely, they became the site of succession initiation.

Lysimachio vulgaris-Filipenduletum. Dominant in this community is *Filipendula ulmaria*, with high proportions of *Lysimachia vulgaris, Lythrum salicaria* and *Carex cespitosa* (Table 2.2). There also occur species usually found in mown meadows, but their contribution to the herbaceous communities is negligible. Often found on the fringes of large *F. ulmaria* aggregations are species such as: *Lychnis flos-cuculi, Ranunculus acris, Cirsium palustre* and *C. rivulare.*

Caricetum acutiformis communities occupy water-logged, peaty substrata where there is a high water-table, with the groundwater level above the surface of the ground throughout the growing season. Over a half of the Reski Range area represents such a habitat type (Table 2.3). The mosaic pattern of the sedge-overgrown patches is closely related to the above-ground water-surface level and its persistence during the growing season. Among large-sized *Carex acutiformis* aggregations there form assemblages of *Filipendula ulmaria, Lythrum salicaria, Iris pseudacorus* and *Caltha palustris*, as well as tree biogroups (Figure 2.1).

Salicetum pentandro-cinereae is found at the forest-meadow interface where it occupies a small swampy area. The main components of this community are: *Salix cinerea, S. pentandra* and *S. aurita.* Solitary *Alnus glutinosa* trees are found dispersed amidst willows. The following woodland species occur there singly or in groups: *Solanum dulcamara, Impatiens noli-tangere, Lycopus europaeus* and *Calamagrostis canescens.* The contribution is also substantial of such species as: *Filipendula ulmaria, Lythrum salicaria* and *Lysimachia vulgaris* which form a tall-herb belt on the margins of willow clumps. In water-filled depressions swamp and aquatic species are found, e.g. *Alisma plantago-aquatica, Cicuta virosa, Oenanthe aquatica, Veronica beccabunga*, as well as *Scirpus sylvaticus, Calamagrostis arundinacea, C. canescens* and *Phragmites australis* (Table 2.4).

Vegetation differentiation in the course of succession

Forest return to fertile and moist meadows, situated close to a forest complex, is distinct for a high rate of vegetation changes, as expressed by the short duration of the successive species combinations, and a relatively early onset of the initial forest phase, e.g., 15—20 years after the cessation of the use of the range. This type of succession is determined primarily by the initial situation of the process, viz.: (a) in meadow *Cirsietum rivularis* phytocoenoses, where the succession is initiated, there occur species that are also components of forest phytocoenoses; (b) the biochores of meadow phytocoenoses are characterized by a considerable microhabitat diversity which increases yet further after the cessation of mowing and favours the mosaic vegetation pattern; (c) due to the location close to a forest complex (the area borders on the Białowieża National Park), forest species seedlings are found in the meadow sward also when the meadows are mown regularly.

The vegetation changes on the range, following the cessation of mowing, vary considerably, and the meadow phytocoenosis decomposition begins and ends at different times. Hence 15 years after

Table 2.4. Salicetum pentandro-cinereae (Almg 1929 and Pass 1961)

Consecutive	1	2	3	4	5	6	7	8	9	10	11	Frequency		
Record number	49	50	51	52	53	54	55	56	59	60	61			
Sector	D_3	D_3	D_3	D_3	D_3	D_3	D_3	D_3	D_4	D_4	D_4			
Coverage of herb layer c%	80	80	80	80	80	80	70	80	70	60	80	number of records	%	constancy class
Record surface m^2	25	25	25	25	25	25	25	25	25	25	25			
Number of species	19	17	16	19	19	14	17	18	16	13	12			
Ch.Ass: *Salicetum pentandro-cinereae*														
Salix pentandra	.	+	.	.	+	.	.	.	+	.	.	3	27	II
Salix cinerea	2	+	.	3	3	3	2	+	3	3	+	10	90	V
Salix aurita	.	.	.	+	+	.	.	.	+	+	.	4	36	II
Ch. *Alnion, Alnetalia, Alnetea*														
Alnus glutinosa	r	+	.	.	+	+	+	.	r	.	.	6	55	III
Ribes nigrum	+	.	.	.	r.1	+.1	r.1	+.1	2.2	+.1	2.2	8	73	IV
Solanum dulcamara	+.1	+.1	+.1	.	.	+.1	1.1	1.4	+.1	1.1	+.2	9	82	V
Calamagrostis canescens	+.1	.	.	.	+.1	.	.	+.1	.	.	.	3	27	II
Thelypteris palustris	+.1	+.1	.	.	.	.	.	.	.	.	.	2	18	I
Lycopus europaeus	.	.	+.1	.	.	.	.	.	.	.	.	1	9	I
Carex elongata	+.1	.	.	.	.	.	.	.	.	.	.	1	9	I
Ch. *Phragmitetea*														
Carex acutiformis	3.5	3.5	3.5	+.1	+.2	.	2.4	.	2.3	2.3	4.5	9	82	V
Phragmites australis	3.5	2.4	+.1	r.1	.	1.2	+.1	4.5	3.5	2.5	5.5	10	90	V
Equisetum fluviatile	+.1	+.1	.	.	.	+.1	.	+.1	.	.	.	4	36	II
Carex appropinquata	.	+.2	.	.	.	.	.	.	.	.	+.2	2	18	I
Poa palustris	.	.	.	+.1	+.1	.	.	+.1	.	.	.	3	27	II
Scutellaria galericulata	+.1	.	.	.	.	.	.	.	.	.	.	1	9	I
Ch. *Molinio-Arrhenatheretea*														
Filipendula ulmaria	2.2	2.2	2.2	2.2	2.2	2.2	2.2	2.2	+.2	+.1	.	10	90	V
Scirpus sylvaticus	2.5	2.5	2.1	.	.	+.2	2.2	+.2	+.1	.	+.2	8	73	IV
Lythrum salicaria	+.2	+.2	2.2	2.2	.	.	r.1	r.1	+.1	r.1	+.2	9	82	V
Lysimachia vulgaris	+.1	+.1	+.1	2.1	.	.	.	.	+.1	+.1	+.1	7	64	IV
Cirsium rivulare	r.1	.	r.1	+.2	.	.	.	r.1	.	.	.	4	36	II
Caltha palustris	.	.	.	.	r.1	.	.	.	.	.	.	1	9	I
Geum rivale	.	.	r.1	.	.	+.1	.	.	.	.	.	2	18	I
Myosotis scorpioides	.	.	.	r.1	.	.	.	+.2	.	r.1	.	3	27	II
Galium uliginosum	.	.	.	.	+.1	.	.	.	.	.	.	1	9	I
Ch. *Querco-Fagetae*														
Impatiens noli-tangere	+.1	+.1	1.1	+.1	2.4	1.4	1.1	1.4	1.1	2.5	+.2	11	100	V
Humulus lupulus	+.1	.	r.1	r.1	+.1	1.4	r.1	r.1	+.1	.	+.1	9	82	V
Fraxinus excelsior	.	.	+	+	.	.	r	r	.	.	.	4	36	II
Galium spurium	.	.	.	+.1	.	.	+.1	+.1	.	.	+.2	4	36	II
Anemone nemorosa	.	.	r.1	.	.	.	.	.	.	.	.	1	9	I
Paris quadrifolia	.	.	.	r.1	.	.	.	.	.	.	.	1	9	I
Accompanying species:														
Urtica dioica	.	.	.	3.4	5.4	2.2	+.1	1.2	r.1	+.1	.	7	64	IV
Cardamine amara	.	.	2.2	2.2	4.4	4.3	.	.	.	.	.	4	36	II
Chrysosplenium alternifolium	.	.	3.2	3.3	+.1	+.1	.	.	.	.	.	4	36	II
Ribes rubrum	.	.	.	.	.	.	r.1	+	2	+	.	4	36	II
Dryopteris carthusiana	.	.	.	.	r.1	.	r.1	.	.	.	.	2	18	I
Viola palustris	.	r.1	.	+.1	+.1	.	.	.	.	.	.	3	27	II
Eupatorium cannabinum	.	r.1	.	.	.	.	r.1	.	.	.	.	2	18	I
Equisetum arvense	+.1	+.1	.	.	.	.	.	.	.	.	.	2	18	I
Rubus idaeus	.	.	.	.	r	.	.	.	.	.	.	1	9	I

the cessation of meadow mowing the range represents 16 different types of phytocoenosis and aggregations of several plant species. At this stage the initial willow-scrub phase *Salicetum pentandro-cinereae* occupies the largest area, while the remainder of the area is grown up with a mosaic of aggregations of species like: *Filipendula ulmaria, Carex cespitosa, C. acutiformis, Lysimachia vulgaris, Salix cinerea, S. pentandra*, and small thistle-meadow patches (Figure 2.1).

Fig. 2.2. Transformations of meadow vegetation for 15 years. Succession phases: (I) meadow, (II) macroforbes community, (III) willow brushwood, (IV) brushwood community.

Floristic composition diversity

The initiation of a succession series from a floristically rich *Cirsietum rivularis* towards a carr (*Circaeo-Alnetum*) is influenced by the species diversity of the initial situation. The phytocoenoses that successively form and replace the thistle meadow, such as: *Lysimachio vulgaris-Filipenduletum* or *Filipendulo-Geranietum*, and willow scrub *Salicetum pentandro-cinereae* include common species. Their contribution to the successive phytocoenoses varies, and so does the manner of space colonization (Figure 2.2).

A characteristic succession of phytocoenoses in a carr sere could be observed in two cases: (1) when, as a result of the cessation of mowing, several or over a dozen species considerably increased in numbers, which led to the formation (within 5–6 years) of a polydominant meadow with the original species composition of the *Cirsietum rivularis* phytocoenosis, but with different quantitative ratios; (2) when in different meadow patches one or two species attained the status of dominants; the original homogeneous phytocoenosis was then replaced with several of its facies slightly differing in their floristic composition (Figure 2.3).

(1) In the former case the changes involved are changes in the quantitative relations between the populations of the species that make up the meadow community. The nature of these changes is manifested by the fact that, generally, the status of dominants is reached by those species which in regularly mown meadows are represented by a few

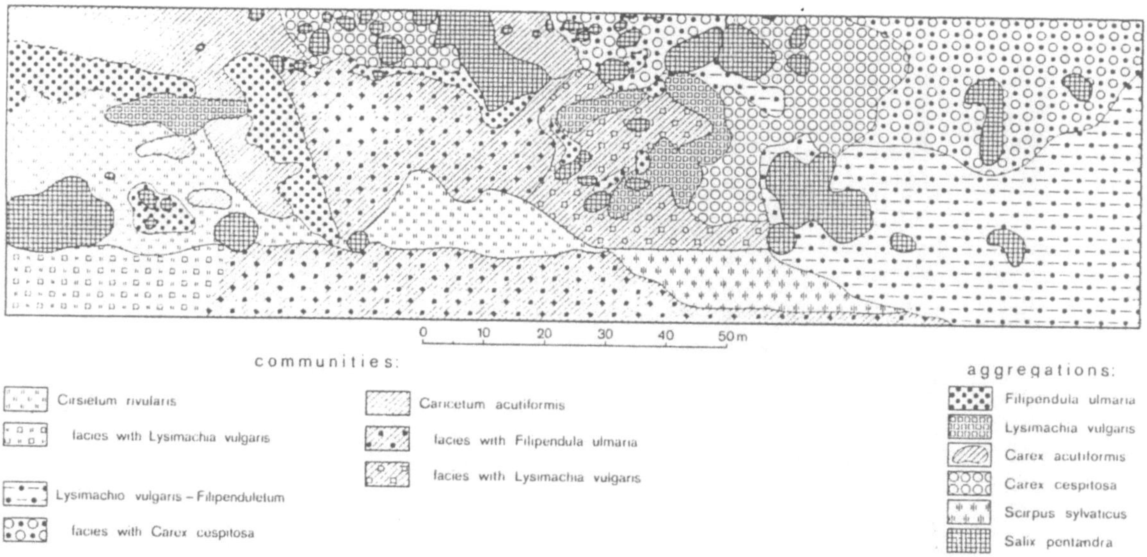

Fig. 2.3. Mosaic of communities and monospecific aggregations formed in the course of 15-year succession in unmown meadows (0.98 ha) (Falińska 1989a).

individuals. Then their proportion in a phytocoenosis, expressed by the percent cover, does not exceed 5—10%, but following the cessation of mowing it comes up to 30—50%. This group includes: *Filipendula ulmaria, Lysimachia vulgaris, Geum rivale, Carex cespitosa* and *C. acutiformis*. If the first two species become dominant and the other two subdominant, then within 7—8 years a meadow is replaced with herbaceous communities *Lysimachio vulgaris-Filipenduletum* or *Filipendulo-Geranietum*. The communities are in fact initiated in the period when *Filipendula ulmaria* increases its dominance, that is, during the first two years following the cessation of mowing. About 9—11 years after the abandonment of a meadow the herbaceous phytocoenosis appears to be fully established, but its durability varies. If during the prevalence of *Filipendula ulmaria* there do not appear *Salix cinerea* and *S. pentandra* individuals, then the herbaceous communities, with a slightly changed spatial- and species-structure, may persist in the same place for 15 years or more. In a situation where the growth in dominance of *Filipendula ulmaria* is accompanied by the appearance of numerous willow assemblages, a herbaceous phytocoenosis does not form, but there develops the initial phase of *Salicetum pentandro-cinereae* with a high percentage of *Filipendula ulmaria*. In such a situation a willow scrub community forms within 10—12 years, that is, it takes as long as the development of a *Lysimachio vulgaris-Filipenduletum* phytocoenosis in other meadow parts. Often both phytocoenosis types develop almost simultaneously in close vicinity. Then it is much more difficult to establish time- and space-boundaries between them, because they form mosaic patterns (Figure 2.1).

(2) If in different places of a meadow biochore different species attain the status of dominants, then there arise several facies of meadow phytocoenoses that occupy comparatively small areas, i.e., from 0.2 to 0.5 ha. Several years later each of them starts the development of a different type of phytocoenosis. For instance, a facies with *Carex acutiformis* develops within 5—8 years into *Caricetum acutiformis* growth, a facies with *Polygonum bistorta* into a *Polygono-Cirsietum palustris* phytocoenosis, a facies with *Lysimachia vulgaris* into a *Lysimachio vulgaris-Filipenduletum*, a facies with *Geranium palustre* into a *Filipendulo-Geranietum*. However, only in the last two is the willow scrub development initiated within a relatively short

Photo 2.1. Temporal changes in the floristic composition and spatial structure at the beginning of *Salix cinerea* expansion. Photo: J. B. Faliński

Photo 2.2. The strongly dominant species (inhibitors of succession): *Carex cespitosa* (top) and *Carex acutiformis* (bottom) in meadow 10 years after abandonment.

Photo: A. Domaszewicz

time (9—10 years). The other facies are not directly replaced with willows scrub, but pass through a number of intermediate stages, e.g. the *Polygono-Cirsietum palustris* is often replaced by *Carex cespitosa* or *Filipendula ulmaria* aggregations, and only some time later (9—10 years) do *Salix pentandra* and *S. cinerea* appear among the aggregations of the dominants. A meadow facies with *Carex acutiformis* and *C. cespitosa* persists the longest time, even when there form willow scrub around it (Figure 2.3, Photos 2.1; 2.2).

Role of tree biogroups in a forest succession

In regularly mown meadows there are patches where a few *Salix pentandra, S. cinerea* and *Alnus glutinosa* individuals are left. After several years they represent large aggregations made up of several tree species (*Salix, Betula* and *Alnus*). Around them there gradually appear forest herbaceous species, e.g.: *Anemone nemorosa, Lycopus europaeus, Solanum dulcamara, Peucedanum palustre, Impatiens noli-tangere* and *Cardamine amara*. A quick growth of those aggregations, due to their vegetative propagation (this particularly applies to the willows), and the appearance of new tree species leads, within a relatively short time (10—12 years), to the union of neighbouring biogroups and initiation of the development of a forest phytocoenosis (Figure 2.1).

CHAPTER 5

SEQUENCE OF SPECIES

Introduction

Floristic-composition diversity and variation with a progressing succession are often interpreted as the result of a dynamic equilibrium between colonization, continued presence and retreat of species. Species-turnover is often related to plant properties as: growth type, reproduction mode and life-history (Drury and Nisbet 1973, Horn 1976, Connell and Slatyer 1977). It is these properties that decide which species will be components of only one or several phytocoenoses succeeding one another in the course of succession. The following species group have been distinguished according to their presence during the succession processes in abandoned meadows (Figure 2.4): (1) species present in the initial or terminal succession phase; (2) species present in all succession phases, but varying in their proportions in the phytocoenoses; (3) species present throughout the succession process as a whole, but not in every succession phase (Table 2.5).

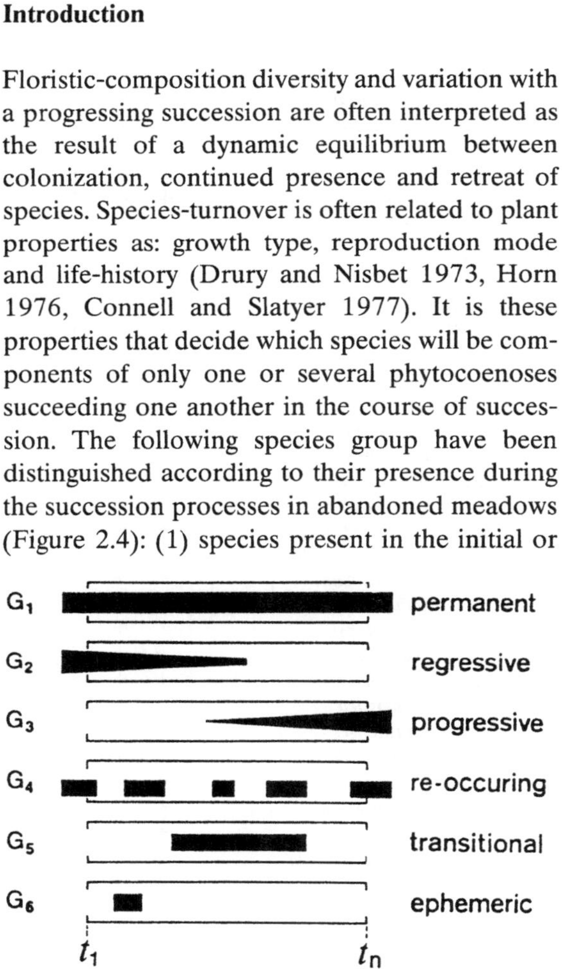

Fig. 2.4. Behaviour of species in a given phytocoenosis or series of successive ones in the course of an arbitrary time interval ($t_1 - t_n$) as a basis for distinguishing dynamic groups of species. The structure and dynamics of each community considered in a longer time interval may be determined according to participation of the dynamic groups of species and the exchange of species in the groups (especially $G_2 - G_5$). (Faliński 1986).

Species contribution and presence durability

Forest return to abandoned meadows is characterized by a considerable species richness throughout the succession series, though varying in abundance and diversity. In the succession process from a meadow to established scrub communities of the *Salicetum pentandro-cinereae* type 235 vascular plant species took part. Only 28 of them occurred exclusively in meadow, and 65 in forest phytocoenoses. The meadow-to-forest seral process is characteristic by the fact that in the *Cirsietum rivularis*, in which it is initiated, there occur a number of species that are also components of forest communities. Many of them are also present during the succession in progress, but their role in the structure of phytocoenoses and in their changes varies (Figure 2.2, Table 2.5).

Plant species respond to the cessation of mowing in different ways, namely: (a) species frequency and percentage in phytocoenoses continue to increase from year to year until willow scrub communities have formed, (b) species frequency and contribution to phytocoenoses drops, which leads on to a complete or partial retreat of populations from a particular area, (c) species frequency and percentage are subject to slight variations, i.e., a particular species is present throughout the succession in progress, but is only represented by a small number of individuals.

Table 2.5. Changes in the species composition in consecutive successional phases

No.	Species	Succession phases I	II	III	IV
1.	*Agrostis stolonifera*	+	·	·	·
2.	*Avenochloa pubescens*	+	·	·	·
3.	*Cardamine pratensis*	+	·	·	·
4.	*Carex canescens*	+	·	·	·
5.	*Carex spicata*	+	·	·	·
6.	*Carex leporina*	+	·	·	·
7.	*Carex pallescens*	+	·	·	·
8.	*Carex echinata*	+	·	·	·
9.	*Carex vulpina*	+	·	·	·
10.	*Leucanthemum ircutianum*	+	·	·	·
11.	*Cynosurus cristatus*	+	·	·	·
12.	*Festuca pratensis*	+	·	·	·
13.	*Hieracium pilosella*	+	·	·	·
14.	*Hieracium umbellatum*	+	·	·	·
15.	*Hypericum perforatum*	+	·	·	·
16.	*Juncus bufonius*	+	·	·	·
17.	*Juncus filiformis*	+	·	·	·
18.	*Juncus tenuis*	+	·	·	·
19.	*Leontodon autumnalis*	+	·	·	·
20.	*Leontodon hispidus*	+	·	·	·
21.	*Lolium perenne*	+	·	·	·
22.	*Polemonium caeruleum*	+	·	·	·
23.	*Polygala vulgaris*	+	·	·	·
24.	*Succisa pratensis*	+	·	·	·
25.	*Trifolium hybridum*	+	·	·	·
26.	*Trifolium medium*	+	·	·	·
27.	*Viola canina*	+	·	·	·
28.	*Veronica officinalis*	+	·	·	·
29.	*Agropyron repens*	+	+	·	·
30.	*Rhinanthus serotinus*	+	+	·	·
31.	*Anthoxanthum odoratum*	+	+	·	·
32.	*Briza media*	+	+	·	·
33.	*Campanula patula*	+	+	·	·
34.	*Cerastium holosteoides*	+	+	·	·
35.	*Epilobium montanum*	+	+	·	·
36.	*Equisetum palustre*	+	+	·	·
37.	*Holcus lanatus*	+	+	·	·
38.	*Lotus corniculatus*	+	+	·	·
39.	*Matricaria discoidea*	+	+	·	·
40.	*Dactylorhiza incarnata*	+	+	·	·
41.	*Parnassia palustris*	+	+	·	·
42.	*Plantago major*	+	+	·	·
43.	*Senecio paludosus*	+	+	·	·
44.	*Sonchus arvensis*	+	+	·	·
45.	*Achillea ptarmica*	+	+	·	·
46.	*Anthemis arvensis*	+	+	·	·
47.	*Bidens tripartita*	+	+	·	·
48.	*Inula britannica*	+	+	·	·
49.	*Polygonum lapatifolium* subsp. *incanum*	+	+	·	·
50.	*Rumex confertus*	+	+	·	·

Table 2.5. Continued

No.	Species	Succession phases I	II	III	IV
51.	*Rumex hydrolapathus*	+	+	·	·
52.	*Rumex thyrsiflorus*	+	+	·	·
53.	*Polygonum bistorta*	+	+	+	·
54.	*Ranunculus acris*	+	+	+	·
55.	*Alopecurus geniculatus*	+	+	+	·
56.	*Carex flava*	+	+	+	·
57.	*Carex hirta*	+	+	+	·
58.	*Carex lepidocarpa*	+	+	+	·
59.	*Glyceria fluitans*	+	+	+	·
60.	*Hieracium lactucella*	+	+	+	·
61.	*Juncus articulatus*	+	+	+	·
62.	*Pontentilla erecta*	+	+	+	·
63.	*Rumex acetosella*	+	+	+	·
64.	*Rumex crispus*	+	+	+	·
65.	*Trifolium pratense*	+	+	+	·
66.	*Achillea millefolium*	+	+	+	·
67.	*Agrostis tenuis*	+	+	+	·
68.	*Alchemilla monticola*	+	+	+	·
69.	*Alopecurus pratensis*	+	+	+	·
70.	*Carex nigra*	+	+	+	·
71.	*Carex gracilis*	+	+	+	·
72.	*Carex panicea*	+	+	+	·
73.	*Carex rostrata*	+	+	+	·
74.	*Cirsium arvense*	+	+	+	·
75.	*Potentilla palustris*	+	+	+	·
76.	*Deschampsia cespitosa*	+	+	+	·
77.	*Epilobium palustre*	+	+	+	·
78.	*Equisetum arvense*	+	+	+	·
79.	*Eriophorum angustifolium*	+	+	+	·
80.	*Galium uliginosum*	+	+	+	·
81.	*Juncus effusus*	+	+	+	·
82.	*Lathyrus pratensis*	+	+	+	·
83.	*Luzula campestris*	+	+	+	·
84.	*Mentha arvensis*	+	+	+	·
85.	*Molinia caerulea*	+	+	+	·
86.	*Phleum pratense*	+	+	+	·
87.	*Plantago lanceolata*	+	+	+	·
88.	*Poa pratensis*	+	+	+	·
89.	*Potentilla anserina*	+	+	+	·
90.	*Prunella vulgaris*	+	+	+	·
91.	*Ranunculus flammula*	+	+	+	·
92.	*Rumex acetosa*	+	+	+	·
93.	*Stellaria palustris*	+	+	+	·
94.	*Taraxacum officinale*	+	+	+	·
95.	*Trifolium repens*	+	+	+	·
96.	*Tripleurospermum inodorum*	+	+	+	·
97.	*Veronica chamaedrys*	+	+	+	·
98.	*Veronica longifolia*	+	+	+	·
99.	*Viola palustris*	+	+	+	·
100.	*Dactylorhiza maculata*	+	·	·	+
101.	*Galium mollugo*	+	·	+	+
102.	*Lysimachia thyrsiflora*	+	·	+	+

Table 2.5. Continued

No.	Species	Succession phases			
		I	II	III	IV
103.	*Dactylorhiza majalis*	+	+	·	+
104.	*Calamagrostis canescens*	+	+	+	+
105.	*Caltha palustris*	+	+	+	+
106.	*Carex acutiformis*	+	+	+	+
107.	*Carex cespitosa*	+	+	+	+
108.	*Carex appropinquata*	+	+	+	+
109.	*Carex vesicaria*	+	+	+	+
110.	*Cirsium palustre*	+	+	+	+
111.	*Cirsium rivulare*	+	+	+	+
112.	*Dactylis glomerata*	+	+	+	+
113.	*Equisetum fluviatile*	+	+	+	+
114.	*Filipendula ulmaria*	+	+	+	+
115.	*Geum rivale*	+	+	+	+
116.	*Lychnis flos-cuculi*	+	+	+	+
117.	*Lysimachia vulgaris*	+	+	+	+
118.	*Lythrum salicaria*	+	+	+	+
119.	*Myosotis scorpioides*	+	+	+	+
120.	*Poa palustris*	+	+	+	+
121.	*Ranunculus repens*	+	+	+	+
122.	*Scirpus sylvaticus*	+	+	+	+
123.	*Scutellaria galericulata*	+	+	+	+
124.	*Urtica dioica*	+	+	+	+
125.	*Anthriscus sylvestris*	·	+	+	·
126.	*Galeopsis bifida*	·	+	+	·
127.	*Galeopsis tetrahit*	·	+	+	·
128.	*Geranium palustre*	·	+	+	·
129.	*Phalaris arundinacea*	·	+	+	·
130.	*Poa trivialis*	·	+	+	·
131.	*Polygonum hydropiper*	·	+	+	·
132.	*Peucedanum palustre*	·	+	+	+
133.	*Galium palustre*	·	+	+	+
134.	*Lapsana communis*	·	+	+	+
135.	*Myosoton aquaticum*	·	+	+	+
136.	*Mycelis muralis*	·	+	+	+
137.	*Polygonum persicaria*	·	+	+	+
138.	*Viola epipsila*	·	+	+	+
139.	*Aegopodium podagraria*	·	+	+	+
140.	*Calamagrostis arundinacea*	·	+	+	+
141.	*Carex elongata*	·	+	+	+
142.	*Crepis paludosa*	·	+	+	+
143.	*Thelypteris palustris*	·	+	+	+
144.	*Galeopsis speciosa*	·	+	+	+
145.	*Iris pseudacorus*	·	+	+	+
146.	*Phragmites australis*	·	+	+	+
147.	*Solanum dulcamara*	·	+	+	+
148.	*Symphytum officinale*	·	+	+	+
149.	*Valeriana officinalis*	·	+	+	+
150.	*Vicia cracca*	·	+	+	+
151.	*Ribes nigrum*	·	+	+	+
152.	*Salix cinerea*	·	+	+	+
153.	*Salix pentandra*	·	+	+	+
154.	*Alisma plantago-aquatica*	·	·	+	+

Table 2.5. Continued

No.	Species	Succession phases			
		I	II	III	IV
155.	*Athyrium filix-femina*	·	·	+	+
156.	*Cardamine amara*	·	·	+	+
157.	*Equisetum sylvaticum*	·	·	+	+
158.	*Eupatorium cannabinum*	·	·	+	+
159.	*Galium spurium* var. *vaillantii*	·	·	+	+
160.	*Geranium robertianum*	·	·	+	+
161.	*Humulus lupulus*	·	·	+	+
162.	*Scrophularia nodosa*	·	·	+	+
163.	*Veronica beccabunga*	·	·	+	+
164.	*Hordelymus europaeus*	·	·	+	+
165.	*Chaerophyllum hirsutum*	·	·	+	+
166.	*Alnus glutinosa*	·	·	+	+
167.	*Ribes rubrum*	·	·	+	+
168.	*Rubus idaeus*	·	·	+	+
169.	*Anemone nemorosa*	·	·	+	+
170.	*Lycopus europaeus*	·	·	+	+
171.	*Adoxa moschatellina*	·	·	·	+
172.	*Ajuga reptans*	·	·	·	+
173.	*Asarum europaeum*	·	·	·	+
174.	*Galium odoratum*	·	·	·	+
175.	*Brachypodium sylvaticum*	·	·	·	+
176.	*Calla palustris*	·	·	·	+
177.	*Carex digitata*	·	·	·	+
178.	*Carex pilosa*	·	·	·	+
179.	*Carex sylvatica*	·	·	·	+
180.	*Carpinus betulus*	·	·	·	+
181.	*Chrysosplenium alternifolium*	·	·	·	+
182.	*Circaea alpina*	·	·	·	+
183.	*Cicuta virosa*	·	·	·	+
184.	*Cirsium oleraceum*	·	·	·	+
185.	*Convallaria majalis*	·	·	·	+
186.	*Corydalis solida*	·	·	·	+
187.	*Crataegus* sp.	·	·	·	+
188.	*Dactylis polygama*	·	·	·	+
189.	*Dentaria bulbifera*	·	·	·	+
190.	*Dryopteris austriaca*	·	·	·	+
191.	*Dryopteris filix-mas*	·	·	·	+
192.	*Dryopteris carthusiana*	·	·	·	+
193.	*Euonymus europaea*	·	·	·	+
194.	*Festuca gigantea*	·	·	·	+
195.	*Frangula alnus*	·	·	·	+
196.	*Fraxinus excelsior*	·	·	·	+
197.	*Gagea lutea*	·	·	·	+
198.	*Lamiastrum galeobdolon*	·	·	·	+
199.	*Geum urbanum*	·	·	·	+
200.	*Impatiens noli-tangere*	·	·	·	+
201.	*Lathraea squamaria*	·	·	·	+
202.	*Lathyrus vernus*	·	·	·	+
203.	*Maianthemum bifolium*	·	·	·	+
204.	*Mercurialis perennis*	·	·	·	+
205.	*Menyanthes trifoliata*	·	·	·	+
206.	*Milium effusum*	·	·	·	+

Table 2.5. Continued

No.	Species	Succession phases I	II	III	IV
207.	*Moehringia trinervia*	·	·	·	+
208.	*Oenanthe aquatica*	·	·	·	+
209.	*Oxalis acetosella*	·	·	·	+
210.	*Prunus padus*	·	·	·	+
211.	*Prunus serotina* (wildering)	·	·	·	+
212.	*Paris quadrifolia*	·	·	·	+
213.	*Gymnocarpium dryopteris*	·	·	·	+
214.	*Poa nemoralis*	·	·	·	+
215.	*Polygonatum multiflorum*	·	·	·	+
216.	*Polygonatum odoratum*	·	·	·	+
217.	*Pteridium aquilinum*	·	·	·	+
218.	*Ranunculus cassubicus*	·	·	·	+
219.	*Sorbus aucuparia*	·	·	·	+
220.	*Stachys palustris*	·	·	·	+
221.	*Stellaria holostea*	·	·	·	+
222.	*Stellaria nemorum*	·	·	·	+
223.	*Viburnum opulus*	·	·	·	+
224.	*Viola reichenbachiana*	·	·	·	+
225.	*Angelica sylvestris*	·	·	·	+
226.	*Equisetum hyemale*	·	·	·	+
227.	*Corylus avellana*	·	·	·	+
228.	*Picea abies*	·	·	·	+
229.	*Quercus robur*	·	·	·	+
230.	*Rubus saxatilis*	·	·	·	+
231.	*Salix aurita*	·	·	·	+
232.	*Salix cinerea* × *S. starkeana*	·	·	·	+
233.	*Populus tremula*	·	·	·	+
234.	*Tilia cordata*	·	·	·	+
235.	*Betula pendula*	·	·	·	+

Species names after Ehrendorfer (1973).

Table 2.5 contains a complete floristic list of the Reski Range. The following groups have been distinguished according to their proportions and duration of presence in the long-term succession process: (1) species that were present during the whole succession process and significantly changed their proportions in the phytocoenoses that were forming; (2) species that were present during the succession in progress, with their proportions in the developing phytocoenoses showing significant changes as their location changed, that is, they retreated from some places and appeared in others; (3) species which occurred in either the initial or the terminal phase of succession, that is, they significantly contributed to the building up of phytocoenoses within a short time, but retreated from the area at an equally fast rate.

Species status during a succession

During an abandoned-meadow vegetation succession, subject to changes is not only the number of species, but also the status of particular species in the successive communities. In a meadow that is periodically mown, some species (Table 2.5, Figure 1.2) represent low proportions in the phytocoenoses (1—5% of the cover), while in the herbaceous and scrub communities of an abandoned meadow they become dominant (50—70% of the cover).

Dominance versus floristic diversity and richness

A characteristic feature of a forest succession course in rich and moist meadows is the fact that: (a) although a dominance increase somewhat reduces the number of species, a floristic richness is maintained throughout the succession process. The number of species varies between 66 and 124, and when the vegetation on the range attains the stage of patchiness the number comes up to 175; (b) 12 of 124 species making up the meadow sward reach, in different succession periods, the dominant or subdominant status. As a rule, these species are common to meadow, scrub and forest phytocoenoses; throughout the succession there persists a species diversity with regard to the biological form (annual, biennial, perennial), ecological amplitude (of forest, meadow, rush swamps, etc.) and to taxonomic groups.

Although in the seral changes of meadow vegetation towards woodland communities dominance is a characteristic phenomenon, a significant reduction in floristic richness (down to just over a dozen or so species) occurred only in some meadow parts, that is, locally, and only as a temporary situation. In general, the dominance of one species eliminates only a fragment of a specified phytocoenosis type, and almost simultaneously initiates the development of another type of phytocoenosis.

Cessation of mowing brings about a rapid growth in numbers of some of the 142 species present. The process begins almost as early as the second growing season following the abandonment of meadows, and within 4—5 years there arises large-sized monospecific aggregations whose area varies between 500 and 2,000 m^2. Distinct for such characteristics are the following species: *Filipendula ulmaria, Lythrum salicaria, Carex acutiformis, C. cespitosa, Lysimachia vulgaris* and *Scirpus sylvaticus*. Several years after the cessation of mowing in different parts of a meadow these species entirely or partly eliminate the characteristic *Cirsietum rivularis* meadow species combination. The later fate of such aggregations is variable. They may continue to occupy their areas for several years (5—9), even when new phytocoenoses are forming around them.

Monospecific aggregations areas are often similar in size to the biochores of the phytocoenoses that arise during a succession (0.2—0.5 ha). Several years later some of them are replaced with the aggregations of another species. The following meadow-to-forest dominant species population sequence has been observed:

(a) *Filipendula ulmaria, Carex acutiformis, C. cespitosa, Salix cinerea*;
(b) *Carex acutiformis, Lysimachia vulgaris, Salix cinerea*;
(c) *Carex acutiformis, Lythrum salicaria, Alnus glutinosa*;
(d) *Filipendula ulmaria, Salix cinerea* and *S. pentandra*;
(e) *Cirsium rivulare, Polygonum bistorta, Scirpus sylvaticus, Filipendula ulmaria, Salix cinerea.*

In the period of successive turnovers of dominants there form combinations, each consisting of several species (2—5), that cannot be classified as belonging to any known types of phytocoenosis. This happens when aggregations of dominants form mosaic patterns that precede the development of a forest phytocoenosis, and are made up of such species as: *Filipendula ulmaria, Carex cespitosa* and *Salix cinerea* (Figure 2.3) or *Lysimachia vulgaris, Carex acutiformis* and *Salix cinerea.* Such combinations last from 8 to 12 years.

CHAPTER 6

COURSE OF SECONDARY SUCCESSION

Succession stages and phases

The initiation of succession in abandoned meadows and the course of its first phases are influenced by the floristic richness and diversity of the phytocoenosis *Cirsietum rivularis*. A diagram of the carr sere (*Circaeo-Alnetum*) has been presented in Figure 2.5, and the species taking part in the succession have been listed in Table 2.5. The diagram has been worked out on the basis of data from researches continued for 15 years in permanent plots located in meadows immediately

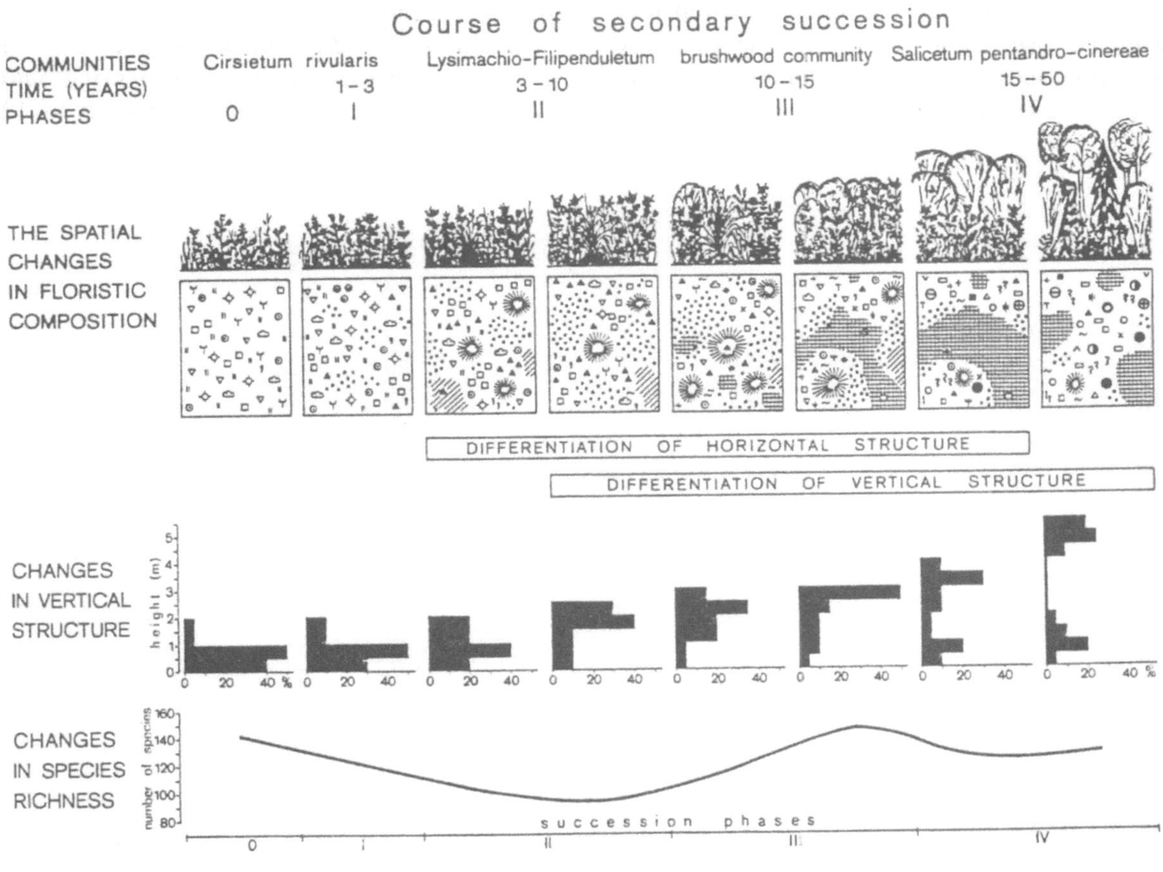

Fig. 2.5. Diagram of vegetation succession in abandoned meadow.

after the cessation of mowing (i.e. in 1974), and in meadows abandoned in different years between 1960 and 1974.

Over the 15 years observation period, data could thus be obtained also on the course of later succession phases. In a situation, for instance, where in 1974 in some parts of the Reski Range willow scrub dominated (a meadow abandoned in 1960), in 1989 willow-alder brushwood community, or *Salicetum pentandro-cinereae* phytocoenosis occurred, e.g. at the woodland boundary (Figure 2.1). From that it has been estimated that the initial phase of the woodland community lasted about 30—35 years at the reserve borderline, and about 40—45 years, at some distance from it.

Applying a comparative approach, 7 phases have been identified in the carr series, and grouped according to three stages: initial, temporary, terminal.

The initial stage comprises the first 2 phases that lasted about 9 years. During that period there occurred significant species-and spatial-structure changes in the *Cirsietum rivularis* phytocoenosis. Of a total of 142 species recorded during the "zero phase", 12 became dominant, half of them retreated, and there arose a macroforb-community (*Lysimachio vulgaris-Filipenduletum*) meadow with a mosaic structure made up of species like: *Filipendula ulmaria, Carex cespitosa, C. acutiformis, Lythrum salicaria* and *Lysimachia vulgaris*. Amidst those aggregations there appeared the first willow individuals.

The temporary stage comprises of two consecutive phases (III and IV) during which there forms a specific spatial complex consisting of a meadow and herbaceous communities and willow scrub aggregations in which the first tree species appear, e.g. *Alnus glutinosa, Frangula alnus* and *Betula pendula*. In the study area this stage lasted about 15 years, counting from the meadow abandonment time.

The terminal stage of succession comprises of two phases (V and VI) during which there form willow-alder brushwoods. The stage can be regarded as the initial stage of the formation of the *Circaeo-Alnetum* woodland community (Phase VII).

Vegetation structure differentiation during succession

A feature noticeable during a progressing succession in abandoned meadows is a gradual increase in the structure complexity of the vegetation; a growth of the number of layers and patchiness. Though much of the original species richness is lost during the first succession phases (Figure 2.5), there occurs a considerable diversification of the vegetation structure (vertical and horizontal).

In the "zero" phase of succession, the meadow spatial structure is as follows. Members of the *Cirsietum rivularis* phytocoenosis make up a multispecific mosaic, with individual plants occurring solitarily or in small assemblages (Figure 2.2). The phytocoenosis consists of three layers. The near-ground layer is made up primarily of seedlings and of the lowermost leaves of many herbaceous plants. The second layer consists of numerous herbs, 30—50 cm tall, like *Ranunculus acris, Lychnis flos-cuculi, Rumex acetosa, Geum rivale, Galium uliginosum, Viola palustris, Myosotis scorpioides, Caltha palustris, Lathyrus pratensis, Festuca pratensis* and *Deschampsia cespitosa*. The third, the tallest layer, is made up of *Cirsium palustre* and *C. rivulare* (1.0—1.2 m).

In the first period (Phase I) following the abandonment of meadows, important changes take place, above all in the third layer of a phytocoenosis. Numerous individuals of *Lysimachia vulgaris, Lythrum salicaria, Cirsium palustre, C. rivulare*, and particularly of *Filipendula ulmaria* attain a height of 1.0 to 1.5 m. In the horizontal structure of a phytocoenosis the proportions also increase considerably of these species in the cover — from 1—5% to 20—30%. The meadow vegetation structure is aggregative.

A further complication follows the formation of large-area aggregations of some meadow herbs (Phase II). The stratification and patchiness then increase considerably of the macroforb-community phytocoenosis. The tallest are then *Cirsium palustre* individuals (about 2 m), especially on the fringes of *Filipendula ulmaria* aggregations. Other herbs reach heights of 1.00—1.40 m. There forms yet another, the third plant layer, of a high density and which reaches a height of 50 cm or less.

The increasing role of tall perennials is followed by a reduction of the number of species making up the phytocoenosis. At that time the species- and spatial-structure of the macroforb-community phytocoenosis is determined primarily by three species: *Filipendula ulmaria, Lysimachia vulgaris* and *Lythrum salicaria*. The tendency of plants to aggregate becomes more intensive (Figure 2.5).

Filipendula ulmaria often forms monospecific one-layer aggregations. The individuals are then 100—120 cm tall. Among them, especially on the margins of the aggregations, there sporadically occur individuals of *Geum rivale, Lysimachia vulgaris, Cirsium rivulare, Lythrum salicaria* and *Carex cespitosa*. The highest density (about 3000 shoots per 200 m^2) and the highest proportions in herbaceous phytocoenoses of *Filipendula ulmaria* can be found 8—10 years after the abandonment of meadows.

The herbaceous phytocoenosis structure is further complicated when there appear willows among the herbaceous aggregations (Phase III). After some time large-area herbaceous aggregations, particularly those of *Filipendula ulmaria*, undergo fragmentation, and the gaps thus formed are inhabited by willows. Willow individuals quickly grow in size, and within several years they form large aggregations (Figure 2.2).

There arises a characteristic spatial complex of herbaceous, sedge-marsh and willow-scrub-phytocoenoses (Phase IV) that can last several or a dozen or so years (Figure 2.3). Vegetation development towards a forest is clearly accelerated if there appear other tree species, e.g. *Alnus glutinosa, Betula pendula* and *Frangula alnus*, among the willows (Phase V).

Almost simultaneously around the tree bio-groups that are forming, woodland herbs appear, like *Anemone nemorosa, Impatiens noli-tangere, Cardamine amara, Solanum dulcamara* and *Lycopus europaeus*.

Species like *Filipendula ulmaria* and *Carex cespitosa* represent comparatively large proportions in the above floristic composition. Besides, species which originally occupied the particular area, that is, *Cirsium palustre, C. rivulare, Lychnis flos-cuculi, Ranunculus acris, Lysimachia vulgaris* and *Lythrum salicaria*, appear again on the tussocks of *Carex cespitosa* (Figures 2.2; 2.3).

A highly mosaic-like multi-layer vegetation pattern arises that initiates the formation of willow-alder brushwoods (Phase VI) or a young forest (Figure 2.1). Further vegetation differentiation leads on to the formation of a storeyed pattern typical of woodland communities, that is, the phytocoenosis is made up of a tree-stand storey, a shrub storey and a herbaceous storey. In the tree storey, three distinct substoreys can be distinguished: the uppermost substorey is made up of *Betula pendula* and *Salix pentandra*, the lower one of *Alnus glutinosa* and *Frangula alnus*, and the lower-most one of *Salix cinerea*. Present in the shrub-storey are juvenile individuals of all the above species, as well as of *Picea abies*.

A fully layered and mosaic-like structure of a forest phytocoenosis (Phase VII) is attained at the end of a succession and formation of a carr (*Circaeo-Alnetum*) community. On the basis of other studies (Faliński 1986) it may be assumed that the formation of such a community takes about 250 years.

PART 3

BIOLOGICAL MECHANISMS OF SPECIES TURNOVER

CHAPTER 7

INDIVIDUAL AND SUCCESSION

Introduction

With a progressing succession in abandoned meadows there appear new species, while those that were present there before retreat. The process is governed by two mechanisms: (a) the free-space colonization mechanism; (b) a mechanism that enables species to survive or leads to their retreat. These mechanisms are closely related to the morphological and biological properties of plants, particularly their growth type and reproduction mode.

Plants possess many traits that enable them to settle in habitats with diverse conditions, and become permanent or temporary components of different phytocoenoses. As emphasized by Harper (1977, 1980), plants can realize individual success in two ways: (1) by growing as tall as possible and shading their neighbours, and thus winning the competition for light, or (2) colonizing as large an area as possible, and thereby obtaining as much food supplies and water as possible. The two strategies depend on the long lasting apical growth, due to which individuals can not only build up their bodies, but also restore their structures lost over their whole life-history. Their specific properties enable plants to quickly and durably colonize free spaces, incorporate themselves into existing species combinations, and displace other species. The manner of colonization depends on the type of reproduction and growth of the particular species. For instance, repent plants can colonize new areas by a close front or by individual rhizomes penetrating into spaces among other plants, whereby they minimize inter-individual and interclonal contacts. A newcomer at first separates the individuals of other species already present in a particular space and then grows and either displaces them or becomes incorporated into the existing combination. The former growth mode is referred to as a "phalanx-" (consolidated), the latter as a "guerilla-" type (exploration) of growth (Harper 1986).

The fact that a plant species, or even an individual can initiate the phytocoenosis-restructuring and species-turnover processes makes one try to answer the following questions: (1) What individual plant properties pre-determine the vegetation-change promoter or inhibitor status of a species? (2) Is the species-turnover affected by: (a) the retreating and space releasing by organisms due to their senescence, (b) an alteration of the conditions of a habitat by one species to such an extent that its individuals can no longer grow in it, but other species can, (c) providing shelter for species colonizing a particular area or returning to it (e.g. forest species)?

The description to follow will take into account the characteristics of those plants whose variation of traits in relationship to age and habitat conditions plays the role of biological mechanisms of species turnover in the course of a succession.

Polycormonal growth and species turnover

The intensiveness and extent of the action of individuals on their habitat and neighbours vary, because over their long life-history their architecture changes, and the area occupied by them increases (Figure 3.1).

Depending on the species biology, individual

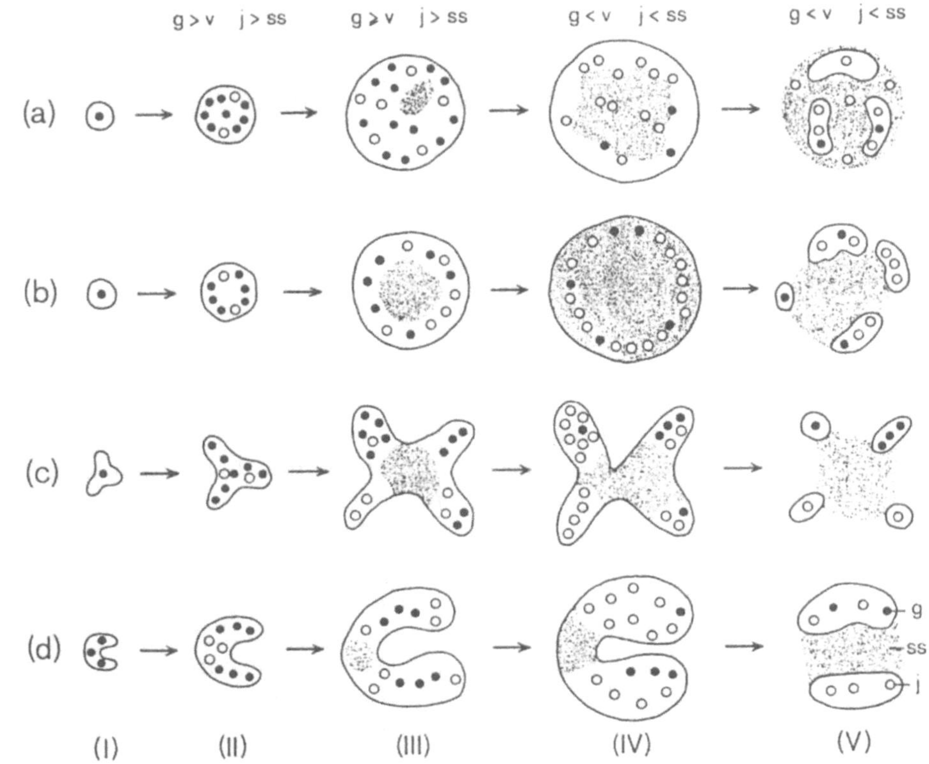

Fig. 3.1. Diagram presentation of growth and senescence of *Filipendula ulmaria*. The shape of the clone, size of the area occupied, the number and distribution of rosettes and shoots change during the 8—12 year growth cycle of the clones. The phases of clone growth (I—V) are described in the text. The degree of modification of clone characters varies depending on the neighbourhood; (a), (b) — without competitors; (c), (d) — with competitors (sedges, willows); g — generative shoots; j — rosettes; ss — senile parts.

age, and habitat conditions, an organism may consist of about a dozen or several dozen, and in some cases as many as several hundred, shoots. The rate of increase in shoot numbers varies, depending on the growth type and habitat conditions. In the first life period it usually follows the geometrical progression rule, and only after several years does it become stabilized at a certain level (Figure 3.2). Changes that occur in later periods are of the type of fluctuations. Though this plant group is characterized by a considerable structure diversity and individual growth plasticity, there is evidence to indicate that a polycormone-type organism possesses certain group features, like those that arise when many individuals co-occur in a particular habitat. The features in question are: spatial structure, age structure, age-stage structure and shoot abundance dynamics (Photo 3.1). The principal characteristics of polycormones can be defined as follows:

(1) they are plant organisms made up of many-rooted shoots whose abundance varies over their long-term life-history, an abundance-growth phase being followed by an equilibrium stage and a decline (Figure 3.2);
(2) the modules that make up a polycormone are usually uneven-sized, uneven-aged, and/or of uneven growth stages. They may consist of shoots representing various forms, e.g. rosettes, upright shoots and repent shoots;
(3) shoot distribution within a polycormonal area can be uniform or aggregative, possibly variable over the long-term growth cycle, most

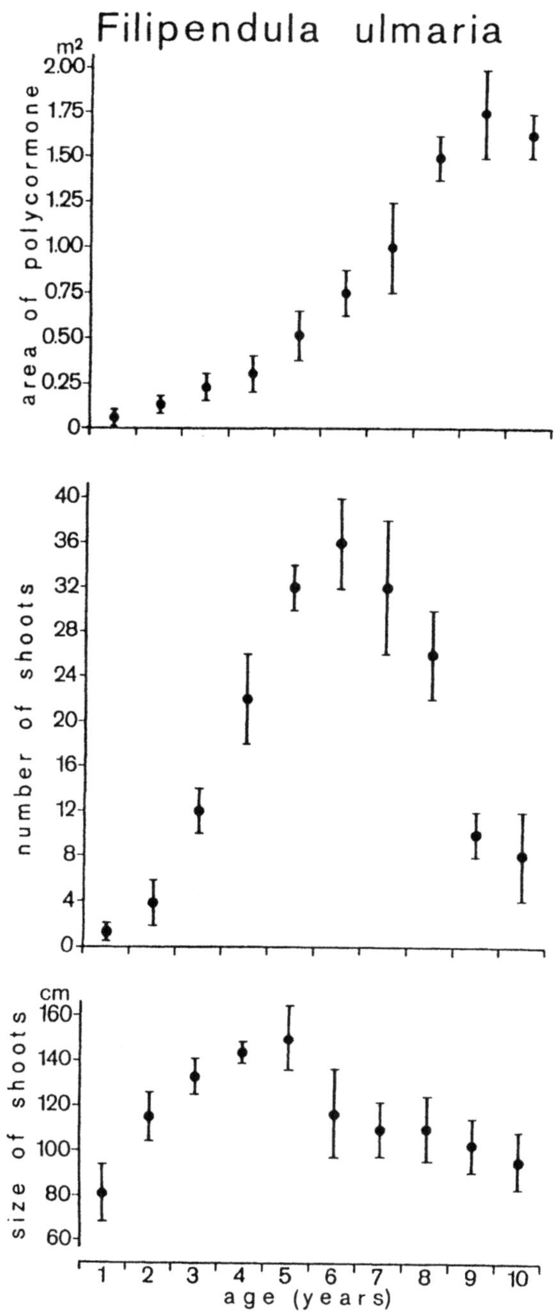

Fig. 3.2. Variation in individual characters with age.

often from an even-distribution pattern to an aggregative one. Shoot density in a polycormone varies, especially between its central and peripheral parts (Figures 1.3; 3.1);

(4) polycormones usually reproduce by two modes (sexual and asexual), their vegetative propagation being affected in many ways: by the separation of only the peripheral fragments or by the disintegration of the whole of a polycormone. The generative-to-vegetative reproduction relationship often changes during the life period of a polycormone (Photos 3.2; 3.3).

Development of polycormonal forms and their permanence

In the history of a polycormone five separate phases can be distinguished, as exemplified by *Filipendula ulmaria* (Figure 1.3):

(I) *Phase of an individual's establishment in a habitat.* The process lasts 2 or 3 growing seasons. During that time the leaf rosette and the root system form. In the second or third growing season a number (10–20%) of individuals produce 1 flowering shoot each, but they seldom reach the fruiting phase.

(II) *Phase of individual growth in size and integration.* Under natural conditions an individual's growth in size by shoot and rootstock-branch multiplication continues up to life year 5 or 6, and in an experimental garden up to life year 9 or 10. At each shoot there appear 2, or up to 5, rejuvescence buds which initiate the growth of new modules in the immediate vicinity. It is at this stage that a polycormone attains a compact structure and exclusive use of its area (Figure 3.1).

(III) *Phase of weakening of an individual's integrity and loss of exclusive use of its area.* Senescent (9 years of age) polycormones still continue to multiply modules, primarily on the periphery of their area. In the middle of the area the inter-shoot distances increase from 5 to 10 cm. New modules appear only in some fragments of a polycormone, where they form assemblages, each made up of 2–6 shoots, among which there appear

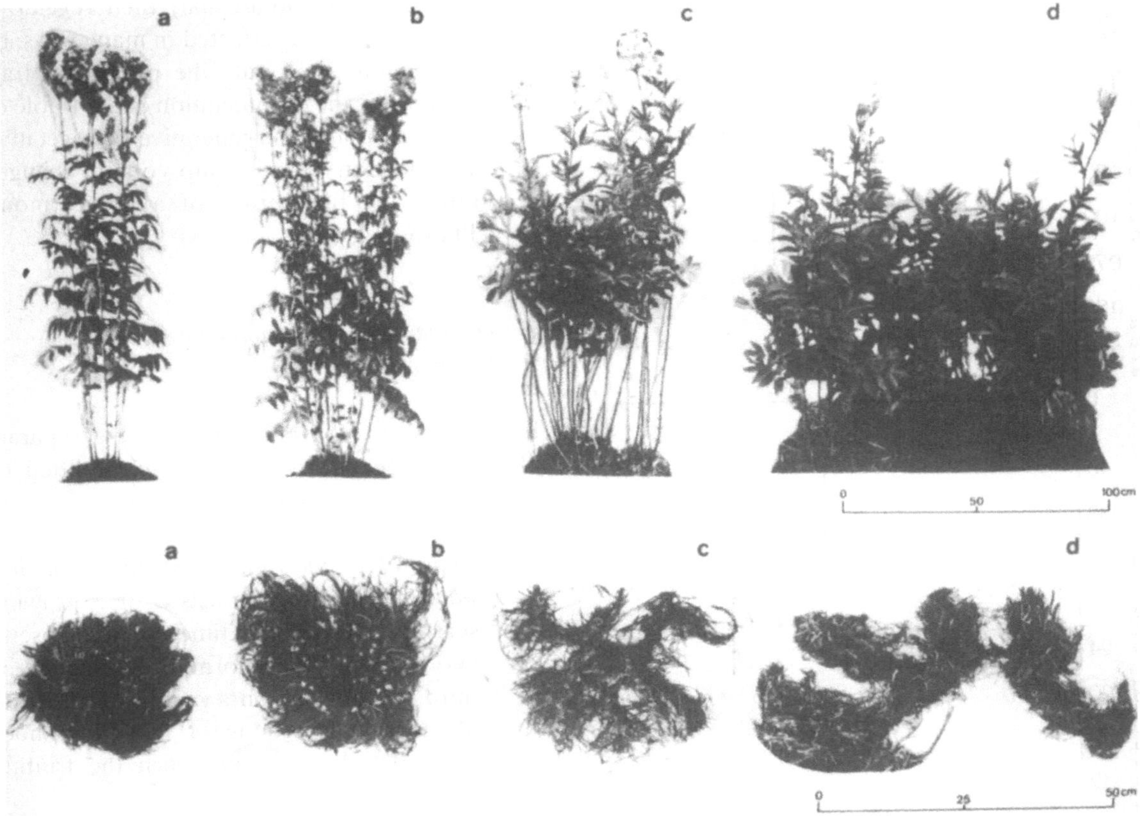

Photo 3.1. Age-related variation in individual characters of *Filipendula ulmaria*: a) 2-year-old individuals, b) 3-year-old individuals, c) 4-year-old individuals, d) 8-year-old individuals.

Photo: J.J. Faliński

empty spaces, due to which the structure of the polycormone loses its compactness. Simultaneously, the rate of production of renewal-buds, near the shoots, continues to decrease. At that time the ratio in the different parts of a polycormone is as follows: 1:1, 1:0.5, 1:0.1, and 1:0.

(IV) *Phase of loss of an individual's integrity.* The proportion of senescent segments increases, and so does the amount of necrotic plant parts in the polycormonal area, which prevents or limits the formation of renewal buds. Even if such buds do appear near some shoots, their chance to initiate the growth of new modules is almost nil, because they are overlain by a thick (10—30 cm) necromass layer (Table 3.1). The restriction of the rejuvenescence processes, particularly in the central portion of an individual, leads on to its gradual necrosis or disintegration (Phase V).

Architecture

Species whose individuals grow in size to form polycormones do not represent a uniform group in respect to their structure. Among them, individuals can be found formed from several or several hundred shoots that develop at some distance from, or close to each other, to form loose or

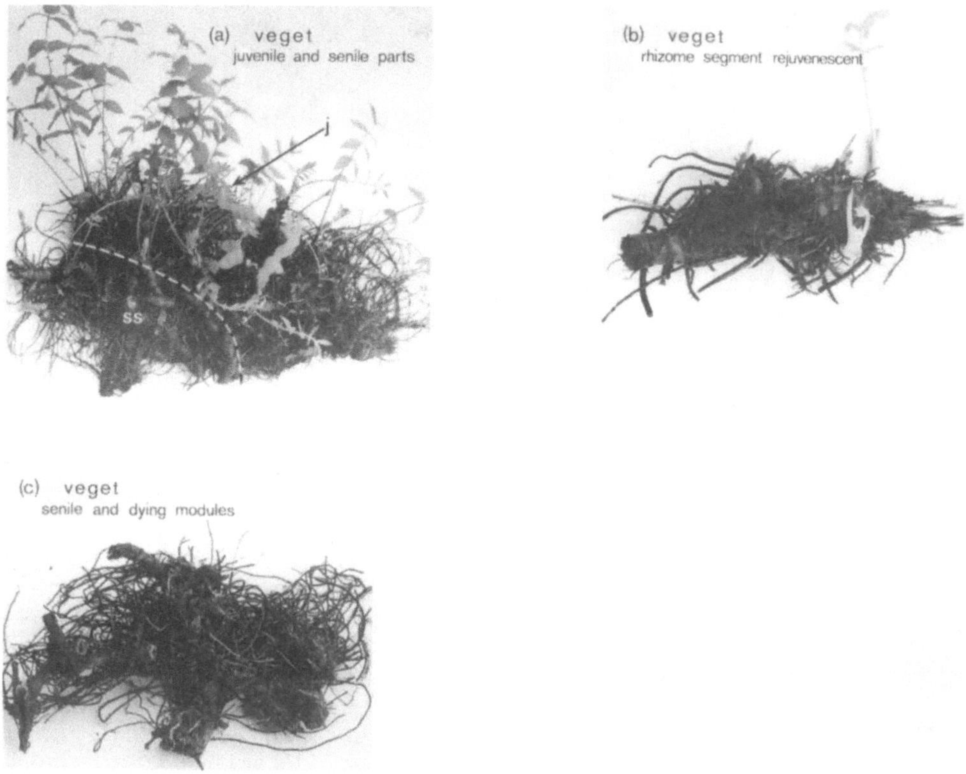

Photo 3.2. Vegets at different stages of renewal, senescence and necrosis: a) a veget in which renewal (j) dominates over necrosis (ss), a high probability of survival and growth; b) a rejuvenescent veget, survival probable at a low competition; c) veget affected by necrosis with a low chance to survive and grow.

Photo: J. J. Faliński

Table 3.1. Variation in individual characters with age in *Filipendula* populations ($\bar{x} \pm SD$)

Phases of growth	*n*	Age (year)	Number of shoots/clone	Height of shoots (cm)	Distance between shoots (cm)	Number of seeds/shoot	Number of renewal buds/shoot	Necromasse tall (cm) within clone
I	50	1–2	1.06 ± 0.48	91.24 ± 4.3	2.41 ± 0.24	319.30 ± 54.20	2.09 ± 0.48	1.50 ± 0.92
II	50	3–4	17.12 ± 0.98	102.35 ± 4.08	4.0 ± 1.16	844.21 ± 55.80	3.80 ± 0.85	6.12 ± 2.40
III	50	5–6	35.28 ± 2.34	122.47 ± 10.21	8.31 ± 1.43	2724.0 ± 529.13	2.09 ± 0.48	16.45 ± 6.28
IV	30	7–8	25.14 ± 9.57	88.25 ± 10.18	19.50 ± 3.12	1570.12 ± 123.62	0.88 ± 0.02	18.92 ± 9.12
V	20	9–10 and older	20.0 ± 10.08	62.0 ± 8.74	25.75 ± 8.80	341.90 ± 13.0	0.25 ± 0.18	15.82 ± 2.91

compact multi-shoot structures, which in turn determines the size of the area occupied and way of its filling (Figures 3.3; 3.4). Though their diversity in this respect is tremendous, they can be included in two types: monocentric and polycentric.

In monocentric polycormones the roots, shoots and renewal buds of an individual are concen-

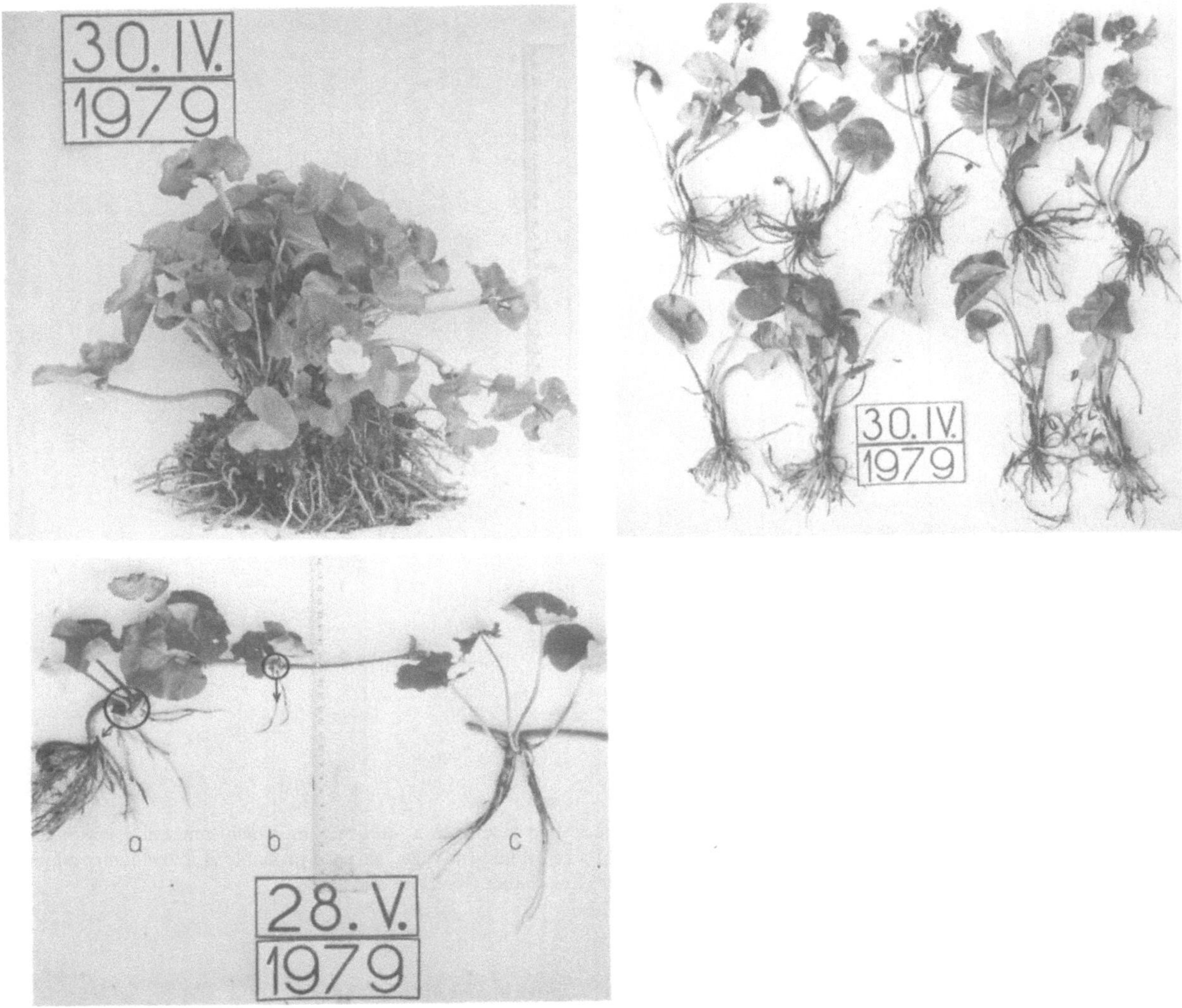

Photo 3.3. A 4-year-old clone of *Caltha palustris* (top) and all its particular units — ramets (bottom). Development of new individuals from different vegetative propagules: gemmae (a, b, c).

Photo: J. J. Faliński

trated in one place, especially in the first ontogenetic period. As time passes, the centre of growth becomes necrotic, while new shoots and roots grow and form peripheral parts of the individual. Each polycentric polycormone possesses several well-outlined growth centres (Photo 3.4). Growth centres represent units, to some extent autonomous within an individual, interconnected by common rhizomes or creeping shoots. Plants with such morphological features show a very wide variation of shape over the long-term individual life-history. In the initial period of their life individuals as a rule represent the monocentric structure type, whereas in later periods each of them forms several visible centres, and become polycentric.

A polycormone that has lived a long time and continued to produce rejuvescente buds may return to the monocentric structure, that is, become secondarily monocentric if some of its centres are affected by necrosis.

Similar processes and their effects could be

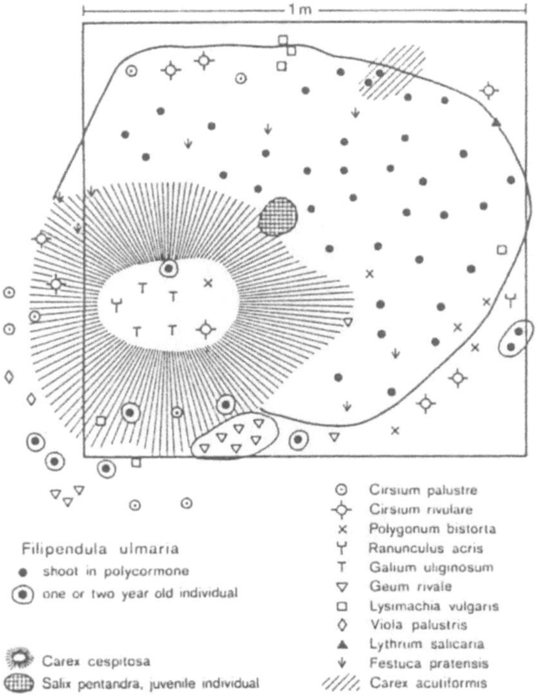

Fig. 3.3. Architecture of an 8-year-old clone of *Filipendula ulmaria*. Within the clone, the distances between shoots and places vacated by dead parts increase with clone age. The loose structure of the clone permits colonization by other species. It is representative for phase IV.

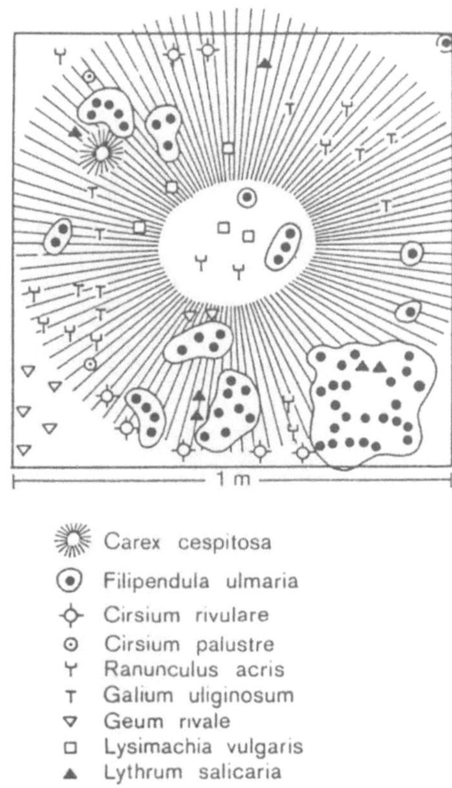

Fig. 3.4. Species turnover 14 years after abandonment of the meadow. *Filipendula ulmaria* returned as a colonizer in the clump of *Carex cespitosa. F. ulmaria* had occupied this place in early successional stage.

observed in populations of species that took part in a long-lasting succession. In the period immediately following the cessation of meadow mowing, populations of *Filipendula ulmaria, Lysimachia vulgaris* and *Lythrum salicaria* were dominated by monocormonal individuals, whereas 3–4 years later most of them formed monocentric entities. With advancing age they multiplied their shoots and grew unevenly in different directions, forming polycentric entities. The same process simultaneously affected individuals of many species, which intensified the mechanical interactions among them, due to which shoot assemblages split up, and those that were most distant from the polycormonal centre became separated. Then again monocentric individuals prevailed (Figure 3.1). The process lasted 12–15 years, and its termination coincided with the initiation of forest phytocoenoses.

The most variable over the long-term growth-history of polycormonal plants are the following characters: the number as well as height of shoots, and the size of area and way of its filling, especially in the case of species like *Filipendula ulmaria, Iris pseudacorus* and *Lythrum salicaria* (Tables 3.1; 3.2).

Depending on their age and habitat conditions, polycormones vary in their shape and in the size of the area occupied by them. They are most often circular, oval or kidney-shaped, and occupy areas of 0.10 to 0.25 m^2, less often 0.5 or 1.0 m^2 (Figure 1.7). Due to this, they can form polydominant communities.

Shoot distribution in a polycormone area

Under favourable conditions individuals of polycormonal species increase the number of shoots

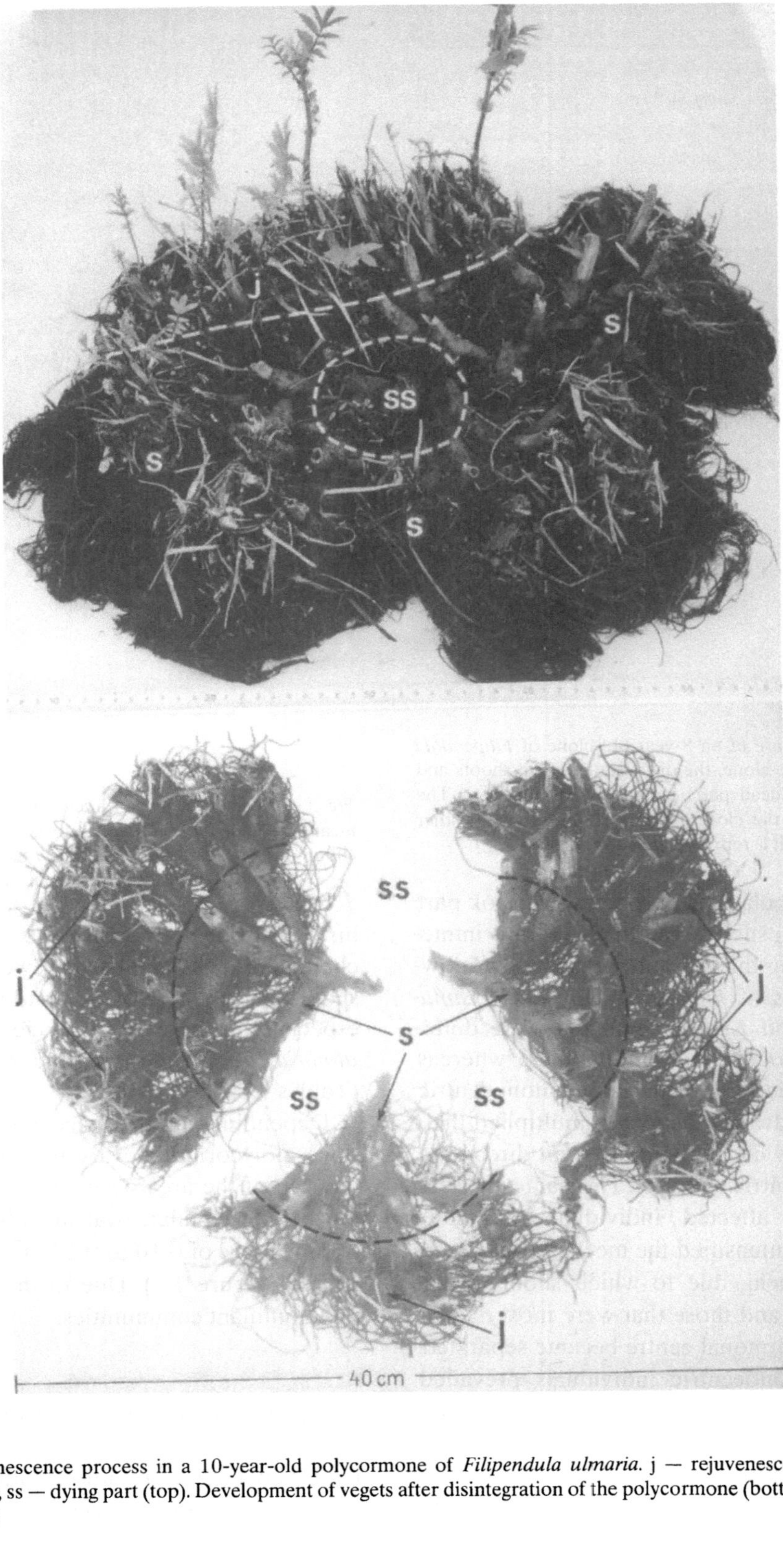

Photo 3.4. The senescence process in a 10-year-old polycormone of *Filipendula ulmaria*. j — rejuvenescent part, s — senil part, ss — dying part (top). Development of vegets after disintegration of the polycormone (bottom).
Photo: J. B. Faliński

Table 3.2. Variation in individual characters with age in population *Iris pseudacorus*

Age (years)	Area (m^2)	No. of rosettes	Height of rosettes (cm)	No. of shoots	Height of shoots (cm)	Biomass (*g*)
1	0.020 ± 0.09	4.57 ± 0.83	59.93 ± 3.46			28.59 ± 12.20
2	0.102 ± 0.82	8.62 ± 4.21	110.41 ± 3.37	1.44 ± 1.09	75.36 ± 7.70	83.55 ± 25.62
3	0.189 ± 0.21	29.00 ± 2.15	115.11 ± 2.44	4.00 ± 2.03	78.21 ± 6.20	795.00 ± 122.05
4	0.328 ± 0.36	109.00 ± 11.00	107.17 ± 1.64	12.50 ± 3.21	65.85 ± 7.52	1334.45 ± 206.20
5	0.928 ± 0.19	211.20 ± 16.21	113.46 ± 2.51	13.00 ± 1.96	79.24 ± 1.37	3488.0 ± 329.67
6	1.250 ± 0.15	149.15 ± 11.60	126.44 ± 4.00	7.62 ± 1.86	83.00 ± 6.46	2556.0 ± 217.90
7	0.586 ± 0.12	93.03 ± 16.28	108.75 ± 6.26	11.29 ± 2.11	43.17 ± 4.43	1918.16 ± 160.05
8	0.252 ± 0.11	62.18 ± 11.20	105.64 ± 3.21	2.61 ± 1.02	55.44 ± 4.0	672.05 ± 190.28

according to the geometric-progression rule, at least up to life year 6. In older, i.e., in age groups 7 to 10 years, further growth proceeds in at least three ways, viz.: (1) a gradual growth of the number of shoots continues until the disintegration phase or death of an individual, (2) the number of shoots becomes stabilized at a level attained in life year 5, usually 18—22 shoots, (3) the number of shoots is reduced to the original level, i.e., 2—4 shoots; in this form individuals can continue to exist for several years, especially when sedge and willow competition intensifies in abandoned meadows.

In very young polycormones, i.e., up to life year 3 and 4, shoots grow close together (2.41 ± 0.24), and they vary little in this respect. In the phase of optimum growth of the polycormones, i.e., between life years 4 and 6, the inter-shoot distance increases at first to 4.0 ± 1.16 and later on to 8.31 ± 1.43. In some polycormones, especially the older ones (7—19 years of age), an aggregative shoot distribution was found in areas where shoots grew very close to each other (1—2 cm). Distances between shoot aggregations ranged from 15 to 30 cm (Table 3.1). They were separated with dead plant remnants, due to which they seemed to be separate entities, but were interconnected by senescent rhizomes that formed a scaffold among the living fragments of a polycormone (Figure 3.1).

Height of shoots in a polycormone

The shoot height of polycormones of every age class varies, e.g. between 60 and 140 cm (Photo 3.1). It has been found that the shoot-size structure of young polycormones varies less than does that of the old ones.

For example, young *Filipendula ulmaria* polycormones produce high (120—140 cm) shoots. An exception in this respect is those individuals that for the first time produce one flowering shoot. Their height then rarely exceeds 80—90 cm. In senescent polycormones most shoots are 80—100 cm tall, while some of them are much taller, i.e., 120—160 cm. They are usually found on their peripheries, (Photo 3.1) where they shade their neighbours and limit their growth, and after the retreat of the latter they extend their individual areas.

Architecture of an individual and the area filled by it

Depending on their structure, individuals occupy their areas exclusively, that is, fill them completely, or share them with individuals of other species (Figures 3.3; 3.4).

The monocentric and compact structure of the polycormones, the result of the close vicinity of rooted shoots, as a rule makes it possible to exclusively occupy an area, for a short or a long time. The exclusiveness lasts the longest time when the inter-shoot distances are not above 2—5 cm, and the spaces are filled by rosettes or basal leaves that tightly cover the area. Endowed with such features are *Filipendula ulmaria, Lythrum salicaria, Geum rivale* and *Cirsium rivulare*

(Photo 3.5). It is only the change of a polycormone's shape from monocentric to polycentric, and the necrosis of fragments in its centre that bring about a loosening of the structure. Within the polycormonal area there form shoot clumps some distance apart (Photo 3.6). This structure of polycormones makes it possible for other species to encroach on their original areas. Then instead of one organism there grow between several and a dozen or so individuals of different species (Figure 3.3).

In tussock plants, e.g., *Carex cespitosa* and *C. appropinquata*, this potential increases only when tussocks have attained 30—50 cm in diameter, and there have appeared empty spots in the middle of each of them (Figure 3.4). They are inhabited by individuals of a fine structure (e.g. *Galium, Viola, Myosotis, Cirsium, Lychnis, Ranunculus*). Individual tussocks then represent "microphytocoenoses" of specified compositions and spatial structure. They can exist in this form for over a dozen years (Photo 3.7). A different fate of tussocks is observed when they are inhabited by species with strong, thick rhizomes e.g. *Geum rivale, Filipendula ulmaria, Lythrum salicaria* and *Polygonum bistorta*. In those species, tussock growth is limited after several years, and tussocks become split up. After 5—6 years tussock parts appear to be completely covered by individuals of species which form polycormones consisting of several shoots with expanded underground organs. Complete necrosis of tussocks occurs when there willows have appeared in a particular area. Their mechanical action eliminates these specific microphytocoenoses. The process lasts between several (7—9) and a dozen or so (12—15) years. Sedge tussocks can be colonized in two ways: they can be overgrown with individuals developed from seeds, and with those developed through the vegetative growth of plants, in which case individuals growing at the base of a tussock are characterized by a strategy that could be called "climbing".

Photo 3.5. The senescence process in a 5-year-old polycormone of *Cirsium rivulare* — short-term promoters. Photo: J. J. Faliński

Clonal growth and species turnover

Loss of plant organism integrity is caused primarily by agencies that disturb the balance between senescence and rejuvenescence. In conditions fully meeting the needs of a particular species (rich soil, sufficient moisture and lack of competitors) senescent rootstocks are continually replaced (overgrown) with juvenile ones. In the experimental garden the process was observed, for instance, in *Filipendula ulmaria* individuals aged 10—13 years, while under natural conditions senescence processes already prevailed in individuals aged 7—10 years, especially when they grew among sedges, or formed monospecific aggregations. The causes of polycormone disintegration can be placed in two groups: endogenous and exogenous. The former includes all those plant characteristics that determine a fast individual growth but do not ensure its integrity. This applies primarily to those plants in which shoots emerge at a considerable distance from a parent centre and sister entities (Figure 3.5). Fragmentation can then occur in almost every place, because the different entities are usually interconnected with delicate rootstocks or aerial shoots. The short duration of a shoot community, and thereby a fast rate of clone

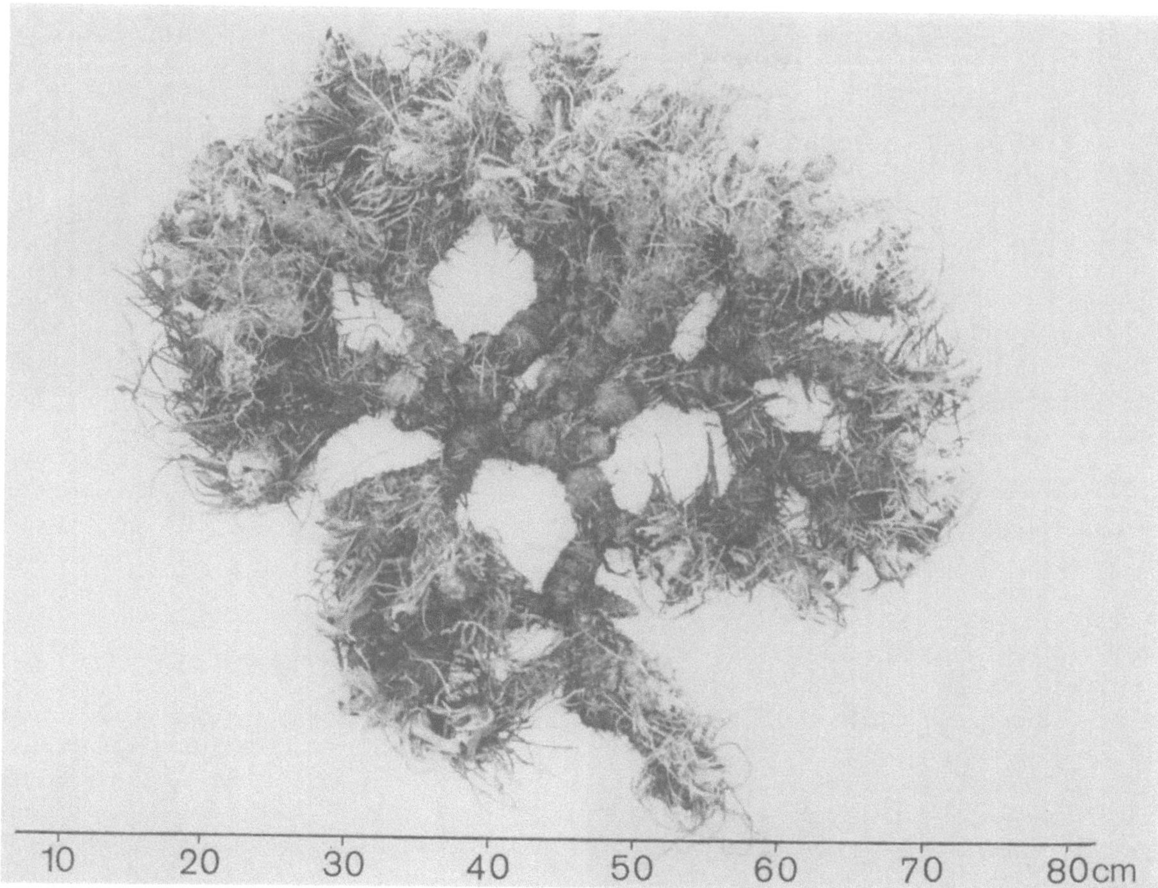

Photo 3.6. Underground organs of a 5-year-old polycormone of *Iris pseudacorus* with distinct "empty space" left by dead rhizomes. The loose structure of the polycormone permits colonization by other species.
Photo: J. J. Faliński

formation is an "inborn" character of stolonate plants, being thus their life strategy. In such cases a plant organism as a whole functions for a relatively short time (only several years), while a clone can exist almost endlessly. A slow rate of clone formation is characteristic of species whose individuals have sizeable and strong underground organs. In such cases the long duration of a shoot community is ensured by many morphological features of the underground organs, e.g.: short inter-node distances, thick rhizomes, long life spans of annual growth of the underground parts (Figure 3.5). Plant organisms of such a kind begin to disintegrate only when there is a joint action of several factors limiting the growth of a perennial plant organism. Though processes of rejuvenescence and senescence occur simultaneously throughout the life-history of an individual, in certain situations the balance between them may be disturbed to the disadvantage of the renewal of old structures.

Clone formation is often evoked by exogenous factors, notably competitors, that is, neighbours. This has been confirmed by experimental-garden studies, where *Filipendula ulmaria* and *Iris pseudacorus* individuals, even those aged 12—13 years, maintained a monocentric structure, with rooted shoots growing close to one another. None of 50 individuals of each of these species initiated clone formation in the garden. Under natural conditions polycormones aged over 6 years were usually subject to disintegration due

Photo 3.7. Species turnover 15 years after abandonment of the meadow. Meadow species return on the clumps of dominating *Carex cespitosa*.

Photo: J. J. Faliński

to the interaction from, e.g., *Carex cespitosa, C. acutiformis, Geum rivale, Lythrum salicaria, Salix pentandra* and *S. cinerea*.

Clone types and specific diversity

Clonal growth to a large extent depends on the nature of the origin of clones, that is, whether they have grown from special vegetative propagules, or have arisen through the fragmentation of a many-year-old plant organism. The simplest clone type is found in species whose individuals produce various types of vegetative propagules such as gemmae, bulbs, tubers. A clone then consists of several members whose life-history and growth make them similar to individuals grown from generative propagules. In the first growth phase, that is, at the stage of rooting and emergence of the first leaves, individuals are similar in their size and habitus to the seedlings of juvenile stages. Clones that have formed that way stand out in the surroundings, as they represent fairly compact assemblages, e.g. those of *Caltha palustris*. (Photo 3.3).

Clones of a similar type also form when annual rootstock growths separate from the parent plant as soon as they are rooted and have produced the first leaves. Individuals which arise in that way, are, from the beginning, in the

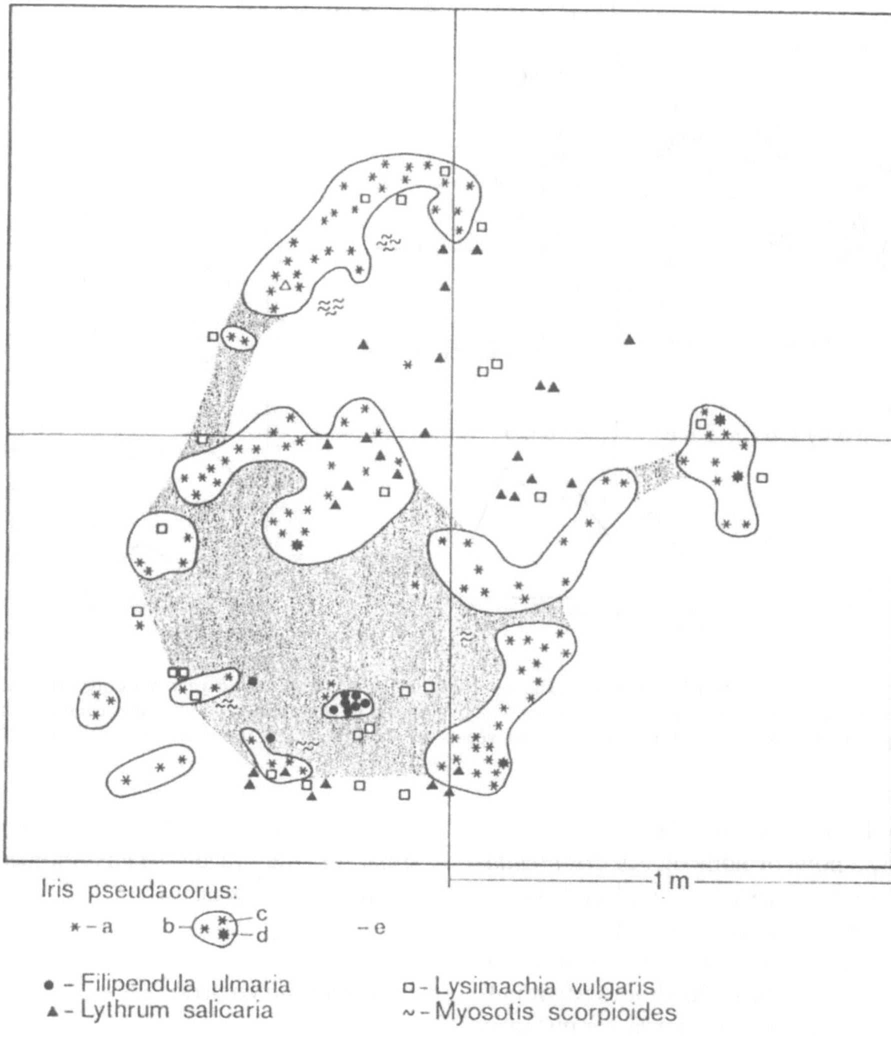

Fig. 3.5. Disintegration of a 7-year-old polycormone and development of a clone of *Iris pseudacorus*, and colonization of its dying fragments by individuals of other species in the sedge communities.

juvenile phase, their age ranging from several months to one year, e.g. in *Myosotis scorpioides*.

Another clone type is formed when an individual disintegrates into several uneven-sized fragments. The clone at first represents a large assemblage with slightly outlined vegets (Figure 3.6). The inter-veget distances increase from year to year because of the necrosis of those vegets in which senescent rhizomes prevailed at the time of separation (Figure 3.5). An opposite situation, that is, a gradual disappearance of veget distinctness in a clone, was less often observed. This occurs when after their separation from the parent plant, individual fragments start growing intensively. Situations of this kind are rare because the growth of individuals within a clone is limited by the species co-occurring in the meadow sward. Individuals of different species usually appear still in the areas of mature, polycormones (Photo 3.3). As the latter disintegrate, individuals of a number of species begin to grow intensively, particularly in the gaps that form

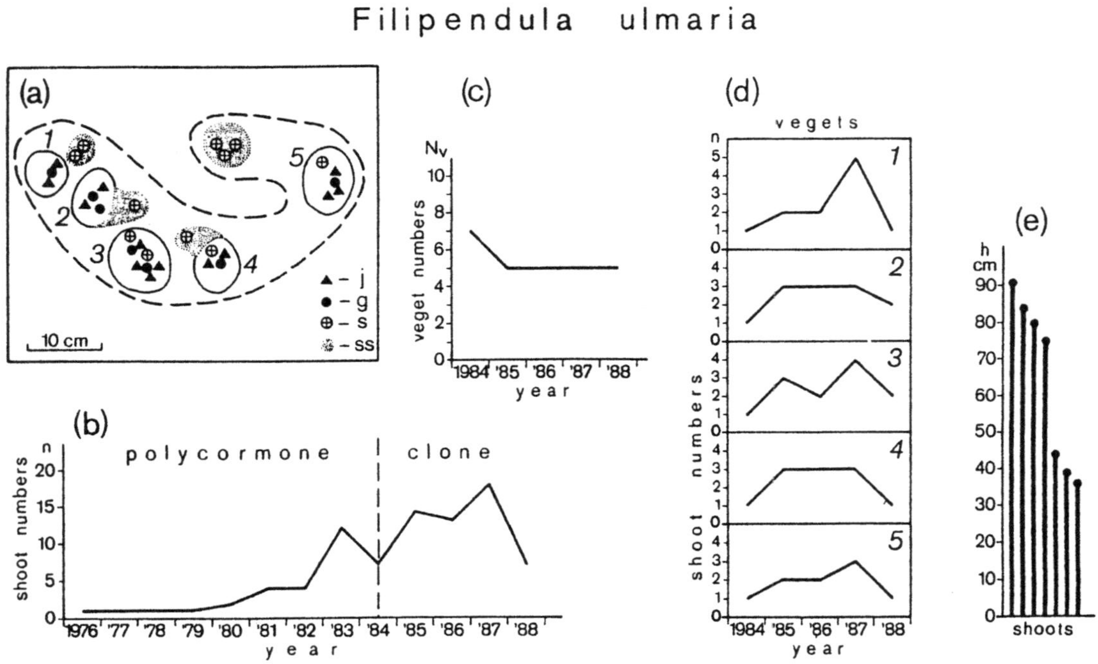

Fig. 3.6. Description of a clone that has formed as the result of the segmentation of a 6-year-old polycormone: a) the size and distribution of vegets (1—5); b) changes in the number of shoots in the polycormone and after its disintegration; c) number of vegets; d) changes in the number of shoots in particular vegets; e) shoot height in the polycormone during the disintegration phase.

among the polycormones that become independent. That way a polycormone of, e.g., *Iris pseudacorus* or *Filipendula ulmaria* is replaced with a mosaic of several species (Figure 3.5).

The largest clones arise, however, when subject to disintegration are multi-shoot individuals with long rhizomes. The growth was, for instance, observed of a *Lysimachia vulgaris* clone which occupied and completely filled 20 m^2, with only small proportions of other species present there.

Age of plant organisms and clones

By observing the life-history of labelled individuals and clones of species taking part in a long-term succession process, it has been found that individuals of such species usually live several, or about a dozen years at the longest, while clones arising as a result of plant organism disintegration may exist in a particular area about 20—30 years. Every year within every clone some individuals die and new ones form due to successive divisions of the vegets. Thus clones consist of uneven-aged vegets.

Intraclonal differences

Clones differ not only in the number of members, but also in the size of the latter, and their age and survival rate. It has been seen that new individuals detached from peripheral fragments of *Iris pseudacorus* and *Filipendula ulmaria* polycormones are aged 1 or 2 years, and are in the juvenile phase. This means that they did not attain the generative phase when they were still parts of a many-year-old organism. Individuals that detach from the central part of such an organism already contain senescent and dying modules at the time of their becoming independent, i.e., fragmentation of the parent organism.

Growth-stage differences among the individuals of a clone make them differ also in respect of their life-history stages. Members of a clone with advanced senescence processes at the time of fragmentation most often become necrotic in the same growing season in which they became independent. If, however, the division of an organism is followed by intensified rejuvenescence processes, then the survival rate of those individuals is the same as that of those detached from young fragments of a many-year-old organism, i.e., 90—100%. Hence clones of the same species may differ from one another by the size of the area occupied and the number of individuals (Figure 3.7). This depends primarily on the size of the individual that has disintegrated, its age and shape, and the morphological and life-history characteristics of the component members of the community (Figures 3.5; 3.6; 3.7).

In fact within clones and in their vicinity there also appear generative individuals (Figure 3.3). The boundaries between different-origin individuals disappear from year to year. After some years a monospecific aggregation will be found to consist of both generative and vegetative individuals (Figure 3.8).

The nature of a clone is, however, determined above all by the biological traits of the species, viz.: (a) how and where necrosis occurs, causing the disintegration of a perennial organism, (b) what the likelihood is of the necrotic structures to be replaced with new ones, and whether this process can stop or retard disintegration, (c) whether individual disintegration is associated with intensified renewal processes or with their regression. In *Filipendula ulmaria*, for example, individuals that have formed from peripheral fragments are usually 2 to 4 years old, that is, in the optimum growth phase, although they are still connected with the senescent and dying fragments of the polycormone. But vegets that form from fragments situated close to the polycormone centre are, as a rule, older, i.e., aged 5 or 6 years. Senescent vegets usually (89%) become necrotic in the growing season in which the polycormone disintegration took place. For instance, in a *Filipendula ulmaria* population, in as many as 52 of 79 vegets, labelled in 1978 and still alive in 1988, senescence signs could be seen: senescent

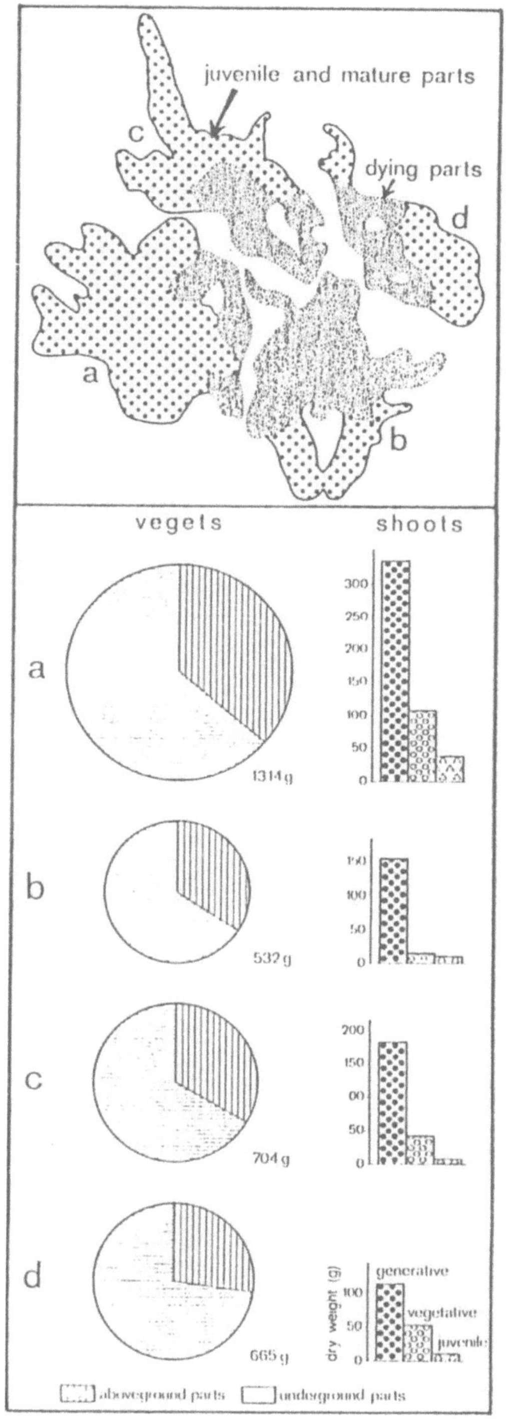

Fig. 3.7. Area size and manner of distribution of necrotic parts in clone of *Filipendula ulmaria*. Description of vegets (a—d) that have arisen as a result of the segmentation of a 10-year-old polycormone.

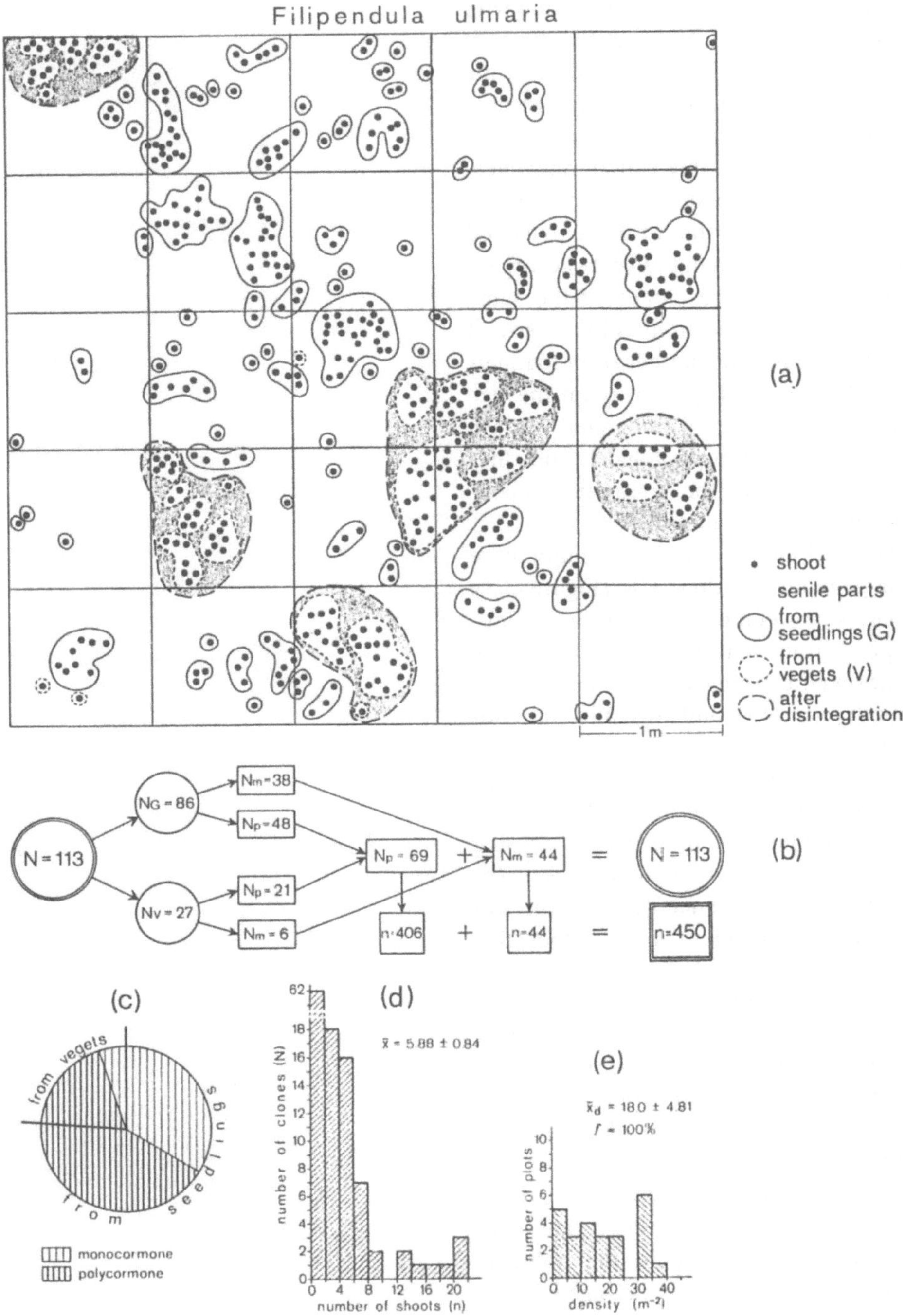

Fig. 3.8. Colonization success of *Filipendula ulmaria* in a 7-year-old abandoned meadow in the early successional stage of vegetation development: a) the spatial organization of population in 5 m × 5 m plot; b) N — number of individuals, n — number of shoots, N_p — number of polycormones, N_m — number of monocormones, N_G — number of individuals derived from seedlings, N_V — number of individuals from derived vegets (independent parts of clone); c) percentage of individuals of different derivation; d) the distribution of number of shoots (n) per polycormone (N), mean number ($\bar{x} \pm$ SD) of shoot per polycormone is indicated; e) frequency of 1 m × 1 m plots with different shoot density, mean ($\bar{x} \pm$SD) density (m^2) is indicated.

or necrotic rhizomes represented 20 to 50% of the underground organs, 50% of the rosettes failed to produce flower shoots, and rejuvenescence buds only formed at 1/3 of the shoots, 1—2 at each shoot.

From known age-related variation of individual characters it was estimated that only 11 of the vegets still alive (1988) had the chance to continue to grow. In the remainder of them, rhizomes senescence and necrosis progressed steadily from year to year (Figure 3.9). Besides, the vegets cannot grow in breadth, that is, spread, because they are surrounded by highly competitive species, such as sedges and willows, which appeared there when the first polycormones had attained the disintegration phase.

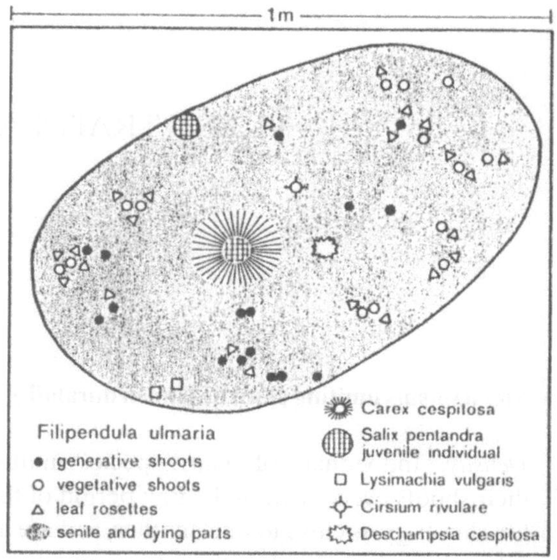

Fig. 3.9. Necrosis of the parts of a 10-year-old clone of *Filipendula ulmaria* and encroachment of individuals of other species on the vacated places. This is representative of phase IV; loss of clone integrity. The senile parts form a permanent framework for the living parts of the clone and for the other indicated species.

CHAPTER 8

PLANT TRAITS AND SPECIES TURNOVER

Mechanisms limiting plant organism durability

Density. Individuals of many species multiply their shoots every year. In the first period of their life shoots grow close to one another, and are as a rule tall and covered with a large number of leaves. Increased shoot densities, and the resultant shading of the polycormonal areas impede the production of renewal buds, and thereby restrict the production of new shoots and rootstock segments. In such situations it is only on their peripheries that polycormones still continue to grow, unless neighboured by species with similar morphological properties. If, however, the spreading rate of polycormones is not limited by competitors, then after some time, fragments that are most distant from the necrosing centre detach and become independent (Figure 3.5). Their structure is at first loose, but several (3—5) years later the area again appears to be filled with numerous shoots, and the same growth type and veget division occurs as those peculiar to the parent organism from which they took their origin.

Necromass. Continued over several (6—7) years, the yearly fall of leaves and aerial shoots of polycormones results in an accumulation of necrotic remnants in large parts of their areas and in their vicinity (Figures 3.8; 3.9). This factor considerably reduces polycormone growth and development, and accelerates senescence. Garden experiments on the growth of *Filipendula ulmaria* polycormones have proved this. Out of 10 polycormones, from whose areas accumulated, biomass was not removed at the end of the growing season, only one reached the age of 10 years, while the remainder of them died between life years 6 and 9. The process started with a decrease in the number of generative shoots, which developed only on the peripheries of polycormones, and were short, 50-70 cm (Figure 3.10), while those of young polycormones are 100—140 cm tall.

Of 10 polycormones, from whose areas falling biomass was removed, as many as 5 reached the age of 15 years, and preserved the monocentric structure. Their progressing senescence was, however, initiated by a large number of rosettes that failed to produce flower shoots. Besides, their underground organs consisted of three layers. The lowest-located of them was made up of the remains of necrotic rhizomes and roots. The thickness of the necromass layer ranged from 5 to 15 cm. It was overlain with rhizomes, partly necrotic and partly living (the second layer), grown up with new segments with renewal buds and above-ground shoots (the third layer). Due to the accumulation of necromass in the underground organs, young shoots lose contact with the soil. A simultaneous action of: shoot overcrowding in a polycormone, considerable shading of its area, and decrease in the decomposition rate of the remnants accumulated in the above-ground and underground parts, causes a gradual necrosis of polycormones (Figure 3.9).

Polycormones from which necromass has not been removed begin to undergo senescence in life year 5 or 6, as under natural conditions. In polycormones from which dead remnants are removed every growing season, the process begins in life year 9 to 10 (Figure 3.10).

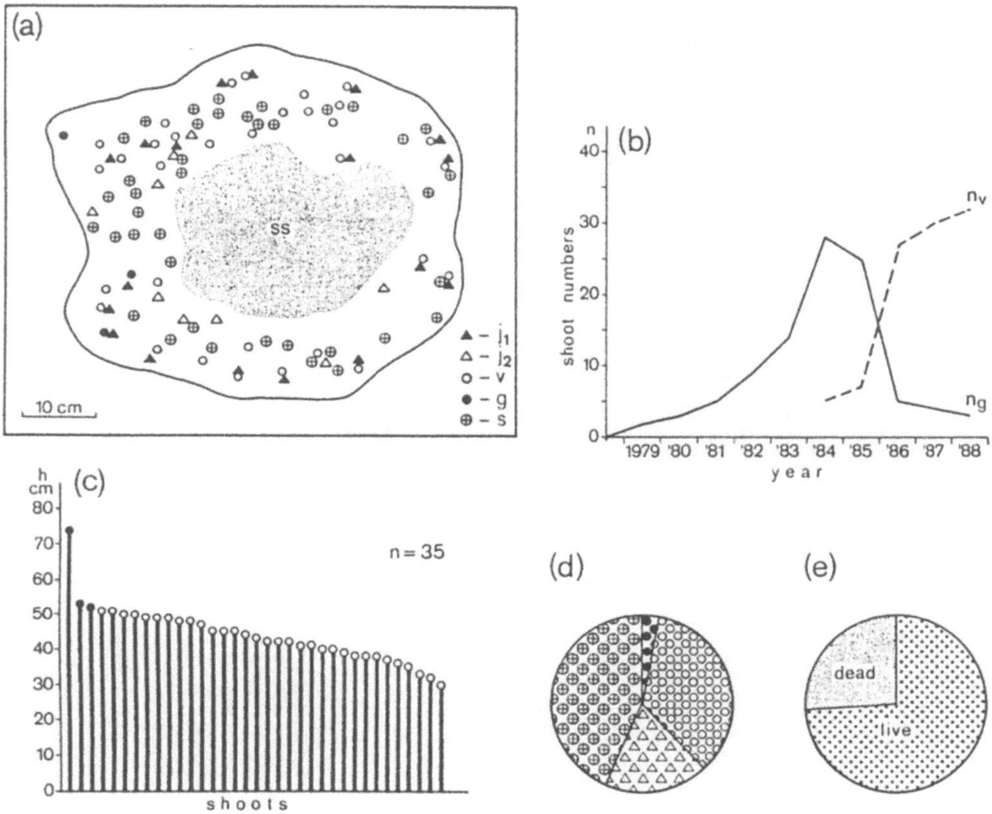

Fig. 3.10. The senescence process in a 10-year-old clone of *Filipendula ulmaria* which was representative of phase IV of clone development; weakening integrity and loss of exclusive occupation of an area: a) the distribution of necrotic parts (ss), senile shoots (s) generative shoots (g) and vegetative shoots (v), rejuvenescent buds (j_1) and juvenile rosettes (j_2); b) number of vegetative (n_v) and generative shoots (n_g); c) shoot-size structure; d) percentages of shoots in different growth phases; e) proportions of live and necrotic parts in under-ground organs.

Following meadow abandonment the percent cover and layer thickness of the surface necromass changes from year to year (Table 3.1). In sedge swamps the highest necromass proportion, expressed as the percent cover, is found between year 9 and year 11 after the cessation of mowing (Figure 3.11). At that stage most of the individuals of polycormonal-species populations attain the senescent phase characterized by a high proportion of dead remnants in the polycormonal and clonal areas (30—60%), which hampers their growth and appearance of new individuals. Following the necromass decomposition, which usually takes 2—4 years, there emerge generative-origin individuals of the species: *Filipendula ulmaria, Lythrum salicaria* and *Cirsium palustre* (Figure 3.11), as well as vegetative-origin individuals, that is those developed from rhizome fragments that survived the adverse conditions period. The phenomenon was observed in the populations of the following species: *Filipendula ulmaria* (Figure 3.8), *Lysimachia vulgaris, Geum rivale* and *Carex acutiformis* (Figure 3.11).

Role of plant-organism senescence and renewal in the species-turnover process

In several-year-old polycormones, renewal and senescence occur simultaneously.

Renewal, which leads to the formation of new

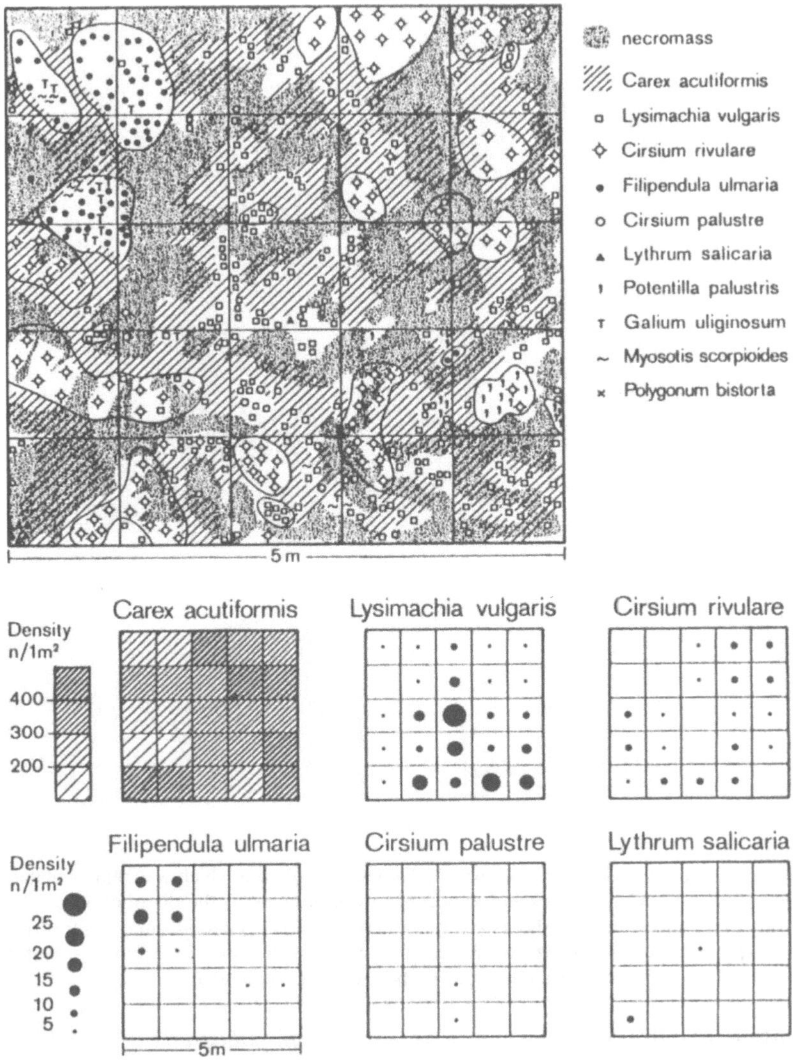

Fig. 3.11. Population mosaic of *Carex acutiformis* and other species in a 10-year-old abandoned meadow. Both high density of *C. acutiformis* and the covering by necromass of surface area limit for many years the growth in numbers of other species.

shoots and rootstock segments, causes plant organisms to grow in breadth. The process favours the formation of monospecific aggregations that quickly eliminate existing species. They locally reduce the floristic richness of meadow communities (Figure 3.11).

Senescence processes usually result in the disintegration or death of plant organisms, or their parts, due to which there appear free spaces in their areas. The spaces are subsequently colonized by various meadow (*Lychnis flos-cuculi, Ranunculus acris* and *Cirsium palustre*) and woodland (*Solanum dulcamara, Lycopus europaeus* and *Chaerophyllum hirsutum*) species (Figures 3.3; 3.4; 3.5).

Plant-organism fragmentation, as well as the loosening of the polycormonal and clonal architecture, thus favour the return of a floristic richness to sites occupied by monospecific aggregations, and make it possible for forest species individuals to grow there. Successive divisions of the different fragments of the parent organism enable some species to take part in a long-term succession.

Polycormone formation and growth, followed by their multiple divisions, play the role of biological mechanisms of species turnover and maintenance of specific diversity.

Role of plant-organism integration and disintegration in the species-turnover process

Plant-organism permanence as integral "communities of shoots" varies between species. Shoot joint existence may continue for several or a dozen or so years, sometimes several dozen years. But even in those species whose morphological characteristics indicate long-durability potentials of polycormones, e.g. in *Lythrum salicaria, Filipendula ulmaria* and *Iris pseudacorus*, a prolongation of their life to several dozen years is rare. This is, in principle, determined by the underground-organ structure and environmental conditions, especially those which lead on to the breaking of the morphological links between rooted units. The cause of this is the presence of individuals of other species with sizeable and strong underground organs. The main factor causing polycormones to disintegrate is, however, the processes that occur over their long-term life-history, e.g. a high rate of growth and spreading, with simultaneous necrosis processes in their senescent parts (Figure 3.7). A fast growth in size of a polycormone usually leads on to a situation where the survival of a whole organism occupying several or about a dozen square metres is impossible. This occurs even when the action of other growth-limiting factors, e.g. competitors, is inefficient. The main factor shortening polycormone durability is the gradual necrosis of the underground organs. This applies particularly to those cases where renewal processes are not fast enough to set off losses caused by the senescence and necrosis processes affecting the durable parts of polycormones. The more so as the renewal processes usually occur on their peripheries, and necrosis in the centre. The processes cause a gradual widening of the "empty centre" (Figures 3.5; 3.7), and a simultaneous narrowing of the outer part, made up of living elements, the growth of which is limited by neighbours. Such a polycormone structure speeds up the disintegration by facilitating the breaking of the morphological completeness of the ring. It has been observed that in individuals of *Filipendula ulmaria* and *Iris pseudacorus* necrosis begins to dominate over renewal from life year 6 or 7, and only then do various actions of neighbours lead to their division. In young polycormones, disintegration can also be brought about by neighbours. This applies to cases where a polycormone occupies a considerable area for 4 or 5 years, but does not fill it completely (Figure 3.6). A fast growth rate of shoots, several dozen centimetres apart, does not favour their joint existence either, e.g. in *Lysimachia vulgaris*.

The following can be included among the characteristics that no doubt determine the length of the functioning of a polycormone as a morphological whole: (a) number of shoots it is made up of, (b) durability of the rootstock segments that connect them, (c) inter-shoot distances, (d) rate and site of necrosis and renewal. Studies carried out so far indicate that the thicker the rhizomes, the shorter the inter-node distances and the faster the renewal filling of spaces relased due to necrosis, the greater the durability of a polycormone. Three polycormone types have been distinguished according to their varying durability:

I — polycormones representing long-lasting "shoot communities" ensured by thick rhizomes, short internodes (5—10 cm) and short inter-shoot distances;

II — polycormones representing short-lived integral "shoot communities" that disintegrate at an early growth stage due to a delicate rootstock structure, long internodes and considerable inter-shoot distances;

III — polycormones representing seasonal (lasting one or two seasons) "shoot communities", found in stolonate plants (with above-ground stolons) and in rhizomic plants if their annual growths do not live longer than one or two growing seasons, e.g. *Myosotis scorpioides*.

Group I includes the polycormones of: *Iris pseudacorus* and *Filipendula ulmaria*. Their high durability result from the thickness and durability of annual rhizome growths and short inter-shoot distances (5—10 cm). Apart from that, their senescent rhizome segments make up a framework on which young polycormone units form. Necrosis and renewal processes proceeding in this way are an efficient mechanism responsible for the permanence of the "shoot communities". It is noteworthy, however, that even with these morphological properties of plants, polycormonal permanence seldom exceeds 10 years (Figures 3.5; 3.7). The oldest (13 years) and the largest *Iris pseudacorus* polycormone, observed in a wet meadow, occupied 2.9 m^2 and was made up of 386 shoots, its biomass amounting to 6103 g. The oldest (15 years) *Filipendula ulmaria* polycormone seen in an unmown meadow, was made up of 287 shoots growing at short distances from one another in an area of 4.5 m^2. Mature polycormones of this species are as a rule made up of 20—40 shoots and occupy areas ranging from 0.25 to 1.00 m^2. Large monospecific *F. ulmaria* aggregations seen in unmown meadows most often consist of several clones.

Long-lived "shoot communities" were observed in species whose individuals produce delicate rhizomes, e.g. 1—5 mm in diameter. Such characters are peculiar to tussock plants, the durability of which comes up to several dozen years (Figure 3.4). In this case it is the result of shoots growing very close to one another and forming durable compact tussocks. This is connected with the fact that individuals produce new storeys of rhizomes and aerial shoots on the necrotic fragments. With passing years there form tussocks, 50—70 cm high, e.g. of *Carex appropinquata* and *C. cespitosa*.

Group II includes species of which the individuals are characterized by a short-lived integrity, due to which they seldom form large-area polycormones. Placed in this group have been *Lysimachia vulgaris* polycormones which produce numerous rootstocks every growing season. The rootstocks are thin, 30 cm to 3 m long. Individual shoots, more precisely clumps of 2—4 shoots, grow at considerable distances from the parent organism and lose the morphological link with it during the same season. Relatively durable shoot clumps also give rise to numerous new shoots at some distance from them. In this way there form, within several years, monospecific meadow *Lysimachia vulgaris* aggregations which are mostly clones consisting of vegets made up of several shoots (2—6) each (Figure 3.12).

Short-lived shoot communities are sometimes also found in species producing thick rhizomes, e.g. 3—4 cm in diameter. This is so when the rootstock structure is spongy, due to which, rootstocks are fragile and breakable, e.g. in *Cirsium rivulare*. Then despite the rhizomes being thick, integrity does not last longer than 5—6 years, whereafter they become fragmented into several smaller entities.

Included in Group III are species of plants that produce, on their above- or underground stolons, new shoots that get rooted and then detach during the same growing season, or during a subsequent growing season, e.g. in *Caltha palustris* and *Myosotis scorpioides*. If individuals of these species grow into polycormones, they, as a rule, last one or two growing seasons, very seldom a longer time. Observed under natural conditions, aggregations of these species are clones, whose members live about 5 years.

Plant longevity or immortality?

A specific characteristic of plants is the active division of the meristem cells throughout the life-history of individuals, making them longaeval, and even immortal. A mature plant organism can exist in a conquered area as long as there continue to arise new structures that not only replace its organs lost, but also lead on to its rejuvenescence. Hence plant immortality has many times been considered to exist (Watkinson and White

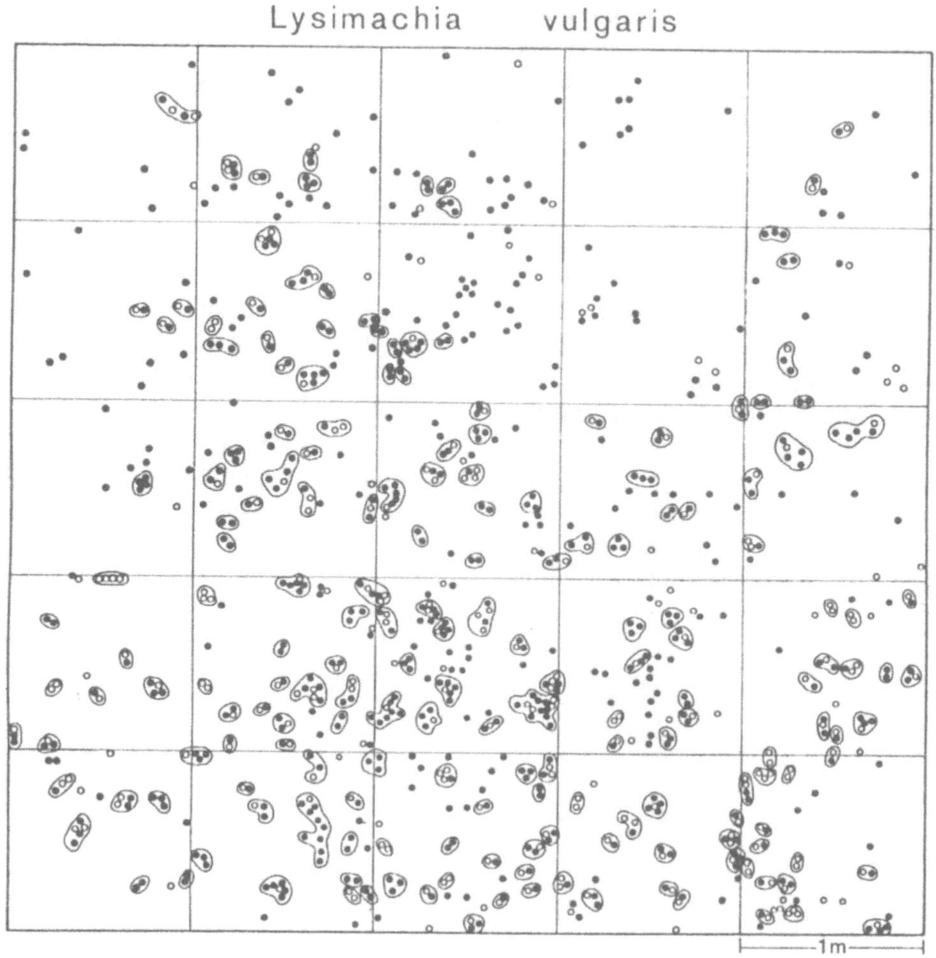

Fig. 3.12. Spatial structure of *Lysimachia vulgaris* in an 8-year-old abandoned meadow (phase II of succession).

1985). Due to the ability of plants to replace old or necrotic body fragments with young modules, many species lack the senescent phase, or it lasts a short time in them. If a plant renews its body vegetatively and multiplies asexually, then it may prove impossible to establish the end of its life as the result of senescence. Harper (1977) reports that a *Pteridium aquilinum* clone found on hills in Great Britain has persisted there since the Iron Age. Harberd (1961) estimates the age of a *Festuca rubra* in pastures in Scotland at 400 or even 1000 years. Similar age has often been reported for *Phragmites australis* clones.

High estimates of plant longevity result from the treating of a clonal growth as the continuation of an individual's life, realized through its numerous replications, and identical genetically. If we assume, however, that a plant that undergoes numerous successive divisions and at the same time renews its vegetative "offspring", then the age, specified for a clone, informs only about the duration of the sojourn of representatives of one genotype in a particular area rather than about the "physiological" age of plants (see Part I).

Individuals arising owing to vegetative multiplication, obviously are, at the beginning, uneven-aged, but as a rule, much younger than the parent plant. Kershaw (1973) reported that peripheral fragments, detached from a 25-year-old *Carex bigelowii* individual, were aged 8 to 12 years. Peripheral parts of several-year-old individuals of *Filipendula ulmaria* and *Iris pseudacorus* are 1 to 3 years old, whereas propagules detaching from *Caltha palustris* individuals are aged 3–4 months (Falińska 1978, 1986a, b, c). In many species the oldest fragments of an individual are usually in its centre (Fiala 1978, Kershaw 1973, Falińska 1982, 1985, 1986b, Kotańska 1986–1987a, b), and the largest number of vegetative offspring become detached from its young peripheral parts. Many studies indicate that the vegetative propagation leads on to the formation of units much younger than the parent plant (Smirnova *et al.* 1976, Szmeja 1987a, b).

Since perennial plants have organs (or parts) of uneven age, the clonal individuals resulting from their fragmentation are already of uneven age at the time of becoming independent. Hence the age estimates of aged plant organisms depend on the criteria used, that is, whether: (1) an individual's age is counted from its birth (seedling emergence) and life years are subsequently added to that date as long as there exist in a particular area plants of a specified genotype, regardless of whether they have survived there in morphological entirety or have been subject to multiple divisions; (2) every division of a plant organism is looked at as its life end heralding the beginning of the life of new individuals to form from its fragments; (3) the fact is taken into account that the entities that have formed as the result of the fragmentation of a many-year-old plant organism are at the very beginning of uneven age and growth phase.

CHAPTER 9

ROLE OF SPECIES IN THE SUCCESSION PROCESS

Introduction

Plant species play different roles in meadow vegetation changes. Some of them significantly contribute to the decomposition of phytocoenoses almost immediately after the cessation of mowing, while others may impede the process for many years.

The adopted criterion for the assessment of the role of species in the succession process was their colonizing capability, that is, the time, rate and manner of settling in an area, and the duration of its occupation.

Following the cessation of mowing, a number of species show a considerable population growth. They usually become dominant within a short time, and reduce the floristic richness of meadows, locally or over large areas (Figures 2.1; 2.2).

Spatial relations among populations of species with high colonizing capabilities are determined by the settling precedence, that is, local differences in their proportions in the meadow communities, as assessed during meadow utilization periods. This applies particularly to species like: *Filipendula ulmaria, Carex acutiformis* and *C. cespitosa*. Due to the fast growth in numbers of these species after the cessation of mowing, in the different parts of a meadow dominance is attained by different species, whereby a mosaic vegetation pattern arises in abandoned meadows (Figure 2.3).

Species with such characteristics were found to differ, the differences being of the following kinds:

(a) size of the area occupied by them (from a dozen or so square metres to several hectares);
(b) duration of area occupation (several to over a dozen years);
(c) manner of space filling (assemblages, aggregations, patches).

After the cessation of mowing, the role of species in meadow vegetation changes, is variable. During a succession as low a proportion as 12% of existing species contribute to the decomposition of meadow phytocoenoses, or initiate the formation of new floristic compositions. According to the nature of the influence, the following species groups have been distinguished:

Species accelerating the meadow phytocoenosis decomposition and initiation of new floristic compositions. Placed in this group have been those species which become dominant in the first years (1—8) in unmown meadows, and significantly affect the numeric ratios among meadow phytocoenosis components. In the later period, i.e. after 8—10 years, they lose the status of dominants and are replaced by woodland species. For those species the term of succession promoters has been adopted. This group includes the following species: *Filipendula ulmaria, Lythrum salicaria* and *Lysimachia vulgaris.*

Species that retard the species and phytocoenosis turnover processes by forming large-sized and long-lasting, almost monospecific aggregations. They have been named the succession inhibitors. This group includes: *Carex*

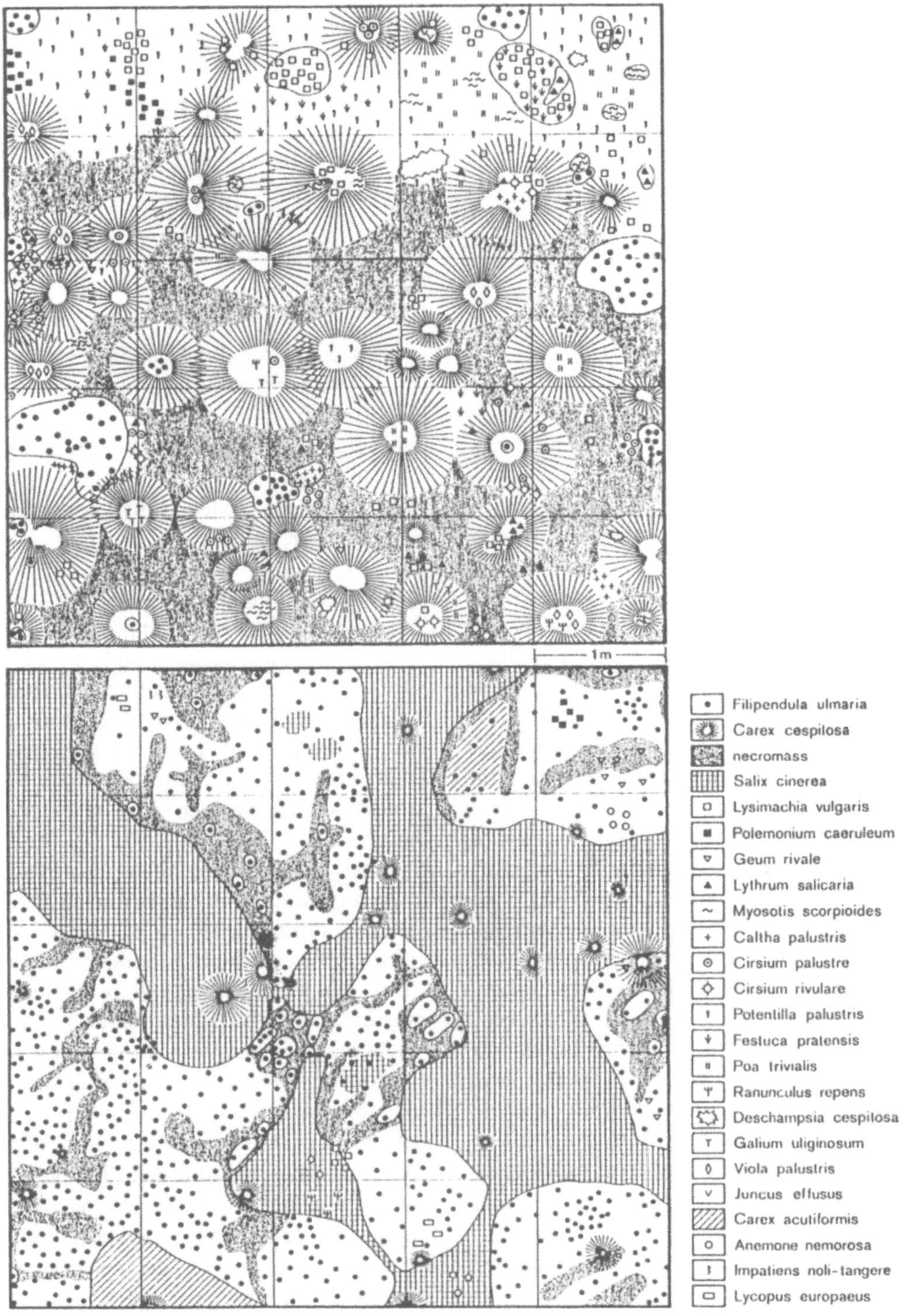

Fig. 3.13. Spatial relation among populations of species with high colonizing capabilities in a 14-year-old abandoned meadow. *Carex cespitosa* as succession inhibitor (top) and *Filipendula ulmaria* as succession promoter — the recovery of woody species (bottom).

acutiformis, C. appropinquata, C. cespitosa, Scirpus sylvaticus and *Phragmites communis.*

Neutral species are present throughout the succession process, but do not significantly affect the species turnover. The following have been included in this group: *Myosotis scorpioides, Caltha palustris* and *Galium palustre.* It means that individuals or populations of these species have not been observed to favour or impede the growth of the co-components, even though in some succession phases they attain a high frequency.

Succession inhibitors

It has been found that only two of the species taking part in the succession process impede vegetation changes. They completely oust, within a relatively short time (i.e. about 10 years), the floristically rich meadow phytocoenoses, and take their place and form large-area stands, poor floristically. For instance, sedge swamps persist even when scrub communities have formed close to them (Figure 2.1).

Species of that group represent various growth types: stolonate — *Carex acutiformis*, and tussock-like — *Carex cespitosa*. They respond in a similar way to the cessation of mowing, that is, they quickly colonize large areas and occupy them for a long time — 15—25 years, nearly as long as it takes for scrub communities to form in areas where sedges have not attained the dominant status (Figure 3.13). Sedge stands almost exclusively continue to hold the areas occupied by them after the elimination of meadow-phytocoenoses until tree species appear in them. Sedge regression starts as soon as the first trees appear. They grow, at first, solitarily and then form groups of several individuals representing different species of the genera: *Salix, Alnus* and *Betula*. After over a dozen years they initiate the formation of forest communities (Figure 2.1). The gradual growth, year by year, of the tree biogroups contributes to the elimination of sedges. This is, however, a long process, and sedges may still continue to exist for many years also on the margins of willow scrub (Figure 2.1).

Carex acutiformis

In regularly mown meadows this species occurs in small numbers of individuals, forming small aggregations among other species (Figure 2.2). Immediately after the cessation of mowing its cover in the meadow communities increases from 5% to 90% (Figure 3.14). This is the result of the fast growth of the individuals that were present when the meadows were still in use, and of new recruitments year by year (Figure 3.15).

In abandoned meadows *C. acutiformis* becomes dominant within 10 years, as illustrated by the vegetation map (Figure 2.1). The fast meadow colonization rate of this sedge species results from the fast growth rate of *C. acutiformis* individuals. Every growing season the number of their modules is multiplied (Figure 3.16), and at the same time long rootstocks are produced which penetrate among the meadow sward plants and become rooted at various distances from sister modules, several to several-dozen centimetres (0.10—1.20 m). This process brings about a growth in density of the modules within the occupied area (Figure 3.15), as well as a fast shifting of the boundaries of individual aggregations (Figure 3.14). This manner of growth in width of *C. acutiformis* organisms leads on to the disappearance of inter-aggregation boundaries and formation of large stands (Figure 3.14). Ten years after the cessation of mowing, meadow phytocoenoses begin to be replaced by sedge marshes which occupy over 50% of a meadow biochore area, and in particular cases even as much as 100%. In some parts of the Reski Range sedge marshes become a permanent landscape element (cf. the map, Figure 2.1). Their characteristic feature is a floristic poverty (Figure 3.17). Individuals of *Lythrum salicaria, Iris pseudacorus, Filipendula ulmaria, Carex cespitosa, C. appropinquata* form small clumps (Figure 3.11) amidst sedges, and sometimes under particular conditions (e.g. local depressions), they occur in small aggregations. However, year by year the number of species and population sizes decrease steadily due to two factors: (1) growth in module density in aggregations (from 30 to 400 modules/m^2; (2) steady accumulation of dead plant remnants on

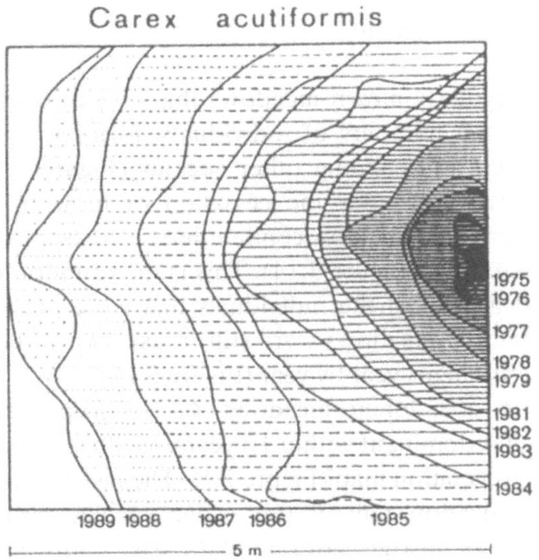

Fig. 3.14. Temporal changes in *Carex acutiformis* spatial structure for 14 years in abandoned meadow.

the ground surface: (a) necromass cover ranging from 15 to 55%, (b) necromass-layer height varying between 5 and 30 cm.

A high density of modules and the covering by necromass of a considerable proportion of the surface area limit for many years the growth in numbers of *C. acutiformis* and other species.

In the initial meadow colonization period (1–5 years) single *C. acutiformis* modules grow dispersed among other meadow species, and several years later they form small aggregations (Figure 3.15). The co-occurrence of sedges and other species in meadow plant communities lasts about 5 years. Then during several subsequent years (3–4) meadow plant species are gradually eliminated and their place is taken by large *C. acutiformis* aggregations (Figure 3.14). A meadow disuse lasting for 8–10 years is followed by a rapid density growth in *C. acutiformis* aggregations, and a significant increase of the sedge-occupied area due to the gradual wedging in of sedges into the meadow sward (Figures 3.14; 3.15).

In certain cases a reverse situation could be seen, e.g. when there appeared in the sedge marshes species of high colonizing capabilities, like *Filipendula ulmaria, Lythrum salicaria* and *Iris pseudacorus*. They are at first represented by a small number of individuals randomly dispersed amidst sedge modules. Several yers (5–6) later individuals of these species form aggregations of varying sizes. In such cases mosaic patterns formed in which 30% of the ground was covered with the necromass of necrosing sedge modules, 40% with living *C. acutiformis* modules, whereas the remainder of the niches were inhabited by species like *Filipendula ulmaria, Lysimachia vulgaris* and *Cirsium rivulare* (Figure 3.11). The latter species often settles in gaps that arise due to the decomposition of accumulated necromass amounts. A sedge marsh holding a small number of species may continue to exist for many years (20–30). In such phytocoenoses population, abundance is subject to only small variations that can be defined as "pulsation" within an aggregation or a stand of uniformly distributed individuals.

Carex cespitosa

During the period of meadow use, this species was represented by small-sized populations, and its proportions in the phytocoenoses (Figure 3.18) were low. For instance, censused in six 200 m^2 plots, the number of tussocks then varied between 1 and 8, and the percent cover between 1 and 3%. Following the cessation of mowing, the number of tussocks increased steadily year by year, and so did the contribution of the species to the phytocoenoses. The maximum number of tussocks recorded during the 15 years' observation in the 200 m^2 plots was 182, the percent cover amounting to 74%. Ten years after the cessation of mowing *Carex cespitosa* covered 1/3 of a 1 ha area, and formed large aggregations there (Figure 2.3). In periods when the meadows were still used *C. cespitosa* tussocks occurred in them randomly, or in small groups of 3 tussocks each (Figure 3.18).

The high rate of abandoned meadow colonization by this species results from the growth in number of the tussocks and their growth in width (Figure 3.18). At the colonization time, for

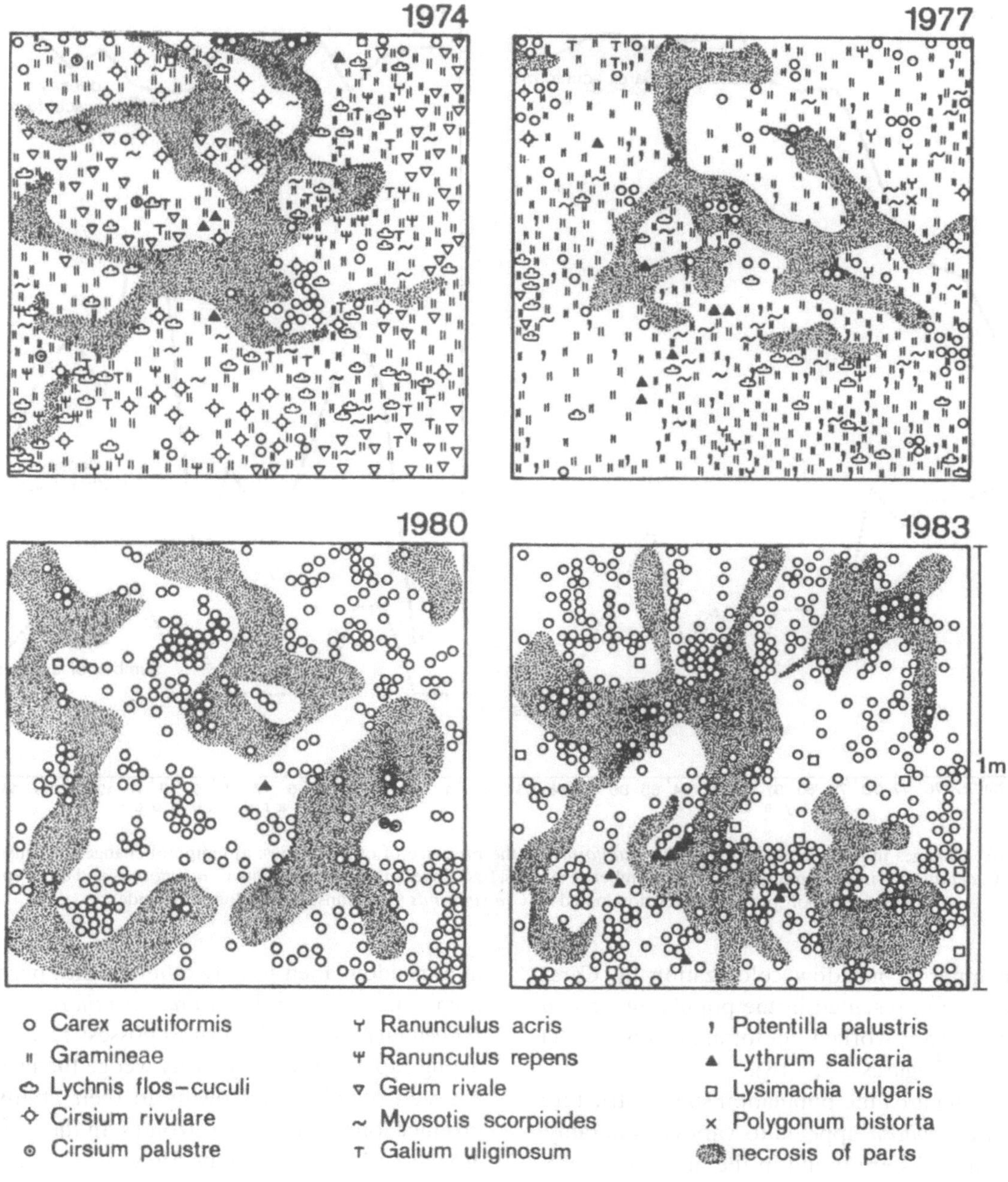

Fig. 3.15. Changes in the spatial structure of *Carex acutiformis* population for 10 years in abandoned meadow, as a result colonization and dying part of sedge in the early successional stages of vegetation development.

example, the tussock diameter rarely exceeds 10 cm, while 5 years later the value comes up to 15—20 cm, and 10 years later, 30—50 cm. Owing to these characteristics, *Carex cespitosa* becomes dominant within 10 years, and almost completely fills the area it occupies. At that time, individual tussocks occupy large areas, 0.50 to 1.0 m^2 in size. In a 25 m^2 plot, for instance, there are found 20, or up to 30, tussocks, 25% of which occupy areas above 1 m^2 in size (Figure 3.18). The rate

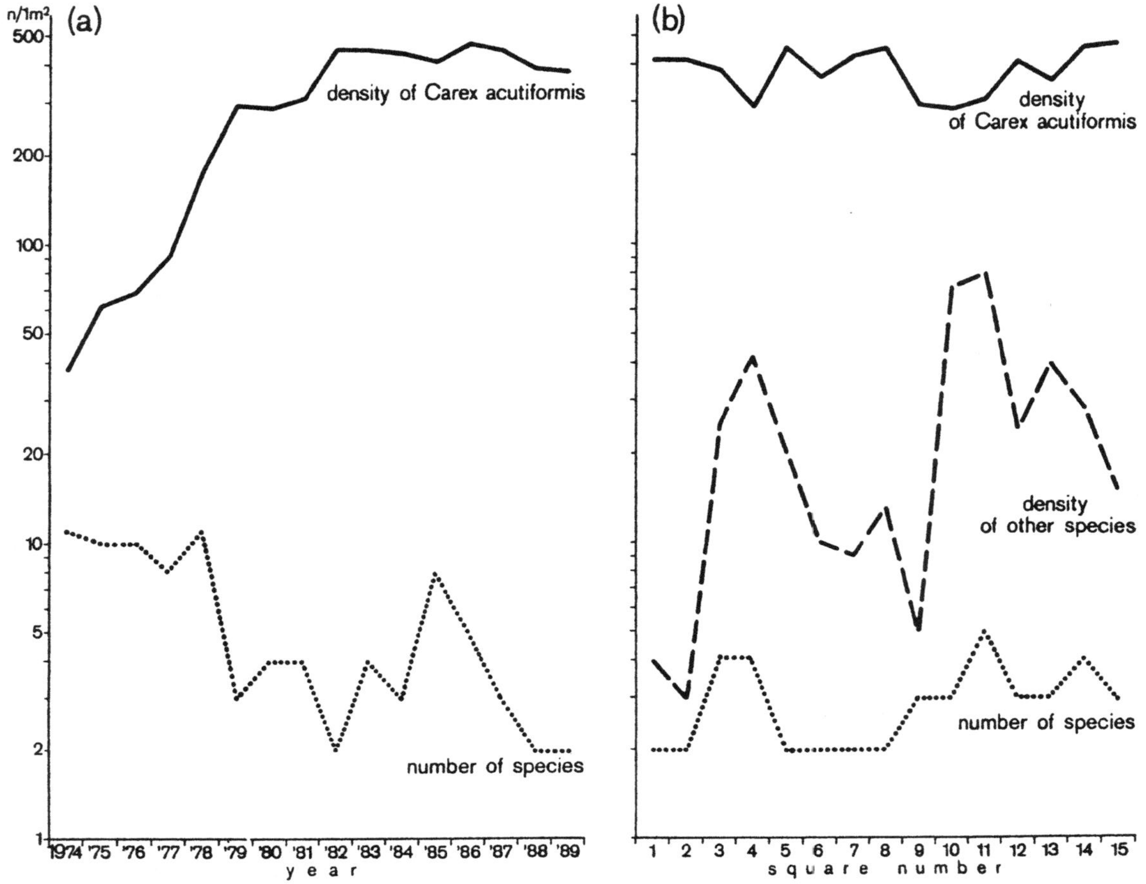

Fig. 3.16. Changes in size of population *Carex acutiformis* in the presence of other species: a) temporal changes in density of shoots *C. acutiformis* in chosen plot (1 m^2) and number of species; b) changes in density of shoots *C. acutiformis* and other species and number of species in patch dominated by *C. acutiformis* (The transect 15 m long, 1 m wide).

and manner of meadow colonization by *Carex cespitosa* are presented in the population spatial dynamics maps worked out for area sizes of 1 ha and 25 m^2 (Figures 2.3; 3.18).

In a 1 ha area the population spatial structure of *Carex cespitosa* appears to vary considerably. Single tussocks are found in the meadow sward, and amidst *Carex acutiformis* and *Filipendula ulmaria* aggregations. They also form almost monospecific stands (Figure 2.3) if in meadow utilization periods there occur, locally, numerous *C. cespitosa* tussocks. The minimum number of tussocks needed by this species to become dominant in an abandoned meadow within a short time is 6—8 years (25 m^2), with a frequency no lower than 30%. If *C. cespitosa* tussocks are rare in a meadow, then usually other species become dominant, as a rule, *Carex acutiformis* and *Filipendula ulmaria*, known to possess high colonizing capabilities. They easily wedge themselves between other species owing to their ability to produce long and strong rhizomes. In their initial growth period, i.e. up to the 5th year, *C. cespitosa* tussocks are small, and are easily eliminated by other species, e.g. *Geum rivale, Lythrum salicaria* and *Carex acutiformis*. After having attained large sizes, i.e., when they already occupy areas of 0.5—1.0 m^2, *C. cespitosa* tussocks in turn begin to eliminate other species. Its large and long-lived tussocks enable the species to remain for a long time in an area of which it takes possession (Photo 2.2).

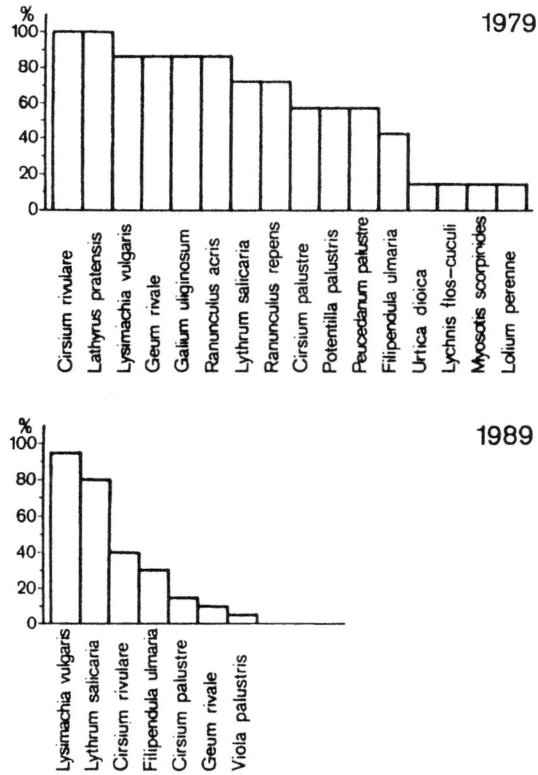

Fig. 3.17. Frequency of species in the patch of *Carex acutiformis* 5 years (1979) and 15 years (1989) after abandonment of the meadow.

Carex cespitosa and other species

In the period immediately following the cessation of mowing, *C. cespitosa* tussocks are dispersed among other meadow species, e.g. *Ranunculus acris, Lychnis flos-cuculi, Lysimachia vulgaris* and *Cirsium palustre.* Through their growth, *C. cespitosa* tussocks, year by year, eliminate their neighbours and take over the space gradually released (Figure 3.18). After a period of 8—10 years, that is, when the tussocks have attained large sizes (0.5—1.0 m^2), there may again appear meadow species in areas released earlier on (Photo 3.7). They grow then mainly on the sedge tussocks (Figure 3.13). The inter-tussock spaces are at that time filled with dead remnants of the above-ground parts that fall every growing season, but their decomposition lasts 2—3 years. The thickness of the necromass layer varies between 10 and 30 cm, and is not usually grown up with vegetation (Figure 3.13).

Found to appear on sedge tussocks are individuals, of only a small number, of the species of the floristically rich meadow that originally existed in the area. The number of species found on a tussock ranges from 1 to 3, and the number of individuals from 1 to 6 (Figure 3.13). They are most often grown up with individuals of *Cirsium palustre, C. rivulare, Filipendula ulmaria, Geum rivale, Lythrum salicaria* and *Lysimachia vulgaris* (Figure 3.19).

Each of the tussocks then represents a "microphytocoenosis". In abandoned meadows, 10 years after the cessation of mowing, the tussocks are found to jointly form mosaic patterns that may persist about 20—25 years, due to the longevity of sedge tussocks estimated at about 50 years.

Succession promoters

Plant species play different roles in the initiation of a succession and in the transformation of a phytocoenosis in the course of it. Some species attain the status of dominants or subdominants in the phytocoenoses that form successively, while others are supplementary elements present only in some succession phases. According to the nature of the relationship between population dynamics and vegetation dynamics, two species groups have been distinguished, namely:

1. *Short-term promoters.* This species involves increase, decrease and dynamic equilibrium of population size in one successional phase. They dominated shortly, especially in the initial phase of succession, whereas later they were represented by single individuals, or even absent individuals (Figure 3.20). This group comprises: *Cirsium palustre, C. rivulare, Polygonum bistorta* and *Potentilla palustris.*

2. *Long-term promoters.* This species involves changes in population size conforming to

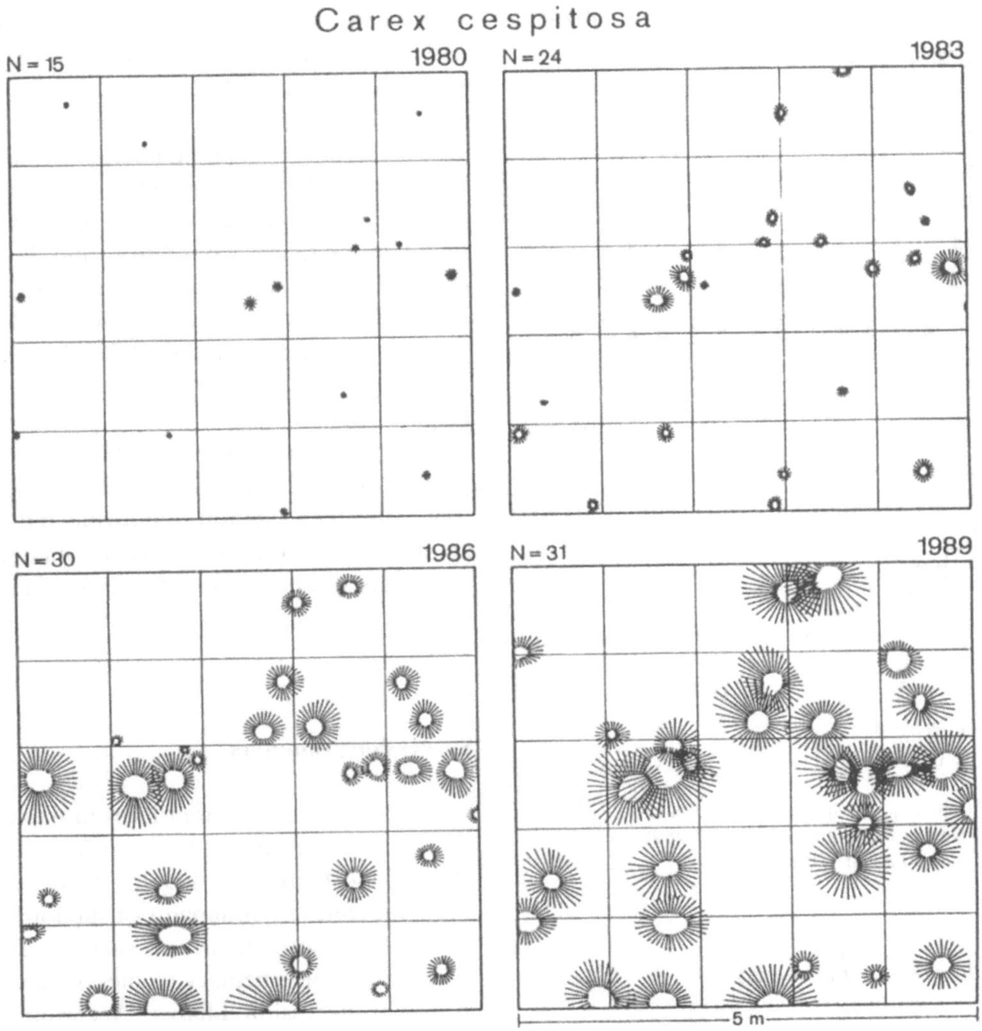

Fig. 3.18. Variation in density and distribution of *Carex cespitosa* individuals in an abandoned meadow.

the directional transformation of vegetation, hence: (a) the population size increased in the initial phase of succession, i.e. in the period of quantitative changes in the structure of meadow communities; (b) population size was maximum in the optimal phase of succession, i.e. in the period of qualitative changes — the formation of new phytocoenoses; (c) dynamic equilibrium of population size was attained in the terminal phase of succession, i.e. in the period of development of forest phytocoenoses. This group includes: *Filipendula ulmaria, Lysimachia vulgaris* and *Lythrum salicaria*. They were present in the whole successional process, although their proportions changed in the consecutive phytocoenoses. These changes either affected the transformation of vegetation or initiated new phytocoenoses.

Short-term promoters' characteristics

Populations of species included in this group

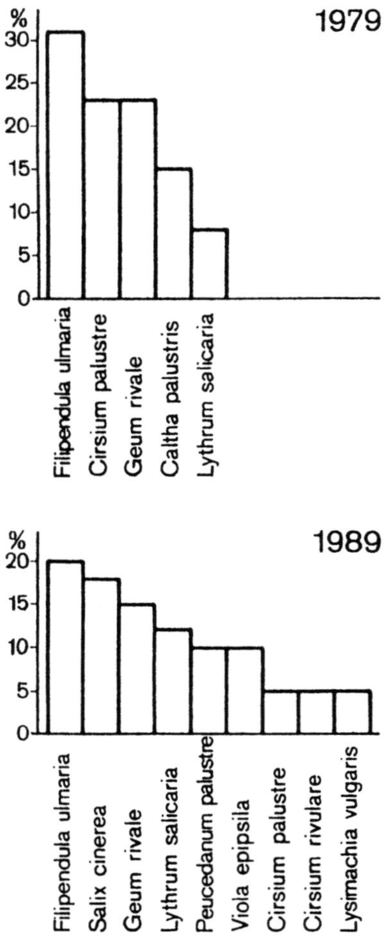

Fig. 3.19. Frequency of species in the clump of *Carex cespitosa* 5 years (1979) and 14 years (1989) after abandonment of the meadow.

undergo the following changes during a succession: (1) in the first 3–6 years following the cessation of mowing their numbers increase considerably, sometimes several-fold. Altered also is their spatial structure — from a random to an aggregative distribution type, with a tendency to form small aggregations (Figure 3.20). Their frequency also increases from 5 to 10%. Estimated 7 years after the cessation of mowing, the population density and frequency are much lower. Then for some time (2–3 years) the populations are characterized by a density and distribution similar to those observed in regularly mown meadows.

In the populations of certain species the growth in density of tall herbs, progressing with an advancing succession, causes an effect similar to that evoked by mowing, that is: (a) a prolonged growth cycle, (b) a reduction of the fraction of individuals attaining the fruiting phase, (c) a decrease in individual fecundity and population reproductiveness.

The short-lived role of some species in the vegetation changes, especially during the initiation of succession in meadows, is primarily the result of the increasing fecundity of individuals immediately after the cessation of mowing, which is in turn related to their high growth rate in that period. In addition, there appear many gaps in the sward of an abandoned meadow. The gaps are colonized above all by plants with a high seed production rate (e.g. *Cirsium*) in the first years of meadow abandonment. Their further persistence in a meadow, especially in periods of increasing population growth and density of many species, depends on a property that may be defined as the "rosette mechanism". In the first period of their life individuals representing these species, particularly those of the genus *Cirsium*, produce rosettes made up of several leaves. The leaves grow in number from year to year and cover larger and larger areas, overshadowing the near-ground layer of the meadow phytocoenosis (Figure 3.21), and thus limiting the growth of individuals of other species. With the progressing succession, however, more and more rosettes die before attaining the fruiting phase. Successively dying rosettes (2–4-year-old) are not replaced by rosettes of the same species, but by polycormonal organisms that at that time colonize meadows, e.g. *Filipendula ulmaria, Carex acutiformis* and *C. cespitosa* (Figure 3.22). This is how the regression begins of some meadow species, terminated by the formation of herbaceous and scrub communities. There sometimes occurs an insignificant new increase in numbers in meadow species populations at the time of willow aggregation formation. Then on the fringes there appear small numbers of *Cirsium palustre, Ranunculus acris*

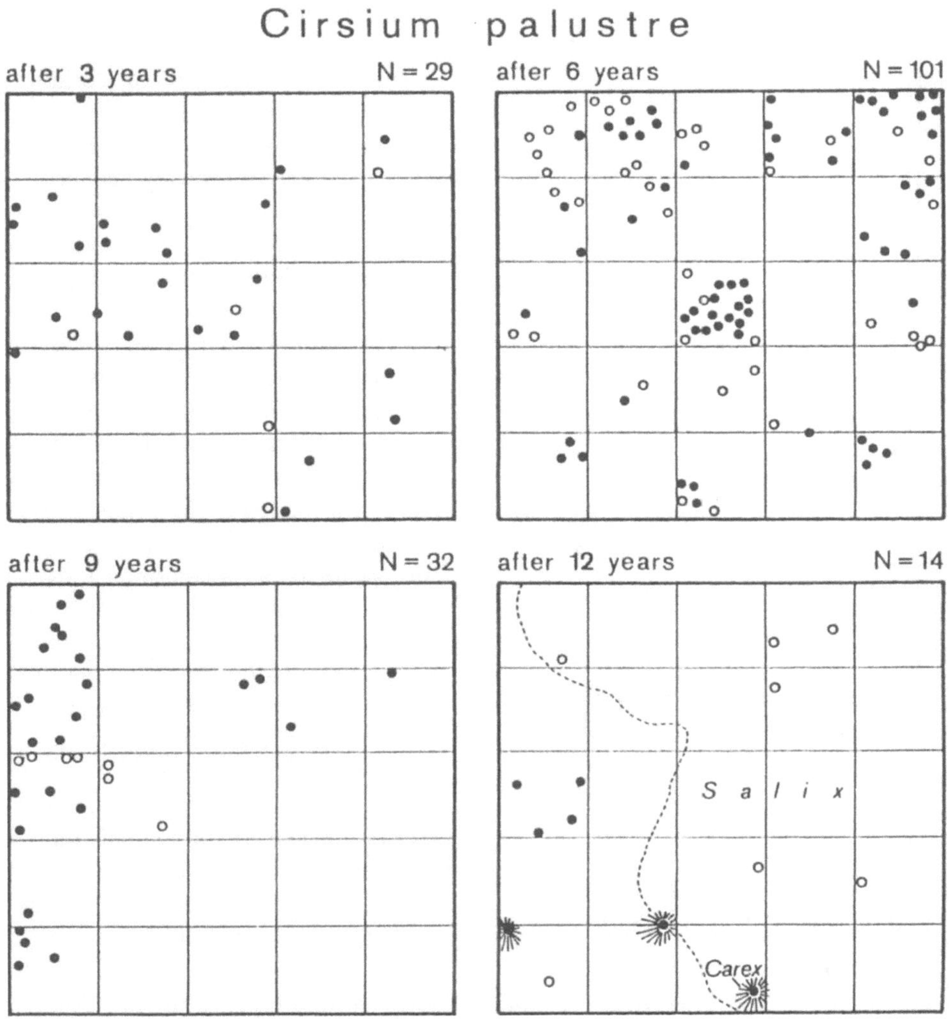

Fig. 3.20. Variation in density and distribution of *Cirsium palustre* individuals in the rosette phase (plain circles) and fruiting phase (shaded circles) for 14 years, i.e. until the formation of willow brushwood.

and *Lychnis flos-cuculi* individuals which may stay there till the return of forest to an abandoned meadow.

Long-term promoters' characteristics

Species included in this group have the following features: (a) ability to form polycormonal entities (multi-shoot organisms); (b) longevity of organisms, acquired owing to the capability of replacing old underground organ structures with young ones; (c) individuals' relocating potential through the growth in size of their peripheral parts and necrosis of their central fragments, in addition, this capability prevents a complete depletion of resources over long periods of area occupation by one organism, and at the same time favours the turnover of individuals in a population and of species in a community (Figures 3.3; 3.5); (d) capability of replacing or supplementing generative reproduction with vegetative propagation in

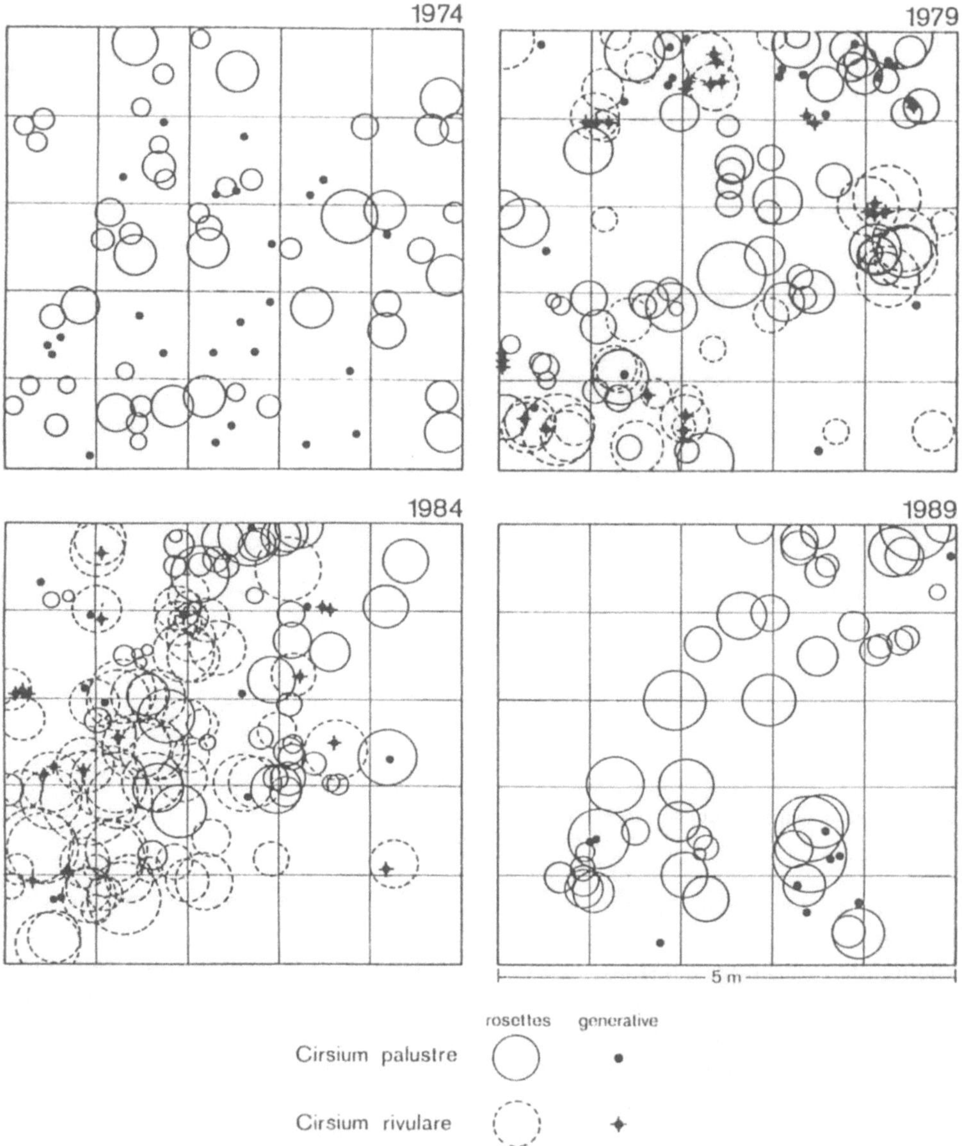

Fig. 3.21. Variation in density and distribution of *Cirsium palustre* and *Cirsium rivulare* individuals as "rosette mechanism" of species turnover in the successional process.

periods of deteriorating habitat conditions and increasing competition during succession (Figure 3.23); (e) growth-cycle modification.

Owing to the above characters, the species attain the status of dominants or subdominants in several successive communities in a progressing succession. Populations of long-standing promoters attain the numeric-balance phase in regularly mown meadows (low and stable numbers), and when there have formed woodland communities (Figure 3.23).

Plant traits and the colonization rate

A fast rate of meadow colonization by some

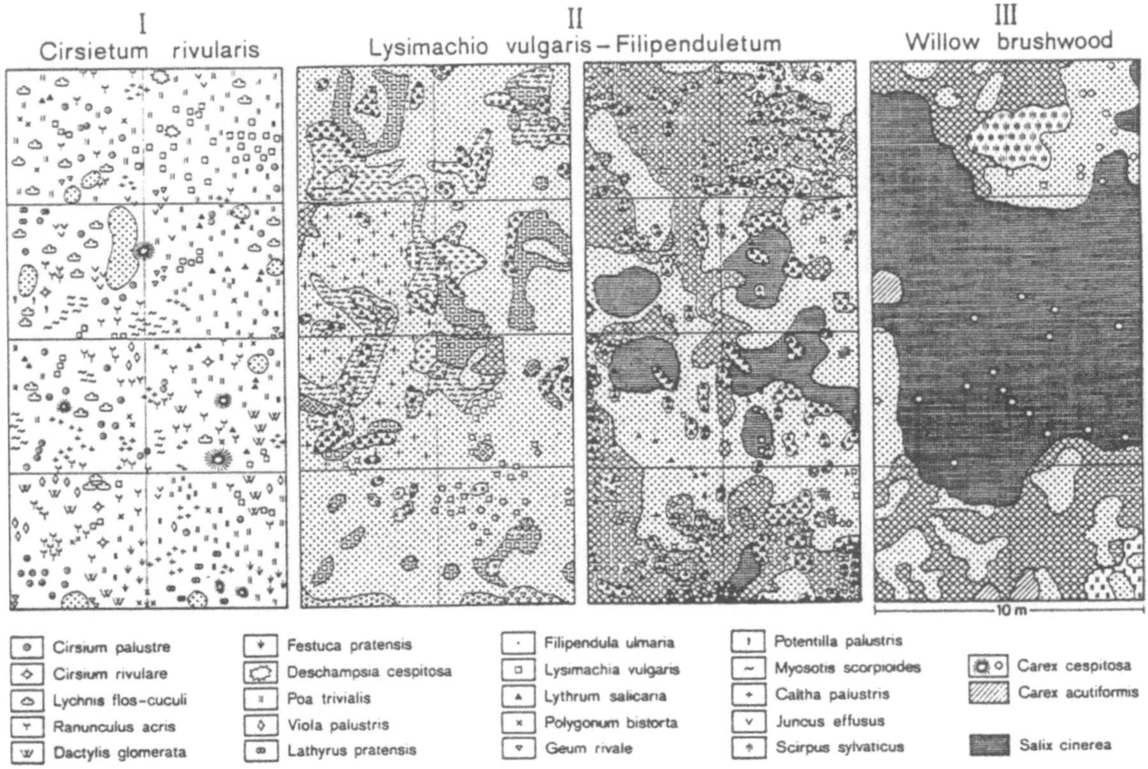

Fig. 3.22. Dynamics of spatial structure of *Filipendula ulmaria* population and its affect on the changes in vegetation during succession. I—III succession phases.

species results from the specific properties of polycormonal organisms, that is, an exponential growth of the number of shoots, a fast increase in area size and rate of its filling, a high biomass production and growth increase with age, as also the occurrence of two reproduction types, especially in periods of deteriorating habitat conditions. Long-term observations of labelled individuals of such species as *Filipendula ulmaria, Lythrum salicaria* and *Lysimachia vulgaris* have demonstrated that some of the traits peculiar to species with the phalanx or "guerilla" colonization strategy (Harper 1986) are noticeable in the same organisms over their long life-history. For example, young individuals produce numerous, intensively branched rhizomes which enable them to become incorporated among the species of a floristically rich meadow and then eliminate those species. Older individuals show a tendency to form compact aggregative structures, each made up of a dozen or so, or several dozen shoots growing close to each other, interconnected with strong, durable rhizomes. With the advancing age of the polycormones, subject to changes is not only their architecture, which significantly affects their relations with their neighbours, but also the quantitative ratios between the generative and vegetative offspring, in favour of the latter. This prolongs the life of some genets in spite of the deteriorating habitat conditions. Thanks to the high plasticity of plants, that is, variable size, biomass and fecundity, the species survive in the course of succession, in spite of the growing competition of trees, e.g. willows. Variation of individual characters of plants during a succession has been reported by many authors (e.g. Knapp 1974, Kershaw 1973, Noble and Slatyer 1980, Fekete and Melkó 1981, Falińska 1986b, c, 1990).

All changes related to individual and clonal

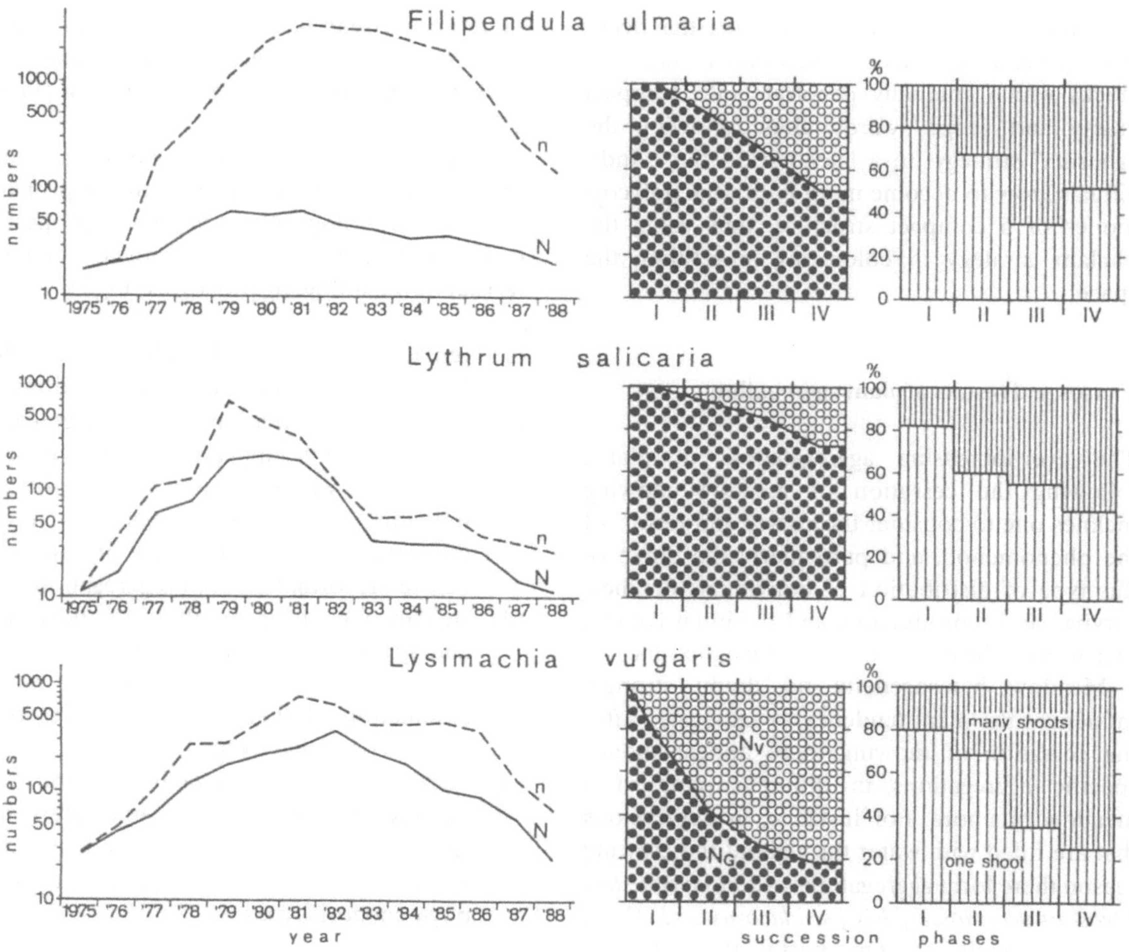

Fig. 3.23. Changes in populations of long-term succession promotors. Changes in number of individuals (N) and shoots (n) for 14 years; proportions of generative (N_G) and vegetative (N_V) origin; proportions of monocormones (one shoot) and polycormones (multishoot) in consecutive succession phases: (I) meadow, (II) macroforbs communities, (III) willow brushwood, (IV) brushwood community.

growth significantly affect the fate of a population. For example, the retreat of *Filipendula ulmaria* several years after the cessation of mowing is often interpreted as the effect of the appearance of competitors (forest species), especially willows. Close studies of the retreat of *F. ulmaria* and its replacement by willows (*Salix cinerea, S. pentandra*) have shown, however, that for such a situation to occur there has to have taken place a process of polycormone senescence and necrosis (Figure 3.13). Only are the areas, thus released, occupied by willows. By shading and mechanical action the latter gradually eliminate all the surviving polycormone fragments of *Filipendula ulmaria, Lythrum salicaria* and *Lysimachia vulgaris*. This way willows contribute to the decomposition of a herbaceous phytocoenosis, the first succession stage.

By observing labelled individuals it has been found that in longaeval organisms age-related changes of characters, especially of the architecture, also involve colonization-strategy changes. For example, in the first period of their life (up to 3 years), individuals of *Filipendula ulmaria,*

Lythrum salicaria and *Iris pseudacorus*, that occur over a long-lasting succession, produce long rootstocks which gradually penetrate into compact grassy and sedge swards, thus following the "guerilla" strategy. In a later period those individuals grow to become multi-cormonal polycormones of a compact structure, thus using the phalanx strategy in colonizing and filling the space.

Causes and results of plant aggregation

The ever-increasing aggregateness of plants following the cessation of meadow mowing inclines one to consider the causes and results of the phenomenon, and particularly the effect of this type of distribution of individuals on their survival and reproduction, and the intensiveness of action on the environment and neighbours.

Meadow heterogeneity no doubt strongly influences the plant tendency to aggregate. After the cessation of mowing there appear monospecific aggregations in different parts of a meadow biochore. For instance, in depressions that are filled with water throughout the growing season there form aggregations of *Caltha palustris, Myosotis scorpioides, Iris pseudacorus* and *Carex vesicaria*, while at temporarily flooded sites there arise assemblages of individuals of *Filipendula ulmaria, Lysimachia vulgaris, Geum rivale, Potentilla palustris* and *Carex cespitosa* (Figure 2.3).

Under natural conditions, however, plants predisposition towards aggregating lie primarily in their morphological and biological properties, owing to which a species turnover is characterised by the following processes (Figure 3.22):

(1) replacement of a floristically rich meadow by large-sized monospecific aggregations;
(2) gradual fragmentation of an aggregation leading on to the appearance of gaps in it;
(3) free spaces within the area are colonized by individuals of various species, as a rule different from those that were present there before the formation of the aggregation;
(4) steadily increasing numbers of individuals and species in the gaps within the aggregation, which results in a total or partial elimination of the dominant species from the area occupied by it;
(5) increasing species diversity; in the area previously occupied by one dominant species there occur forest and meadow species, as also those common to both phytocoenosis types.

CHAPTER 10

RECAPITULATION: SUCCESSION AS A PLANT-BY-PLANT REPLACEMENT PROCESS

Some perennial species are present in the whole successional process, from meadow to forest, but with different proportions (5% to 80% area) in consecutive communities. The species exhibits high colonization ability, increases rapidly in number in a few years and maintains its dominance for some time. Crucial in the colonization and establishment of promoters is their specific clonal growth. Geometric increase in the shoot number and longevity of organs determine the role of clonal species in the dynamics of populations and communities, and determine their competitive ability. Only senescence and disintegration of polycormones give the chance to other species to appear within their area. Senescent polycormones are replaced by willows that initiate the development of brushwood. This indicates that competitive ability is closely related to individual age, which was also confirmed by Kershaw (1962) for *Carex bigelowii*. From earlier studies on promoter populations it can be inferred that it is the age structure which is responsible for the course of many population phenomena and determines the role of species in the structure and dynamics of vegetation (Falińska 1989a, b, 1990). Clumped spatial structure is mostly interpreted as population response to habitat microstructure and as the effect of uneven dissemination (Harper 1977, Connell and Slatyer 1977). However, an increase in the clumping-degree in populations results mainly from the growth of individuals into multishoot polycormones and clone formation. The division of polycormones and senescence of their oldest parts leads to the comeback of random spatial structure of the population, associated with decreasing proportions of these species in the structure of consecutive communities (Falińska 1985).

The architecture of clones determines their relationship to space and neighbours. The range and rate at which an individual can draw on the habitat resources depends on the size of its area and on the way it is filled (White 1979). Often a plant growth model is analysed as a way of space occupation (Buss 1979) and thereby as one of the main factors shaping the spatial structure of populations (Kershaw 1973).

For instance, young individuals can produce numerous highly branched rhizomes, enabling them to penetrate the space between species of a floristically rich meadow and to eliminate them. On the other hand, older individuals tend to form compact structures, composed of nearly a hundred shoots, growing in close vicinity and connected with long-lived and strong rhizomes. Then, the population has a pronounced clumped spatial structure, which results not only from the architecture of polycormones that changes with their age, but also from the fractions of generative and vegetative offspring. This considerably prolongs the life of some genets in spite of the worsening environmental conditions (Grime 1979).

When conditions are generally favourable for growth, the clone size increases progressively with genets age (Watkinson and White 1985). Clones may form large strands of a single genotype (Silvertown 1987).

Due to the high plasticity, variable size, biomass and fecundity of individuals the species are present throughout the succession, despite increasing competition from, e.g. trees.Variation of individual properties in the successional process has often been encountered (e.g. Holt 1972, Kershaw 1973, Knapp 1974, Fekete and Melkó 1981, Falińska 1986a, 1989a, b, 1990).

Data from studies of the fate of individuals over the long-term cycle of their life indicate that their architecture and the intensity of the renewal and ageing processes change significantly with advancing age. It has been found that the above properties play the role of a biological mechanism responsible for the exchange of species during succession. This can account for the similarity between the population age structure and the phases of population and succession. The clonal growth form of many species contributes to its success in successional series from the floristically rich *Cirsietum rivularis* meadow towards the flood-plain forest of *Circaeo-Alnetum*.

Changes that take place over the long-term growth cycle of polycormones, culminating in their disintegration, can hardly be unequivocally defined as senescence processes. The problem is, among other things, that the causal relationships between particular age-related signs and the size of individuals are not yet understood well enough. Besides, it is not always known which of them are primary and which are secondary. It seems, however, that the formation of necrotic zones in polycormones is a sign of senescence. The phenomenon is irreversible, at least in polycormone fragments, and becomes noticeable at a certain age. If, however, this necrotic part performs a reinforcing function (as a framework preserving the entirety of an individual) and does not cause gradual deterioration in the functioning of a polycormone – can it be accepted that is the sign of senescence? Even if senile rhizomes occupy 50% of the area of a polycormone, active growth continues on its periphery, while its death or disintegration is forced by external factors.

It has often been discussed whether the dying of modules and other parts of long-lived organisms results from their senescence, hence whether it is an irreversible process of the worsening of their habitat conditions (cf. Turner 1950, Hamilton 1966, Bell 1984, Watkinson and White 1985). Reproduction has frequently been associated with the age of an organism and its modules (Rabotnov 1985, Sarukhan 1980, Charlesworth 1980, Callaghan and Emanuelsson 1985). A population rejuvenates due to the ability of plants to regenerate lost structures, and there is a constant emergence of new structures with a high reproductive potential. Moreover, migration of new modules outside the area occupied by the oldest parts of an organism contributes to the specific "movement" of individuals in space. The process prevents total resource depletion, caused by the long-term occupation of one site by a single genotype. Parallel processes of senescence and rejuvenation, characteristic of perennial species, prevent the stagnancy in natural populations, and at the same time enable them to be permanently present in a site, e.g. in the course of long-term succession.

In many plant species, in perennials in particular, evolution went towards increasing the number of rooted shoots and enlarging the underground organs, that is, towards polycormonal forms (Penzes 1960). Organism enlargement in width considerably increases a plant's competitive power in communities, but at the same time it threatens its existence if the physical limits of material strength, or the possibility of functioning of joint parts, are exceeded. If we assume that a plant organism in the form of an integrated "assemblage of shoots" has a better chance to survive and produce offspring, it is worth considering the question as to what role disintegration processes play in its life, and in the dynamics of populations and vegetation, and whether they result from the action of competitors, deterioration of habitat conditions, or are caused by the senescence of the plant, that is, deterioration of the quality of its functioning.

Everything indicates that in polycormones, cyclic senescence and rejuvenescence take place. It is not known, however, whether this is proof that senescence processes are reversible, or that ageing is prevented.

It is probable that in the case of many modular plants the essence of old age is a gradual disintegration of the whole individual, which may result from disturbances in the relationships between the above-ground and the underground parts (Woolhouse 1972). For instance, disturbed biomass proportions between the underground and the above ground organs in 10—12-year-old *F. ulmaria* polycormones impede the functioning of the plant body, i.e., reduce the uptake of water, mineral salts, and hamper assimilation, which results in the death of an individual or necrosis of its part (Falińska 1989a, b, 1990).

Over the long-term life-history of clones there take place changes which can, following Noodeén (1980), be defined as growing age in the sense of elapsing time (ageing), and not in the sense of becoming old (senescence).

Maybe clone growth accompanied by clone disintegration provides a mechanism ensuring a long-lasting presence of a genet in the form of replications, and at the same time it is a factor preventing resource depletion. New entities as a rule develop on the periphery of the area occupied by the parent plant. Thus, disintegration causes the formation of many independent genetically identical entities, and may in the end result in a genetical monotony in a population. In fact the vegetative propagation is very common, but the genetical monotony does not occur in natural populations as often as it might be expected. For instance, in a 1 m^2 meadow there can be an average of 44—49 white clover genets (Trathan 1983), with clones intermingled and not forming pure patches (Cahn and Harper 1976). Even in the case of *Mercurialis perennis*, a species of very low generative reproduction efficiency, an aggregation covering a 400 m^2 area developed from 8 individuals of generative origin (Falińska 1982). In the population of *Filipendula ulmaria* the genetic diversity decreased almost by 50% within 15 years (Part 4).

Genet-level stuides on clonal plant demography thus support the view that vegetative reproduction is part of an integrated growth process (Caswell 1986, Harper 1986, Cook 1983, DeSteven 1989).

Integration within clones probably prolongs the life of some genets in spite of the worsening habitat conditions. Disintegration within clones probably facilitate clonal expansion and maintains the dominance of *F. ulmaria* in abandoned meadows.

In studies of population dynamics, much attention is paid to density which is often treated as one of the agents that exert an influence on the demographic processes. The effect, often confirmed, of density on individual size and habitus, and on the ability of individuals to attain the reproductive phase, provides the basis for considering the phenomenon of autoregulation in plant populations (White and Harper 1970, Harper and White 1974, Lee and Bazzaz 1980, Silvertown 1987). Simple relationships between density and reproduction and mortality were seen primarily in annual species. Less unequivocal results have been obtained in the case of perennial species. During their life cycle individuals of some herbaceous perennials increase the number of rooted shoots, which indicates that density increase and decrease are closely correlated with changes in population age structure. The low density of polycormonal species in the first colonization period results from the growth of polycormonal individuals randomly distributed in the meadow sward. A high rate of shoot multiplication leads on to the formation of polycormones and a gradual filling of space by them, which significantly limits the development of juvenile individuals and seedling emergence. Here the following question arises: Does this result from population density, or from specific interactions between mature individuals and the offspring? Hartnett and Bazzaz (1985a, b, c) found that interconnected ramets of *Solidago canadensis* were less affected by interspecific competition at high density than were individual shoots grown from seed.

Distances between individuals determine the relations between them, which vary with their age, because during the life-history of plants their demand for space varies (Mithen, Harper and Weiner 1984). However, the effects of density are usually considered in relation to the number of individuals or shoots in a particular area, not in relation to the distances between them. In *F. ulmaria* populations it is difficult to separate the effect of density from the effect of polycormone senescence (Figure 3.13). For example, in the period of maximum density in a population, the largest number of polycormones attain the age at which there occur signs of ageing, e.g. a decrease in the number of generative shoots and renewal buds.

The tendency towards a reduction in the number of genets in a population, accompanied by an increase in the number of ramets per genet was observed in *Solidago canadensis* (Hartnett and Bazzaz 1985a, b, c).

PART 4

PLANT DEMOGRAPHY AND POPULATION DYNAMICS

CHAPTER 11

SUCCESSION AS THE RESULT OF DEMOGRAPHIC PROCESSES

Introduction

The processes of phytocoenosis formation, changes in it, and its passing, no doubt result from intra-population processes and inter-population interactions against the environment of living, broadly understood. The range of inter-individual and inter-population reactions within phytocoenoses can be explained by the life strategy of species. Individual characteristics as the size, habitus, organ durability, manner of vegetative renewal, and reproduction mode, all shape the population properties, and thereby determine the role of a particular species in a phytocoenosis. The role is expressed by the species being a member of one or several phytocoenoses during a progressing succession, where its status is that of a dominant permanent component, or a temporary complementary element.

The appearance of new species, their reaching a certain level of numbers, as well as the retreat of various species lead on to floristic-composition changes with a progressing succession. Important for the course of succession are two fundamental demographic mechanisms, viz.: a) a plant's ability to colonize free spaces, b) a plant's capacity to continue to exist in the colonized area, or their lack of such a potential (Gray, Crawley and Edwards 1987).

The plant population's ability to change is an imminent character of its existence, being the resultant of demographic processes and at the same time a response to the varying environmental conditions. The rate and intensiveness of the changes in the population properties, and thereby the nature of plant communities depend primarily on the life strategies of species, and particularly of characteristics such as growth types, life forms and reproduction modes (Horn 1976, Drury and Nisbet 1973, Connell and Slatyer 1977, Noble and Slatyer 1980).

Individual life, reproduction and death, and thereby population size, are effected by many abiotic, biotic and intra-population agents (Harper and White 1971, Harper 1977). At the successive stages of their growth, the survival and further development of individuals, and thereby the fate of a population, are decided by various factors (Grime 1979). The latter may include shading (Barkham 1980a, b), plant competition (Sarukhan and Harper 1973, Sharitz and McCormic 1975, Gross 1980, Turkington and Harper 1979), closed vegetation cover (Platt 1975, Grubb 1977, Gross and Werner 1982), as well as weather conditions.

Studies of the population dynamics of the species taking part in a succession have demonstrated that all the above factors affect the demographic processes of plants (Falińska 1989a, b), but the effects of the particular agents on the fate of individuals vary with species and succession phases (cf. Part III).

In this book population dynamics is understood as a series of changes in population properties with time. Descriptions of the populations of the species participating in the succession processes take into account the direction of the seral changes, their intensiveness, long duration, as well as their causes, especially the population size in relationship to space.

Population abundance dynamics types during a succession

According to the direction, intensiveness and

duration of population size changes during a progressing succession, three species groups have been identified (Figure 4.1):

1. Species whose populations attain high levels of numbers within a short time after which a decline follows. As a rule, most intensive abundance variations occur at one succession stage, usually the initial one (Figure 4.2). The individuals of these species have short life-histories. This group includes *Cirsium palustre, Ranunculus acris* and *Lychnis flos-cuculi.*

2. Species with population-abundance variations concordant with seral vegetation changes, that is: a) the growth in numbers begins at the initial succession stage, that is, immediately after the cessation of mowing; b) the maximum abundance coincides with the optimum succession stage, that is, the time when there form new floristic compositions, e.g., herbaceous and scrub communities; c) the abundance equilibrium stage is reached at the terminal stage (in forest phytocoenoses). Included in this group are *Filipendula ulmaria, Lysimachia vulgaris* and *Lythrum salicaria.*

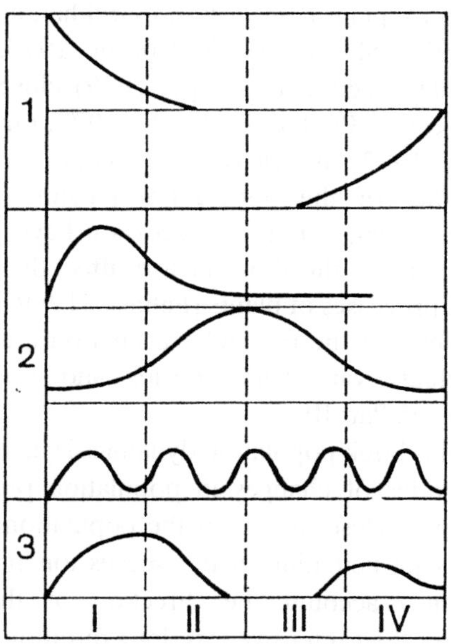

Fig. 4.1. Model of population dynamics types in the successional process; succession phases: (I) meadow, (II) macroforbes community, (III) willow brushwood, (IV) brushwood community; 1—3 — species types distinguished with respect to their role in the succession: 1 — the development of populations accompanied one successional phase; 2 — the development of populations lasted as long as forest comeback; 3 — the population size fluctuates, more rarely changes in cycles.

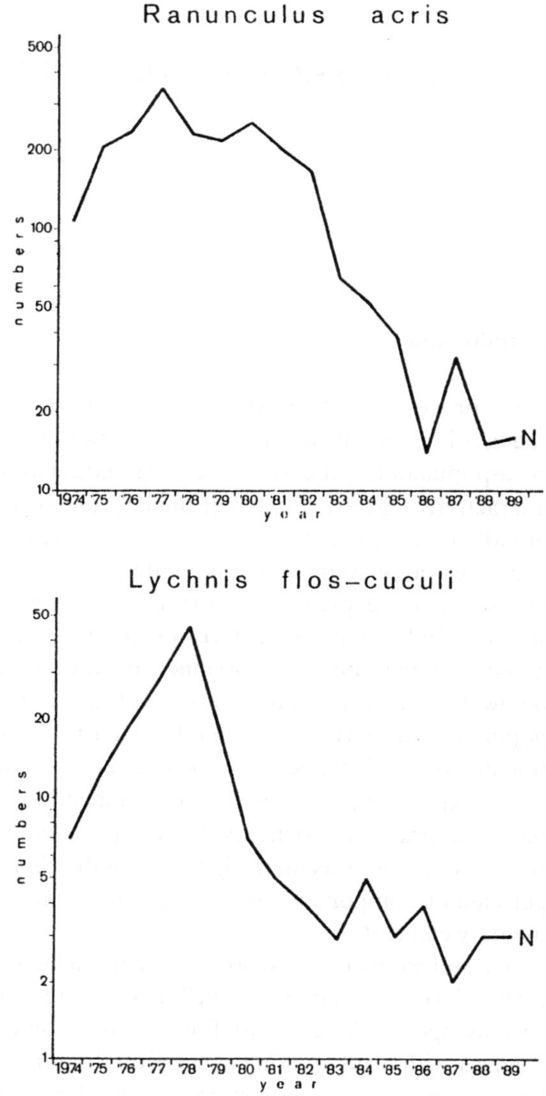

Fig. 4.2. Population dynamics of two species succession. A fast rate of abundance growth and decline after abandonment of the meadow.

3. Species with population-abundance changes of the type of fluctuations. A growth or reduction in numbers is not clearly correlated with the

sequence of vegetation communities. The following species have been included in this group: *Caltha palustris, Myosotis scorpioides* and *Iris pseudacorus.*

The course of abundance variations and the demographic processes of the particular groups have been exemplified by selected species to which the role of demographical models has been assigned.

A demographic model — *Cirsium palustre*

Abundance dynamics

In meadows that are still used, *Cirsium palustre* individuals occur in small numbers and are distributed randomly. Mowing considerably reduces their growth because it is done during their flowering. Only some individuals do then develop fruit, but even this ensures a sufficient recruitment of new individuals every growing season, and thereby a continued presence of the species in mown meadows. In the last mowing year (1974) a total of 128 individuals were found randomly distributed over a 0.6 ha meadow, and an aggregation made up of 140 individuals, occupying a 124 m^2 area. The cessation of mowing was followed by a considerable growth in numbers and aggregateness in the *C. palustre* populations. Five years after the cessation, 4 large aggregations were found in a 0.6 ha meadow area, consisting of the following numbers of individuals: 88, 95, 101, 141, and smaller aggregations made up of different numbers of individuals from 12 to 48. There were 138 individuals dispersed among the aggregations.

The course of abundance changes in the *C. palustre* populations was not the same in the different meadow parts. Three types of population abundance dynamics have been distinguished (Figure 4.3).

1. A fast rate of abundance growth and decline when a particular species inhabits meadow parts in which *Salix cinerea* attains the status of dominant within a short time following the cessation of mowing.
2. A gradual or rapid growth in numbers followed by a stabilized low level of numbers. A short phase of equilibrium is followed by regression and a new stabilized state at low numbers. This type of population dynamics was observed in meadow fragments in which after the cessation of mowing macroforbes, aggregations formed.
3. A high level of numbers during meadow usage, and after meadow abandonment a population decline preceded by fluctuations lasting several years. This situation was seen when on poorly-mown sites *C. palustre* formed small aggregations in which willows or sedge appeared immediately after the cessation of mowing. They gradually eliminated *C. palustre* individuals which for some time still persisted on the margins of willow aggregations, or in gaps in them.

Fecundity and survivorship

Cirsium palustre is a facultative biennial. In the first growing season, each of its individuals develops a rosette, and in the second growing season it forms a flowering shoot. Under unfavourable environmental conditions rosette growth lasts 2 or 3 growing seasons, and therefore the first individual flowering is considerably retarded. The growing herbaceous species density that follows the cessation of mowing brings about an increase in the fraction of individuals "held up" in the juvenile phase, from 25% to 78% (Figure 4.4). Over a half of the individuals held up at the rosette stage never reach the flowering phase. This is so because the flowering phase is attained by individual rosettes that are 25—30 cm in diameter, and several-year-old *C. palustre* rosettes growing amidst sedges and herbs seldom exceed 20 cm in diameter (Figure 4.4). With the growing herbaceous and scrub communities there is little chance for *C. palustre* to complete their growth cycle. They are then found only on fringes of *F. ulmaria* and *Salix cinerea* aggregations or in *Carex ces-*

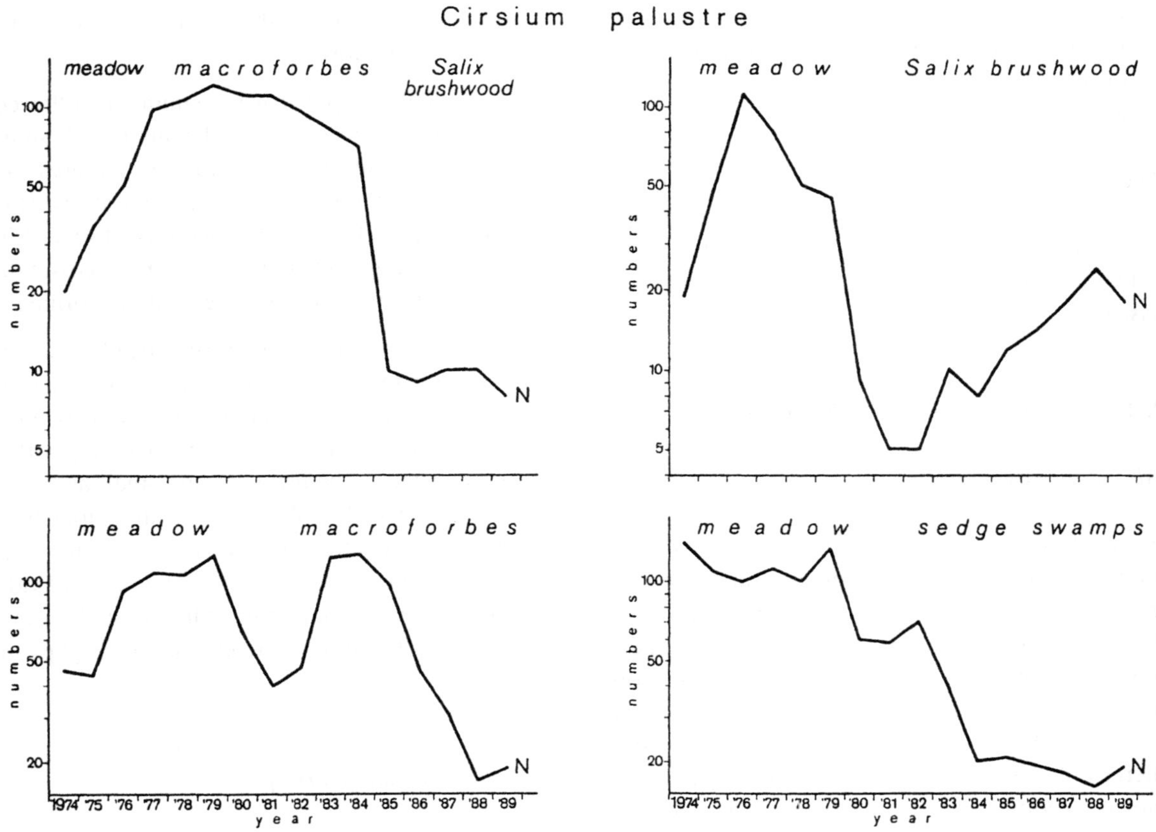

Fig. 4.3. Population dynamics short-term promoters of succession in different patches in abandoned meadows.

pitosa tussocks (Figure 3.13). This situation may stay until the return of forestation (Photo 3.7).

A demographic model — *Filipendula ulmaria*

Abundance dynamics

In meadows that are mown regularly *Filipendula* occurs in small but stable numbers. For instance, during a 5-year period the number of individuals per 1 ha varied between 32 and 36, and the number of shoots between 93 and 107. Immediately after the cessation of mowing there begins a rapid growth of the population. The rate of growth in number of the individuals (N) and shoots (n) is not, however, the same in all parts of a meadow abandoned at the same time (1974). A comparison of N and n in two 200 m^2 plots located in a 1 ha meadow shows that they differed in this respect for 14 years (Figure 4.5).

The fastest abundance growth rate was recorded when *F. ulmaria* was the first of many meadow species to begin to colonize a particular area. In every situation N increased 10-fold within 5—6 years, and n 100-fold (Figure 4.5). During the same period, however, the number of individuals (N) increased only 3-fold, and the number of shoots (n) only 50-fold when a particular area was colonized also by *Carex acutiformis* simultaneously with *F. ulmaria* (Figure 4.5).

The mechanism of the changes is as follows. In the first *F. ulmaria* population growth period the increase in the number of the individuals is accompanied by shoot multiplication within polycormones, which results in a complete or partial filling of an area (3000—5000 n/200 m^2) (Figure

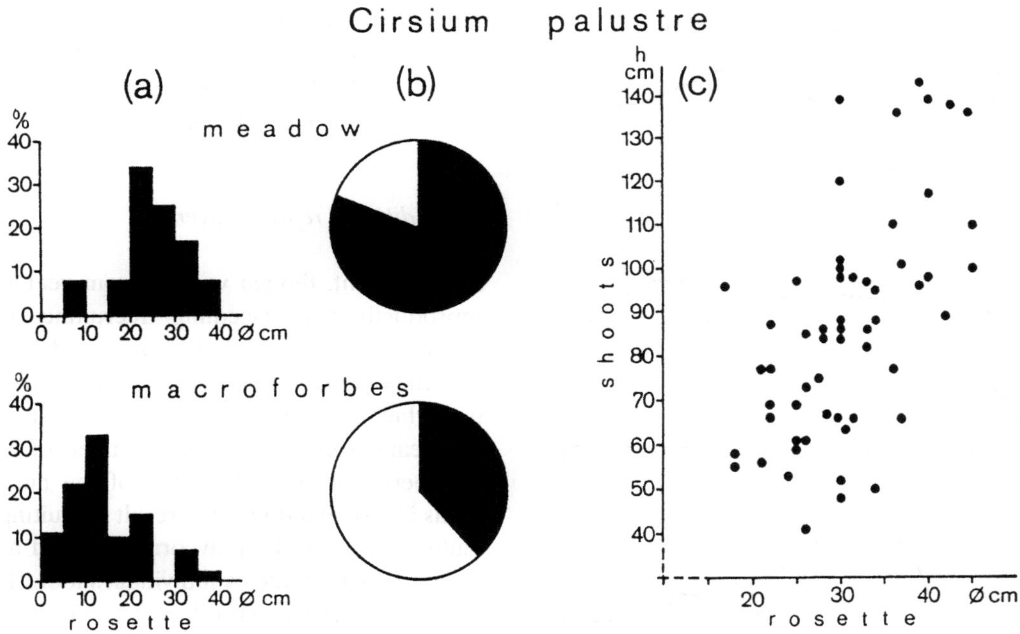

Fig. 4.4. Changes in *Cirsium* populations in the course of succession: a — differentiation of rosette size in meadow and macroforbes communities; b — the fraction of fruiting individuals (shaded); c — shoot height versus rosette size (Falińska 1989a, b).

3.22) and that in turn impedes the growth of seedlings, and contributes to a mortality increase of juvenile (the rosette stage) and monocormonal (one-shoot) individuals. Then, i.e., between 7th and 9th population-growth years, the first abundance decrease is noticeable. At the same time, however, some of the polycormones reach the disintegration stage, giving rise to vegetative-origin individuals (vegets), and thereby bringing about a growth in numbers. During the subsequent 2—3 growing seasons part of the vegets die, which leads to another reduction in numbers between 12th and 14th population growth years (Figure 4.5).

In areas where there occurs *Carex acutiformis* the number of *Filipendula ulmaria* individuals doubles within the first two growing seasons following the cessation of a meadow, and 7 years after, its gradual decrease begins (Figure 4.5). Within a sedge aggregation, *F. ulmaria* individuals do not grow in size to form multi-shoot polycormones, and they seldom disintegrate. For this reason, the role of vegetative propagation is less important than in the former situation. After 14 years, 18 in 25 individuals (in a 200 m^2 area) were of generative and 7 (28%) of vegetative origin, the latter having arisen as a result of the fragmentation of 3 polycormones. If, however, *F. ulmaria* is the first colonizer, then among 60 individuals 22 are of generative and 38 of vegetative (63%) origin.

A comparison of the variation in numbers of *F. ulmaria* in 6 areas in a meadow abandoned in 1974 shows that the first type of population growth (Figure 4.5a) occurred in two, the second (Figure 4.5b) in one of them. In the remaining three areas the abundance growth rate was much faster which resulted in the formation of monospecific *F. ulmaria* aggregations (Figure 3.22). In as many as 17 of the 24 plots set up in the Reski Range for studying phytocoenotic population dynamics *F. ulmaria* aggregations formed within 6—8 years. Since this is the typical mode of abandoned meadow colonization by this species, the demographic analysis has been made on the basis of data from a selected plot (25 m^2), situated in an *F. ulmaria* aggregation (2310 m^2).

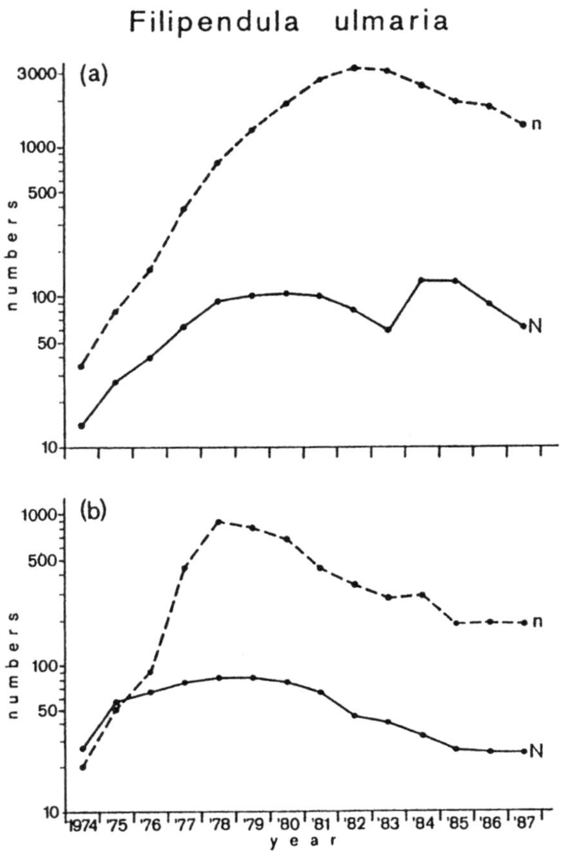

Fig. 4.5. Changes in the number of *Filipendula* individuals (*N*) and shoots (*n*) in 200 m² study area: a — *F. ulmaria* aggregations; b — *Carex acuttiformis* aggregations (Falińska 1989a, b).

Density

During a 14 years' population growth the shoot density changed significantly from 5 to 82 per m², i.e., from 45 to 1406 per 25 m². It was mainly the effect of shoot propagation by individuals over the long-term growth cycle, and to a lesser extent of a growth in number of monocormonal organisms. Even at peak abundance their density is small, ranging from 1 to 20 per 1 m², while at the same time the density of the shoots produced by polycormones ranges from 30 to 82 per m².

At maximum shoot numbers, that is, at a density of 1406/25 m², only 128 shoots had been produced by monocormonal organisms. All the remainder were members of 108 polycormones including varying numbers of shoots, between 3 and 60. Polycormones consisting of 6 to 22 shoots constituted the larges group (Figure 4.6).

Fecundity and reproductiveness

A permanent, though varying from year to year, reproductiveness (seedling recruitment) of a population (Figure 4.7) is maintained by a high fecundity (seed production) of *F. ulmaria* individuals which varies with their age. Individuals aged 3 to 7 years produce the largest numbers of seeds. In that period a gradual growth of the number of seeds in a population is the result of fruiting-shoot multiplication in the polycormones, and a higher seed production rate by individual shoots. Individuals of this age group produce 5 to 36 shoots each, the average number of seeds per shoot coming up to 425 ± 156. In this case high variation could be seen, for example the number of seeds per shoot could vary between 12 and 3200.

Depending on the population growth phase, the number of fruiting shoots in polycormones, and the production of seeds vary considerably also within the same age group of individuals. In the colonizing phase, for instance, individual shoots produce on an average 844 ± 554, and in the abundance-growth phase (6 years after the cessation of mowing), 1893 ± 1256 seeds. During that time in a 25 m² area there develop a total of 243 to 364 fruiting shoots. There is no relationship between the number of seedlings appearing during that period and individual fecundity. In 1 m² plots there appear on an average 2 to 16 seedlings.

Noticeable over the 14-year population-growth cycle is a relatively high proportion of juvenile individuals (Figure 4.6). Their number at first varies between 15 and 62, and estimated after 10 years, it appeared to have doubled. This was due to the successive growth in number of the offspring whose development was retarded, i.e., rosettes that did not attain the flowering and fruiting phases. *F. ulmaria* individuals as a rule flower and fruit in the second or third growing season, but from year to year there are more and

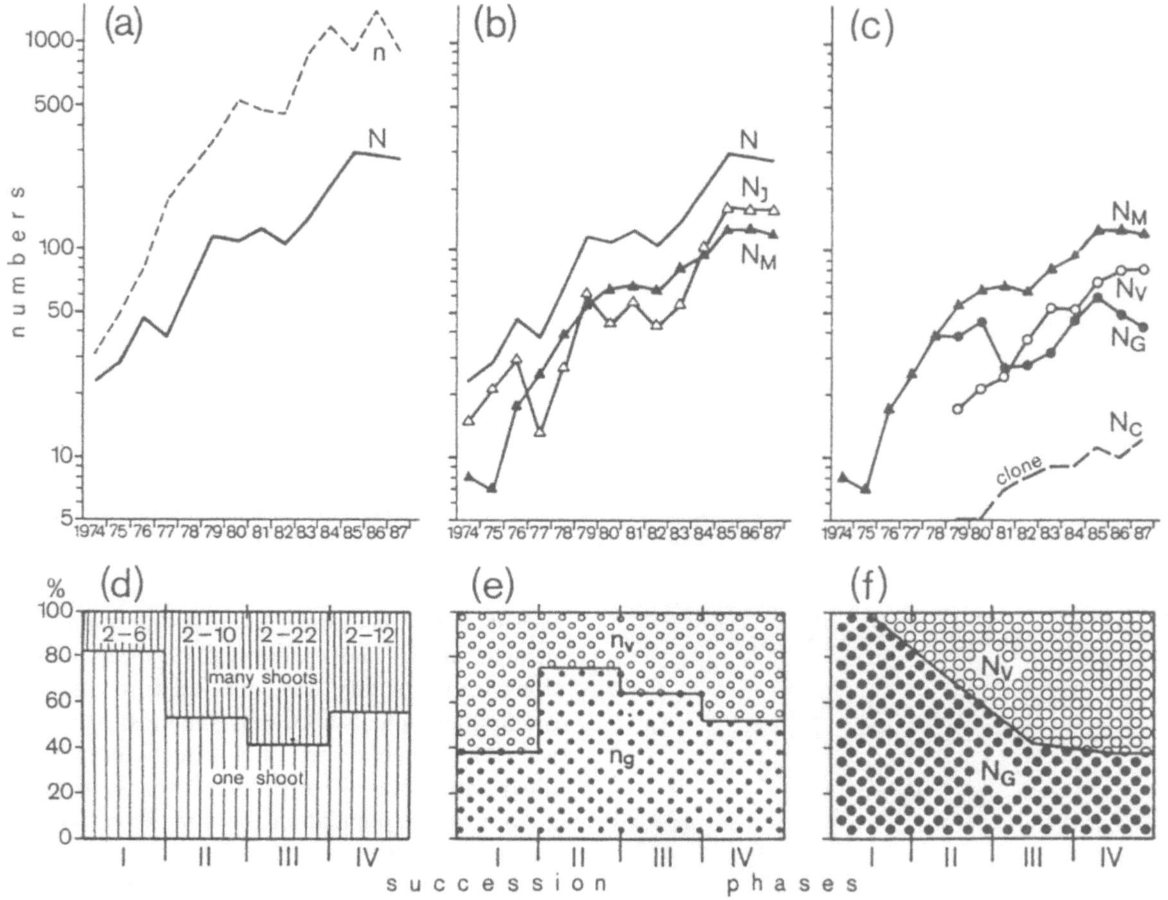

Fig. 4.6. Changes in the population of *Filipendula ulmaria* in consecutive successional phases: (I) meadow, (II) macroforbes community, (III) willow brushwood, (IV) brushwood community; a — number of individuals (N) and shoots (n); b — number of juvenile (N_j) and mature (N_M) individuals; c — number of mature individuals of generative (N_G) and vegetative (N_V) origin, ($N_M = N_G + N_V$) and clone number (N_C); d — the proportions of monocormonal (one shoot) and polycormonal (multi shoot) individuals in a population (the numbers indicate variation range of the shoot in polycormones); e — variation in the proportions of generative (n_g) versus vegetative (n_v) shoots in population; f — variation in proportions of individuals of generative (N_G) versus vegetative (N_V) origin. (Falińska 1989a, b).

more individuals that stay in the rosette phase (Figure 4.8). The process coincides with the period of an intensive growth in size of the polycormones, and clone development (Figure 4.7). Uneven-aged (from 2 to 5 years) juvenile individuals form a "bank of generative offspring" which plays a significant role in the population renewal during the formation of high-density aggregations.

The balance of numbers indicates that individuals developed from (N_G) seeds appear every growing season. Over the 14 years of observations the number of seedlings per 10 m^2 ranged from 11 to 160. A total of 680 of them appeared during that period, and 59, i.e., 6.2% survived till the juvenile or mature phase. Seedling survivorship varied between years: in four cohorts 20 to 36% of them survived, and in five, 0.8 to 5%. In four cohorts the mortality rate came up to 100% (Table 4.1).

Vegetative-origin (N_V) individuals begin to appear only when a particular population has grown for 6 years, i.e., when the first polycormones have attained the phase of disintegration. Between 1979 and 1987 there arose a total of 95 vegetative individuals (vegets), and 79, that is,

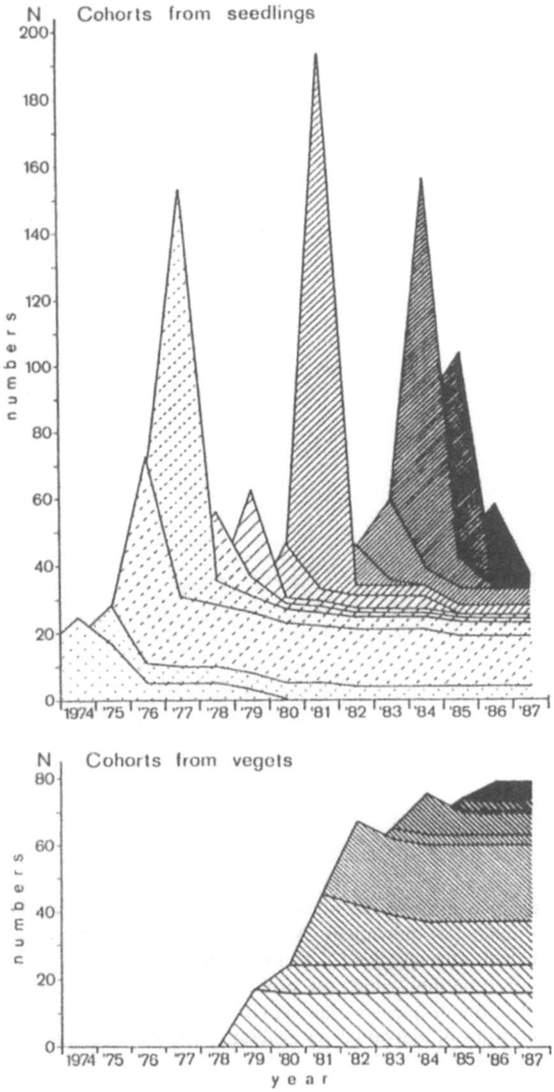

Fig. 4.7. Number and survival of individuals of generative (genets) and vegetative (vegets) origin in various cohorts (Falińska 1989a, b).

83% survived till 1987. Those vegets took origin from 12 individuals initiating 12 clones (N_C).

A demographic model — *Iris pseudacorus*

Abundance dynamics

Iris pseudacorus populations that grow under

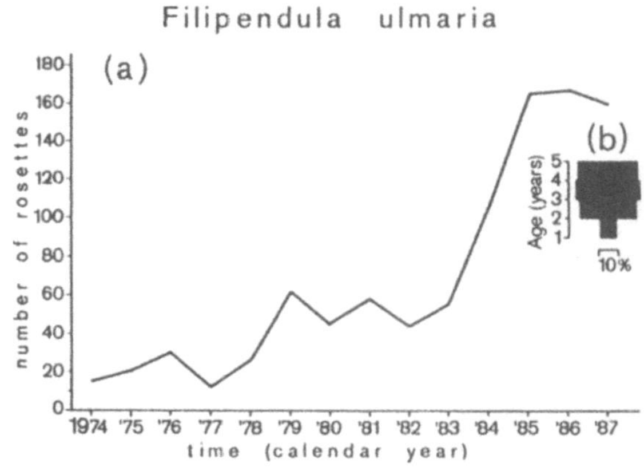

Fig. 4.8. Changes in the number of marked rosettes in a 25 m² plot in the period 1974—1987 (a) age structure of rosettes; (b) 14 years after abandonment of the meadow (n = 140).

varied conditions differ much in the abundance growth rate (N and n) and size. Of three populations whose history was followed from the time the first individuals appeared, the most numerous was that which occupied a trough temporarily filled with water (Figure 4.9). After an 8 years' growth of the population the number of individuals in a 400 m² area (2 × 200 m²) came up to 126, and the number of rosettes and shoots to 2240, while at the same time the respective quantities for a population in an alder-scrub were 95 (N) and 1806 (n). The lowest numbers were recorded for a sedge-swamp population where in 1984 $N = 20$, and $n = 899$. At the initial stage the populations under observation already differed from one another in abundance. The group that initiated the population growth in the trough included 15 individuals — the one in the alder scrub, 12, and that in sedge swamp, 5 — counting from the juvenile phase onwards. In all the populations, the highest abundance increase rate was recorded in the first period following the cessation of the use of the range.

A comparison of the numbers curves (N) indicates that the populations studied, except the one in the trough, attained a relative-equilibrium phase (Figure 4.9). When compared, the abundance curves for rosettes and shoots (n) show that these units still continue to grow in number in all

Table 4.1. Survivorship of *Filipendula ulmaria* individuals of different origin in abandoned meadows. Survival of cohorts of plants marked as seedlings (G) and independent parts of clone (V) in the field on the plot 25 m^2. Between August and October of each year (1974—1987) new plants were marked and their development observed.

(G)	Number of plants which survived to next year														Survivorship of cohorts until 1987
	1974	1975	1976	1977	1978	1979	1980	1981	1982	1983	1984	1985	1986	1987	%
	25	17	5	5	5	2	0	0	0	0	0	0	0	0	0
		11	6	5	5	5	5	4	4	4	4	4	4	4	36
			61	21	18	15	18	17	17	17	17	15	15	15	25
				122	7	5	4	4	4	4	4	4	4	4	3
					20	6	2	1	1	11	1	1	1	1	5
						25	1	1	1	1	1	1	1	1	4
							14	4	4	4	4	0	0	0	0
								160	3	3	3	3	3	3	2
									12	1	0	0	0	0	0
										25	5	5	5	5	20
											118	10	1	1	1
												61	1	0	0
													14	3	21
														12	100
Total in each	25	28	72	153	55	58	44	191	46	60	157	104	49	49	
(V)	Number of parts of clone														
						17	16	16	16	16	16	16	16	16	94
							8	8	8	8	8	8	8	8	100
								21	18	15	13	13	13	13	62
									25	23	23	23	23	23	92
										3	3	3	3	3	100
											12	7	7	7	58
												4	4	4	100
													5	5	100
														0	0
Total in each year						17	24	45	67	65	75	74	79	79	

the populations. According to the number of individuals (N) and units of which they are made (n), three phases have been distinguished in the growth of *I. pseudacorus* populations.

In the first population-growth period the rates of the increase in numbers of the individuals (N) and rosettes (n) are similar. This phase lasts 2, or up to 4, years, i.e., until part of the monocormonal individuals have developed into polycormonal forms. Only in the first year of new area colonization by *I. pseudacorus* can the number of individuals (N) be equal to the number of rosettes and shoots (n). With the passing population-growth years the difference between the number of individuals and the number of rosettes increases considerably due to the multiplication of rosettes and shoots in polycormones.

The second population-growth period is distinct for a high growth in numbers of rosettes (n), with a much slower numeric growth of individuals (N). It happens so when some individuals of a population attain the age of 3 to 5 years and start a geometrical-progression increase.

The third peirod is distinguishable for slight variations in the number of individuals (N), and a slower numeric growth rate of rosettes (n), although growth still continues in all populations. The changes indicate that the particular popula-

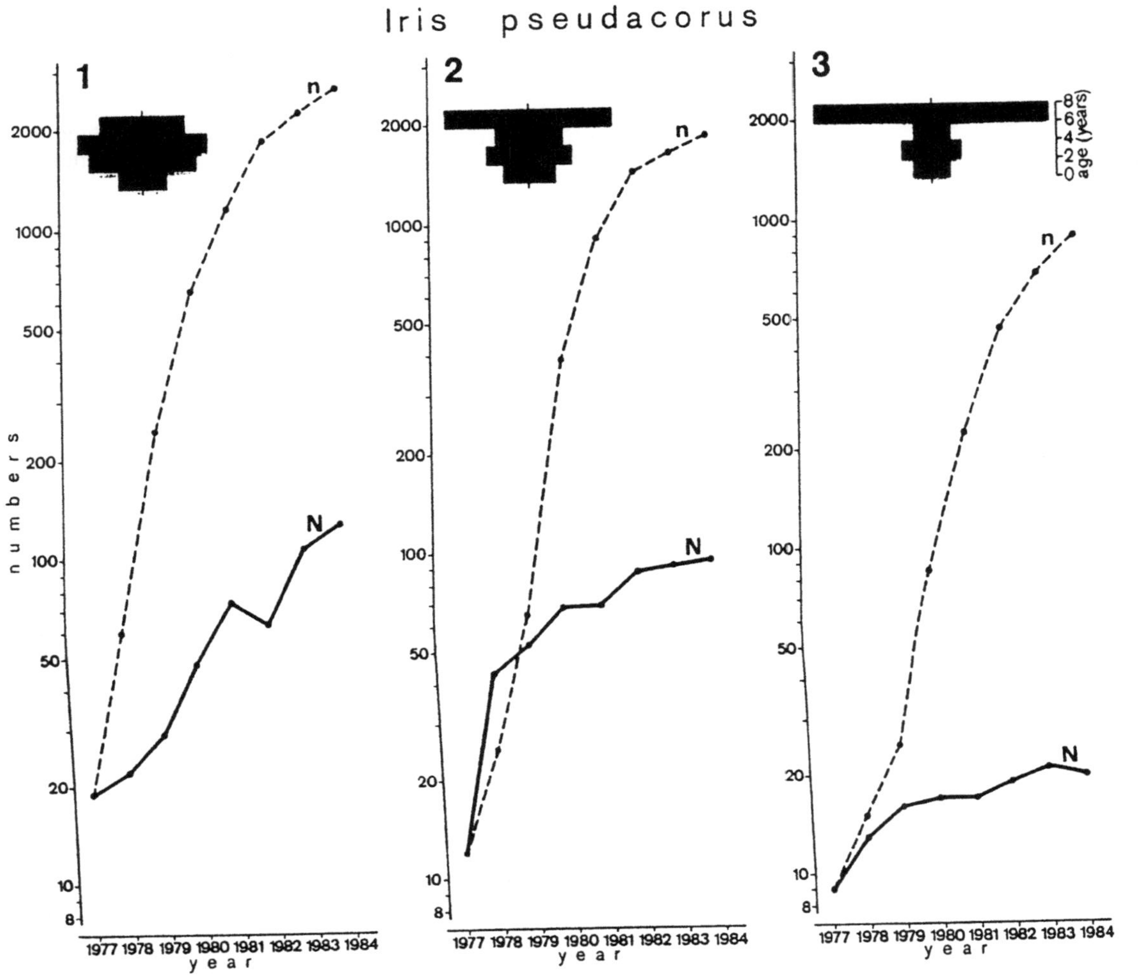

Fig. 4.9. Curves of numbers of individuals (*N*) and shoots (*n*) and age structure in the populations inhibiting the muck (1), brushwood (2), and sedge communities, (3).

tions reached the abundance-equilibrium phase. The phase starts in population-growth year 6 or 7.

Fecundity and reproductiveness

A permanent, though varying between years, seedling recruitment is the result of individual fecundity (Table 4.2). *Iris pseudacorus* individuals reach the generative phase in the first or second year of their life. Each of them produces then 1 or 2 flowering shoots. Generally, the number of generative shoots increases with the age of an individual. Individuals aged 3 to 5 years produce the largest numbers of them (2—14 shoots). The number of shoots and seeds varies, depending on the hibitat conditions. For example, individuals growing in sedge swamps produce on an average 3.42 ± 1.92 shoots, and in a trough 12.12 ± 4.01. The average number of seeds per shoot is 282.76 ± 75.5 in a trough, and 157.30 ± 44.50 in a sedge swamp. If a mature polycormone produces several or over a dozen generative shoots, then its seed production in different seasons ranges from 2200 to 3048.

Though the number of generative individuals in a population increases from year to year, and so does the number of fruiting shoots, and thereby the seed pool, the number of seedlings has not been found to follow a similar regularity. For

Table 4.2. Demographical description of the *Iris pseudacorus* population in the muck (1), alder scrub (2) and sedge communities (3). G: individuals of generative origin from the juvenile phase; V: individuals of vegetative origin

	Populations																	
	1						2						3					
Year	birth		death		net	N	birth		death		net	N	birth		death		net	N
	G	V	G	V	increment		G	V	G	V	increment		G	V	G	V	increment	
1977	19		0		19	19	15		3		12	12	9		0		9	9
1978	5		2		3	22	42		11		31	43	4		0		4	13
1979	10		3		7	29	12		2		10	53	7		4		3	16
1980	21		2		19	48	18		2		16	69	4		3		1	17
1981	28	2	5		25	73	17	4	18	2	1	70	0		0		0	17
1982	10	6	22	5	−11	62	7	12	1	0	18	88	2	2	2	0	2	19
1983	11	38	0	2	47	109	9	11	7	9	4	92	5	6	8	1	2	21
1984	14	12	4	5			125	37	50	17			41	12	27	6		
Total	176		50				162		67				53		33			
Mortality (%)	28						41						60					

instance, in a population with the highest numbers in a 400 m^2 plot in the successive study years (1977—1984) there emerged: 22, 138, 268, 468, 566, 242, 142, 169 seedlings, whereas in populations with the lowest abundance the respective quantities were as follows: 12, 69, 312, 110, 420, 92, 52, 12. The following example illustrates the seed-seedling relationship: in 1981 in a 400 m^2 trough almost 1/4 individuals attained the generative phase, i.e., of 73 individuals present 19 (22%) fruited. They jointly produced 108 generative shoots, from which 24992 seeds were collected, counted and left on the ground. The following year 566 seedlings emerged. So the seed fraction that developed into offspring was small, 2.3%.

A low seed germination capacity under natural conditions is compensated for by a high production rate — 62 to 95 seeds per 1 m^2 in the colonization phase, and 224 to 900 in the later period. Though varying from year to year, seed density ranges from 5 to 32 per 1 m^2 plot, the frequency in the study plots being 5 to 38%.

The number of individuals that arise as a result of polycormone fragmentation depends primarily on the rate at which polycormones reach the disintegration phase. The largest number of vegetative individuals, i.e., 58, formed in a mid-meadow trough, and in 12 sedge swamps.

The differences were the result of uneven proportions in those populations of polycormones which gave to vegetative individuals. In sedge swamps, for instance, the fraction of polycormones which disintegrate into several parts each, or from which peripheral fragments separate, comes up to 12%, and in a mid-meadow trough, to 28%. This is due to uneven polycormone growth rates, dependent on habitat conditions.

The first vegetative individuals appeared there in 1982, that is, in the 6th growing season after the colonization of the trough by *I. pseudacorus* (Figure 4.10). The disintegration phase was then reached by one polycormone aged 5 years (No. I). From that year onwards senescent polycormones disintegrated every growing season. In 1983, two polycormones aged 5 years (III and IV), and one aged 6 years (II) attained the disintegration phase. In 1984, three polycormones split up: two 5-year-old ones (V and VI), and one 6-year-old (VII). The number of fragments that arose due to the splitting of 7 polycormones amounted to 60, of which 18, i.e., 30%, were dead or undergoing necrosis.

Survivorship of different-origin individuals

The varying quantitative relations among generative and vegetative individuals over the 8-year population-growth cycle resulted from uneven

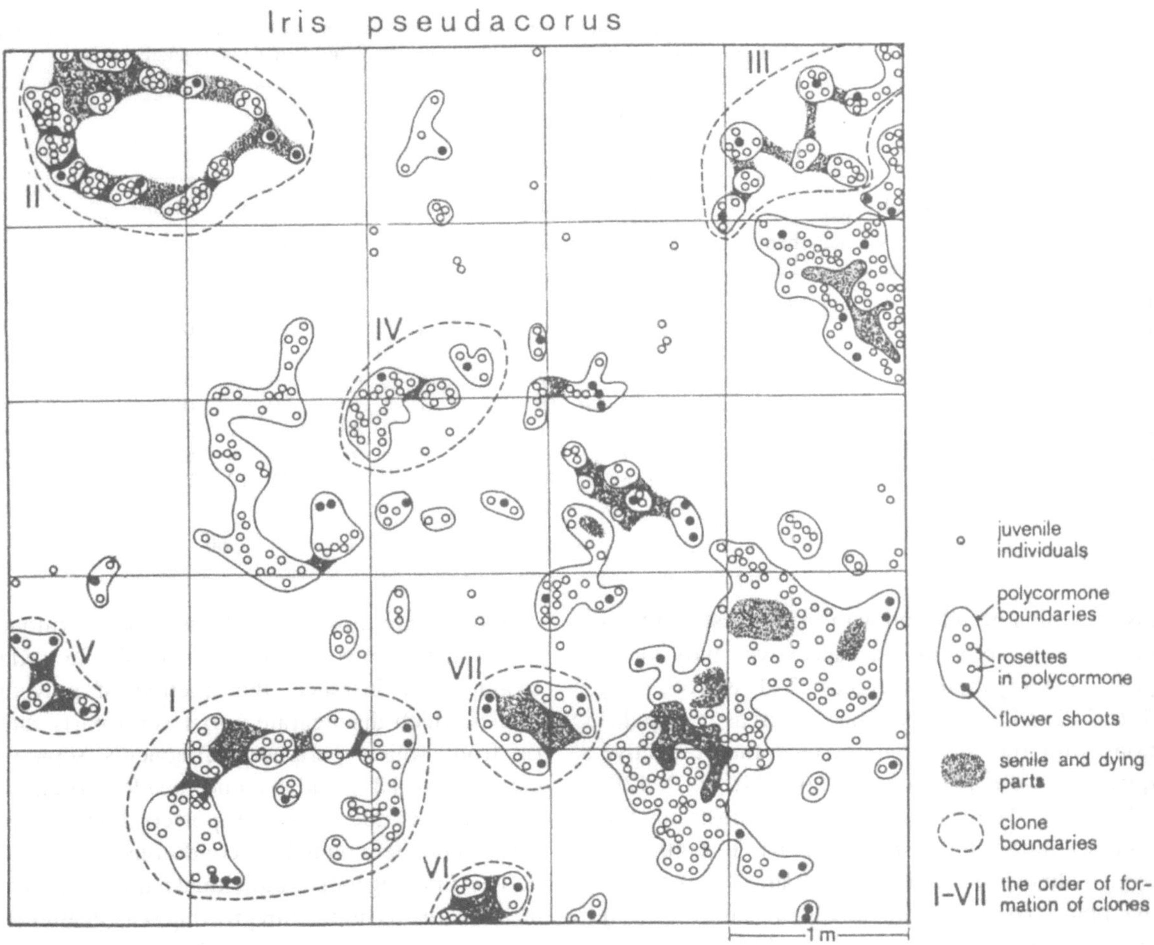

Fig. 4.10. Spatial structure of population as a result of development of polycormones and clones in the much 8 years after abandonment of the meadow (see text). (Falińska 1986b, slightly changed).

rates of recruitment of individuals of different origin and their uneven survival rates. Generative individuals as a rule appear throughout the population-growth cycle, their abundance being higher and their survival rate lower than those of vegetative individuals. For instance, during a period of 8 years in a population that inhabited a trough, 2015 seedlings emerged, 3.9% of which survived until the mature phase. During the same period there formed 58 vegetative individuals of which 79.4% survived. In sedge swamps the respective proportion for seedlings was 1.2%, and for vegetative individuals, 50%.

The high survivorship of vegetative individuals results from the fact that at the time of their "birth" they are already in the juvenile or the mature, seldom in the senescent, phase. The growth phase and age of these units depend on the polycormone part from which they have taken their origin. New individuals that have taken their origin from the peripheral portions of a polycormone are as a rule at a growth stage similar to the juvenile stage of generative individuals, however, those formed of the central parts of a polycormone are similar in habitus and size to the mature-phase individuals (aged 3—5 years).

Percentage of different-origin individuals

Iris pseudacorus populations growing under different environmental conditions differ not only in size, but also in the proportions of different-origin individuals. Throughout the population-growth period, however, individuals developed from seeds prevail, whereas the proportion of individuals formed through polycormone splitting-up does not begin to increase until the population has grown for several years (Table 4.2). Eight years after the species had colonized new sites, the proportion of generative individuals was 63%, that of vegetative individuals 37%, and in the least abundant populations: 70% and 30%, respectively.

Population age structure

Eight years after the colonization of unmown meadows by *Iris pseudacorus* the population age structure shows that the stabilization stage has been attained. Populations that have grown under different conditions but during the same period (1977—1984) differ primarily in that the proportions of the oldest individuals are uneven. The largest number of polycormones aged 6 to 8 years is found in a population inhabiting sedge swamps (65%), and the smallest in a mid-meadow trough (Figure 4.9). It is the result of an uneven course of the growth cycle of individuals growing in different conditions. In the trough, individuals attain the successive growth stages faster than elsewhere, e.g. they start the senescence and the disintegration phase as early as the 4th and 5th years of their life, while those in the sedge swamps, 6th and 7th life years. The course of the senescence processes in polycormones determines the presence or absence of 6—8-year-old individuals, as well as the rate of recruitment of vegetative individuals. Depending on the intensiveness of these processes, individuals aged 5 to 8 years either die or disintegrate. The effect of vegetative individuals on the level of numbers becomes noticeable only in the second population-growth period when some of them are 4 or 5 years old. In that period also their mortality increases, but it does not cause a reduction in numbers. A reduction in abundance, due to the necrosis of several-year-old polycormones is offset by the recruitment of new vegetative individuals. The gradual increase in a population of the percentage of individuals that form from senescent-polycormone fragments accelerates its senescence. The new individuals that arise through fragmentation continue the senescence processes of the parent organism. The phenomenon can be seen in the age pyramids prepared in two ways. If only generative individuals have been taken into account, a flat pyramid is obtained (G). If vegetative-origin individuals were also taken into account, and for their age assessment, successive years were added to the age of the parent organism, from which they had separated, then a slender pyramid was obtained (G + V). Then a population contained similar proportions of individuals representing different age classes (Figure 4.11).

Equally significant population age-stage structure differences were noticeable if generative and vegetative individuals were treated separately. Among vegetative individuals (V) mature individ-

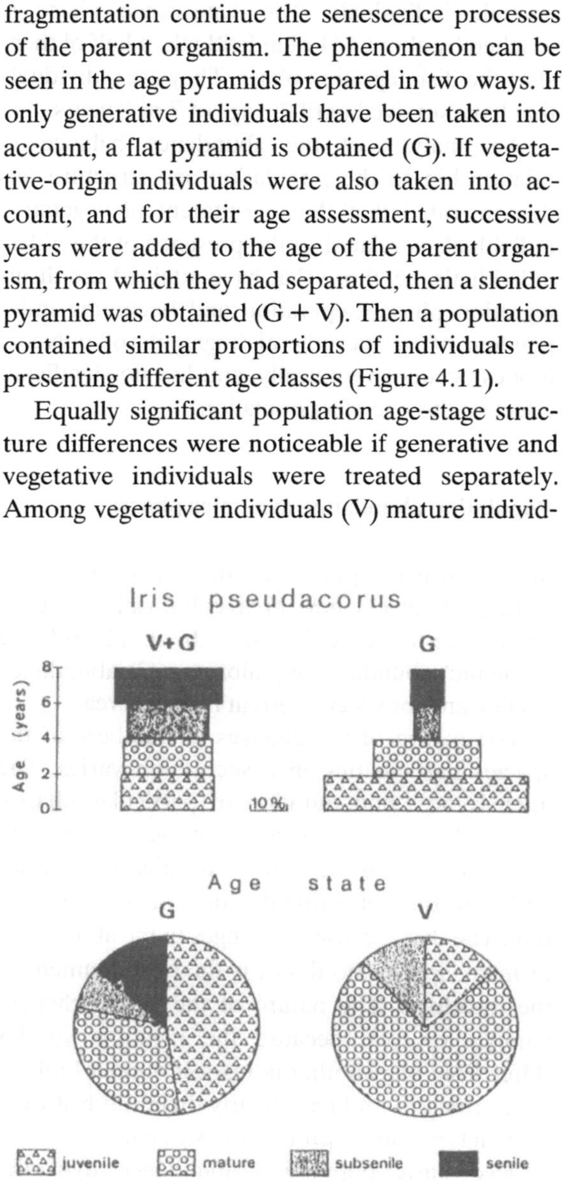

Fig. 4.11. Age structure and age state of the same population. Individuals from vegets (V) and from seedlings (G). (Falińska 1986b; see text).

uals prevailed, with a low percentage of juvenile individuals. The latter form only from polycormone peripheral portions. At the time of becoming independent the latter are similar in their growth stage to juvenile (one-year-old) generative individuals. Apart from this, the group does not include polycormones in the disintegration phase. About 50% of the generative individuals (G) are juvenile individuals. The senescence phase is reached by about 1/4 of individuals, a half of them start the disintegration phase. They are individuals that have survived until the age of 7 or 8 years.

An analysis of the abundance balance in relationship to the population age structure has demonstrated that the recruitment of vegetative individuals is related to the presence of the oldest individuals, i.e. those that have attained the disintegration phase. Separation and becoming independent of the peripheral fragments of polycormones occurs sporadically, and has no significant effect on the size of a population.

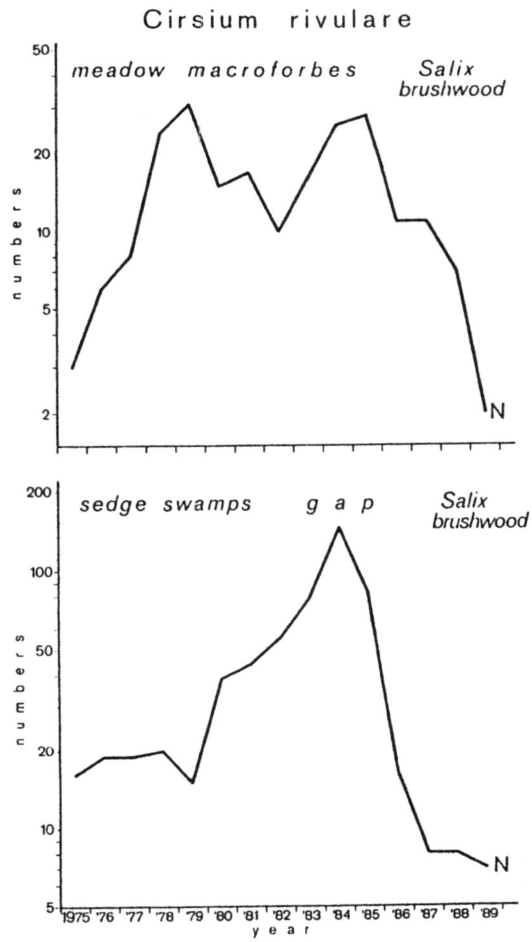

Fig. 4.12. Population dynamics of short-term promoters of succession in different patches in abandoned meadows.

Population abundance dynamics phases

In the growth of a population three main phases are distinguished: (1) new area colonization and/or abundance increase in the habitat; (2) relative (dynamic) abundance equilibrium; (3) abundance decline and/or species retreat from its area.

The nature of the changes in numbers of the species participating in a succession varies. The intensiveness and duration of particular phases, that is the time of the whole population growth cycle, depend on the biological traits of a species and conditions affecting the numbers of a population. The direct cause of changes in the abundance of many species no doubt is the abandonment of the meadows. The nature of the actual changes varies, however, because the meadow species differ from one another in respect of their biological characters, and particularly their, life-histories, growth types and reproduction strategies.

Two basic population abundance dynamics types have been distinguished (Figure 4.12): (1) Abundance changes include a period of fast increase, usually in geometric progression, terminated with peak numbers followed by an abundance decline and retreat of the population from its area, or stabilization of numbers at a very low level; (2) Growth in numbers continues until the growing population fills all free-meadow sward niches and spaces released by species retreating due to the cessation of meadow mowing. Having taken possession of a habitat, the population reaches the stability (equilibrium) phase. In certain situations the attainment of the peak level is followed by a short period of decrease in numbers, whereafter the numbers become stabilized at a lower level.

Type 1 of population abundance dynamics was

observed primarily in all those species whose individuals live a short time, that is 2, or up to 5 years. Besides, they are of a delicate body structure, monocormonal growth type, and their reproduction is primarily, or exclusively, generative.

Populations of species like *Cirsium palustre, Ranunculus acris* and *Lychnis flos-cuculi* attain the abundance growth and decrease phases within a relatively short time, 5—6 years, whereafter the populations retreat or stay at a stable, low level of numbers (Figures 4.2; 4.3). For example 10 years after the meadow abandonment these species were represented in the 200 m^2 plots by small numbers of individuals, e.g. 1 to 20.

Cessation of mowing was immediately followed by a 3—5-fold , sometimes even 10-fold, growth in numbers of these species. The rapid growth in numbers in the first years of meadow disuse was the result of the presence of a bank of seedlings and juvenile individuals which had not reached further growth phases before the cessation of mowing. Within the first three years they filled all the gaps in the fairly loose structure of the meadow sward. The growth in numbers of these populations did not cease until the growth in numbers had started of species with a stolonate or clump growth type. These are highly competitive species whose competition could be defined as an "ousting competition". It causes a physical limitation of the growth of neighbours by the competing species through their own growth. This type of competition is peculiar to species like *Filipendula ulmaria, Lythrum salicaria, Carex acutiformis, C. cespitosa, Geum rivale* and *Potentilla palustris.* They significantly contribute to the reduction in numbers of those species which had considerably increased their abundance in the first period following the abondonment of the meadows (Figure 4.3).

When dominance has been reached by species with a polycormonal structure type, then species representing Type 1 of abundance dynamics are in the regression, or equilibrium phase, at a low level of numbers (Figures 4.3; 4.12).

It has also been seen that within about a dozen years *Cirsium palustre* and *Lychnis flos-cuculi* populations could twice attain an abundance equilibrium phase preceded by a considerable reduction in numbers. This occurs when after a short-lived growth in numbers, immediately after the cessation of mowing, a species begins to retreat during the decomposition of the meadow phytocoenoses, and then it appears again as soon as there has formed a new floristic composition (Figure 4.3). A second abundance growth and fall of *Cirsium palustre* can, for instance, be caused by the expansion of *Carex cespitosa.* As a result of the growth in width of *C. cespitosa* tussocks, *Cirsium palustre* individuals become partially or totally eliminated. When a *C. cespitosa* uniform stand has formed, *C. palustre* individuals begin to appear, among other species, on the tussocks of the latter sedge species. Then there again occurs an abundance increase of *C. palustre*, sometimes to a level typical of the first phase. A second reduction in numbers of the population is also caused by the same species which caused the first abundance decrease several years earlier. *Carex cespitosa* tussocks that have attained large sizes, and occupy areas of 0.5—1.0 m^2 gradually become grown up with species like *Filipendula ulmaria, Lythrum salicaria, Lysimachia vulgaris, Geum rivale* and *Polygonum bistorta*, and individuals of a delicate body structure become elimated (Part III).

Type 2 of abundance dynamics was observed in species whose individuals, of a polycormonal structure, are known to be longaeval. In such situations the populations remain a long time in a stable-abundance phase. The following species have been included in this group: *Filipendula ulmaria, Lythrum salicaria, Lysimachia vulgaris* and *Iris pseudacorus*. Within several, a dozen or so (or more) years of their life individuals of these species can build up their underground organs to considerable sizes and multiply the number of shoots, forming assemblages of a variable size and permanence. Such plant organisms show a seasonal and long-term inner dynamics manifested by an increase or decrease in the number of renewal, juvenile and generative shoots. Depending on the individual age and habitat conditions, variations in the number of shoots can follow the exponential or logistic model. In an individual that has reached a certain degree of equilibrium of growth and development, these variations are usually of the nature of fluctuations.

In the analyses of the polycormonal species population dynamics Harper's (1980) concept of a two-level population organization was adopted. Harper assumes that the growth of modular organisms is itself a demographic process. Therefore to understand the population dynamics of species whose individuals are made up of many shoots, we need data on the recruitment and loss of individuals, and on the increase and decrease in the number of shoots.

This view is supported by the fact that variations in the number of individuals and of shoots in the same population are different. These differences are primarily variations in the number of individuals (*N*) and shoots (*n*) in the particular population growth phases (Figure 4.9).

Studies carried out so far indicate that differences in the variation of shoot (*n*) and individual (*N*) numbers over a long-term population growth history follow the following rules: (1) The individual-shoot numeric ratios are similar throughout the period; (2) The individual-shoot numeric ratios vary with the population growth in the following way (Figure 1.3):

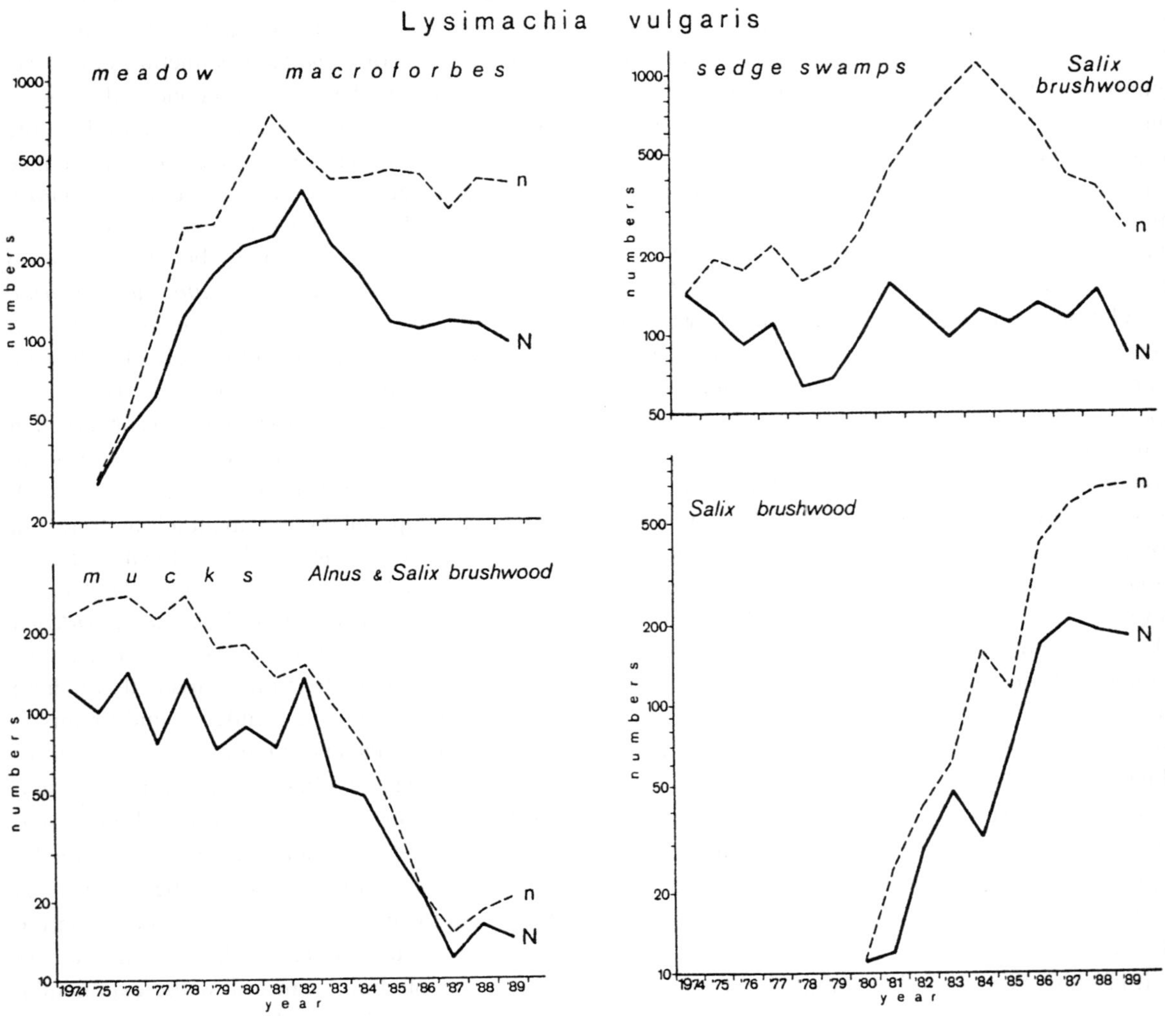

Fig. 4.13. Population dynamics of long-term promoters of succession in different patches in abandoned meadows.

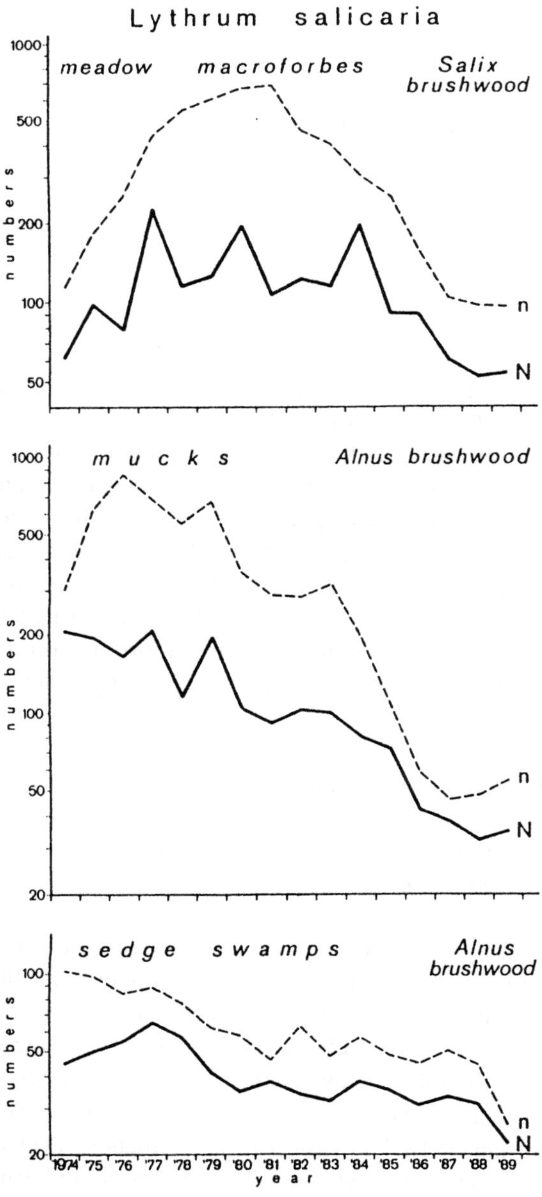

Fig. 4.14. Population dynamics of long-term promoters of succession in different patches in abandoned meadows.

a) In the initial phase, the number of individuals is almost the same as the number of shoots, since each juvenile individual is usually made up of one shoot. With the elapse of time the proportion of polycormonal individuals increases, so the values of *N* and *n* begin to differ.
b) A growth in number of the individuals is accompanied by a growth in number of the shoots, the increase of the value of *n* being usually faster than that of *N* due to the growth biology and life-histories of perennial individuals.
c) Related to the numeric equilibrium of individuals are fluctuations of the value of *n*, and a considerable increase of that value, since some plant organisms are able to multiply their shoots (Figures 4.13; 4.14).

These three possible situations result from the following causes: (1) A decrease in numbers of the individuals may be accompanied by both a decrease and growth, as well as a relative equilibrium of the number of the shoots; (2) if multi-shoot individuals necrose completely and rapidly, then the decrease of the value of *N* is similar to that of *n*; (3) if a part of individuals die, and the remainder increase the number of shoots, then with a decreasing *N* value the value of *n* increases considerably; (4) if some individuals become necrotic, and others use the released area only to some extent by growing in size, that is, they only replace the shoots lost, then there occurs an equilibrium of *n* with a reduction of *N*.

In many herbaceous populations the growth in number of the individuals was found to follow the logistic model, whereas that of the shoots, the exponential model (Figure 4.9). The beginning of population stabilization is as a rule accompanied by a rapid growth in the number of shoots. In many species the phenomenon results from an intensive enlargement of several-year-old polycormones that often represent the most abundant group, when in a newly colonized area the population attains the abundance-equilibrium phase.

CHAPTER 12

REPRODUCTION AND SUCCESSION

Introduction

The effect of habitat supplies on the survivorship and reproductive success of plants has been fully understood (Harper and Ogden 1970, Harper 1977, Silvertown 1983). It has also been found many times that the proportions of species with different reproduction strategies vary with the succession phases (Horn 1974, Newell and Tramer 1978, Noble and Slatyer 1980). It has been demonstrated, for instance, that species with a high production rate of small seeds (r — selection) dominate in early succession phases, and those with a low production of large seeds (K — selection), in later phases (Platt 1975, Grubb 1977, Grime 1979, Gross and Werner 1982).

Recently, much attention has been given to species taking part in a long-term succession. Their individual fecundity and survivorship, as well as the amount of biomass allocated for reproduction usually change with the progressing succession (Fekete and Melkó 1981, Falińska 1979, 1989a, b).

The populations of some species respond to habitat conditions, varying with the succession progress, by gradually increasing the role of vegetative reproduction. The number increases, for example, of vegetative propagules in a population, this being usually accompanied by a decreased abundance of seedlings (Figure 4.7). The varying relationship between the efficiency of generative reproduction and that of vegetative reproduction causes variations in the proportions of different individuals of a population taking part in succession processes. The proportion of vegetative individuals as a rule increases from year to year (Figure 4.6, Tables 4.1; 4.2).

This chapter describes the efficiency of reproduction of the populations of selected perennial species which take part in a succession from the initial to the terminal phase. The efficiency has been estimated on the basis of the following quantities: a) proportions of individuals participating in generative and vegetative reproduction; b) number of generative and vegetative propagules.

Generative propagation

Proportion of individuals participating in reproduction

During a succession in progress the proportion of individuals taking part in reproduction in the populations of many plant species changes significantly (Figure 4.15). The cessation of mowing is usually followed by an increase in the number of fruiting individuals, particularly in the populations of tall herbs. They form then the highest layer of the meadow phytocoensis, and flower and fruit intensively (Photo 2.1). This situation persists until the formation of willow scrub, i.e., until Phase IV. Only then does the fraction drop of fruiting individuals in the populations of many perennial species to several or about a dozen percent, as opposed to 50 or up to 70%, in the early succession phases (Figure 4.15). At the time of willow scrub formation *Filipendula ulmaria, Lysimachia vulgaris, Lythrum salicaria* and *Cirsium rivulare* populations show a clear stratification: a) individ-

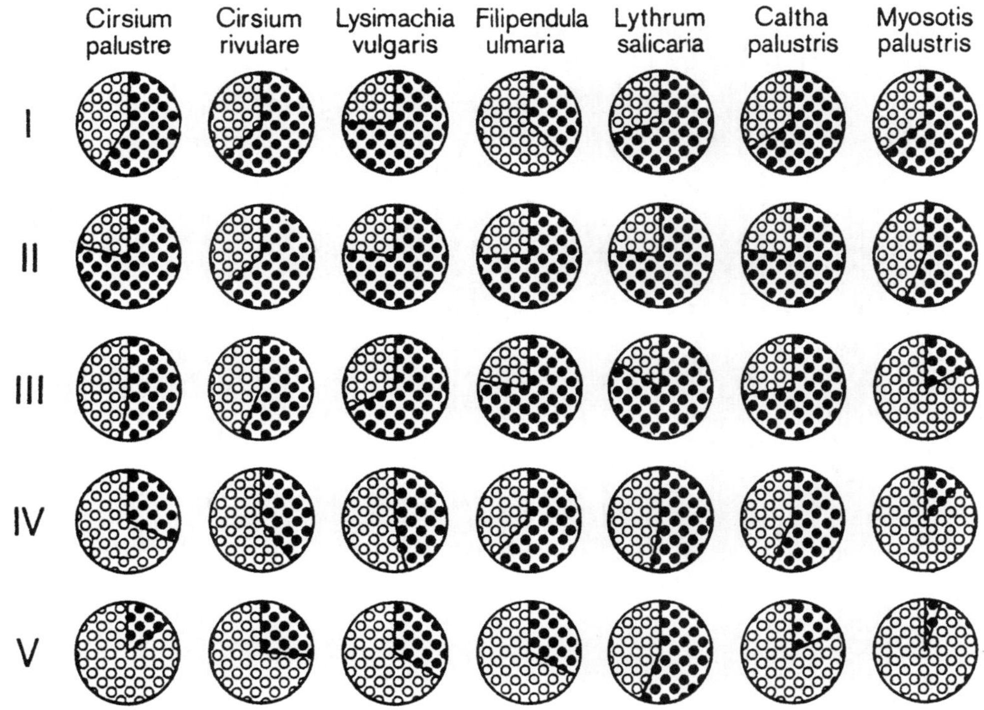

Fig. 4.15. Participations of flowering individuals in populations in consecutive successional phases: (I) meadow, (II) macroforbes community; (III) willow brushwood, (IV) brushwood community, (V) *Salicetum pentandro-cinereae.*

uals growing under the canopy of willows represent a group of a low flowering and fruiting fraction; b) individuals found on the edge of willow scrub represent groups with high fruiting-plant fractions (over 50%). This regularity has been recorded for tall-herb populations (Photo 4.1).

In the populations of those species whose individuals make up the lower layer of a meadow phytocoenosis, the fruiting-individual fraction drops as early as Phase III, that is, when there forms a polydominant meadow. This type of response has been found in *Myosotis scorpioides* (Figure 4.15). The flowering effect of tall herbs on these species is similar to that of willows on tall herbs.

Seed and fruit production

Changes in the proportions of fruiting individuals are followed by variations in seed production rates, per individual and per unit area. In the first period following the cessation of mowing some species increase their seed production, e.g. *Cirsium palustre* (Table 4.3). The individuals of some species produce most fruit in Phase II and Phase III of succession. In those communities also the individuals attain the largest sizes (Figure 4.16).

Wide seed production rate variation was observed in the populations of succession promoter species, that is, *Filipendula ulmaria, Lysimachia vulgaris* and *Lythrum salicaria*. The individuals of these species produce most seeds in Phase II and Phase III, that is, when there form macroforb and willow-scrub phytocoenoses. The individuals of succession promoters found on the scrub edge then attained large sizes and produced 3—5 times as many seeds as did those seen in Phase 0 and Phase I. In Phase 0, for instance, *Filipendula ulmaria* individuals produced 500 to 800 seeds, while in Phase III about 4000. A similar response has been found in *Lysimachia vulgaris* individuals which produce the largest

Photo 4.1. "Edge effect" on the brushwood of *Salix cinerea* boundaries in meadow 15 years after abandonment of the meadow. Development of tall herbs on the edge of aggregation of Salix.

Photo: J. B. Faliński

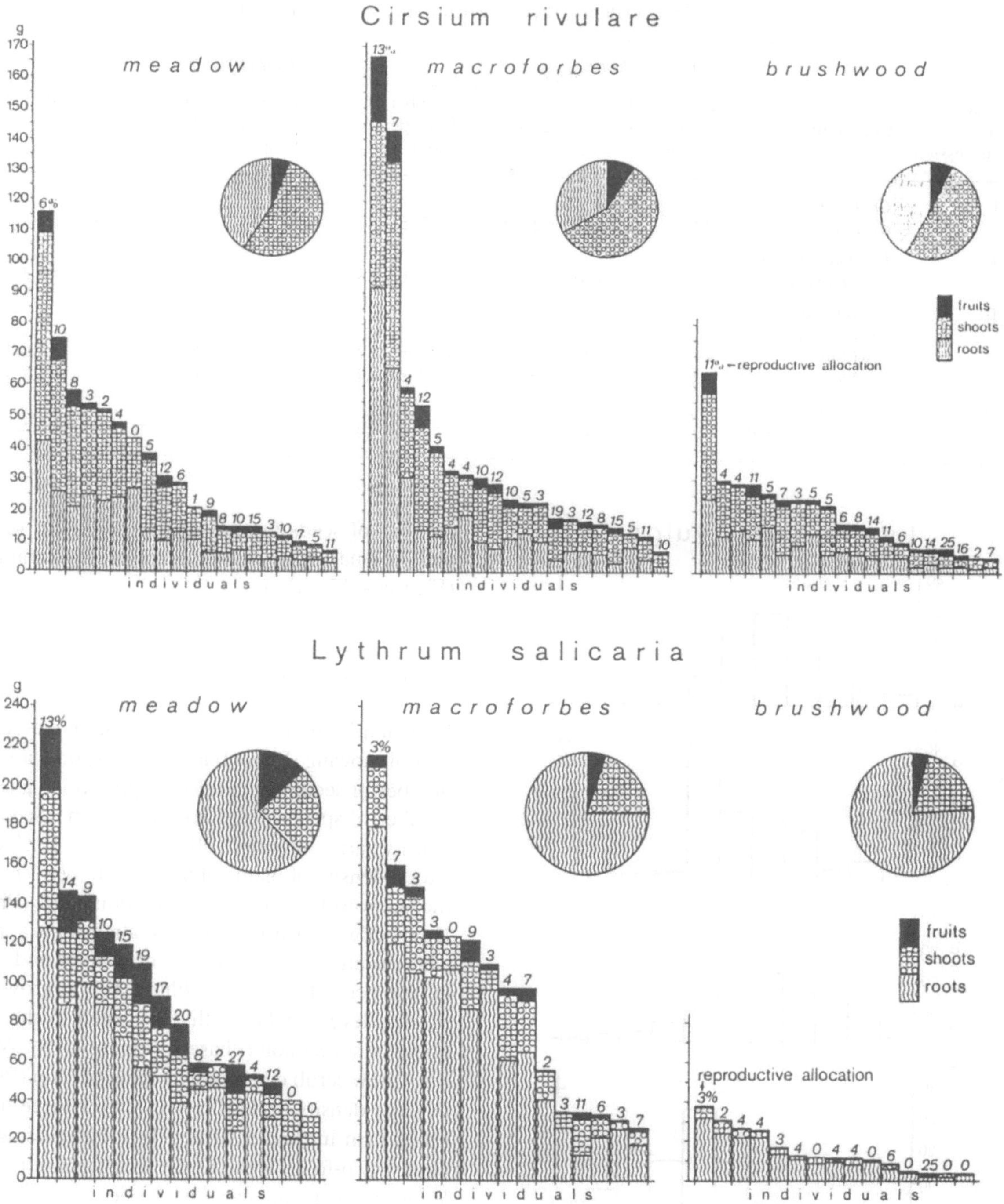

Fig. 4.16. Differentiation of individual biomass in the successional process in populations of short-term (top) and long-term (bottom) promoters of succession.

Table 4.3. Differentiation of shoot size of few species in various communities

Phases of succession	Communities	*Cirsium palustre* Number of individuals /25 m²	Height of shoot (cm) $\bar{x} \pm$ SD	Number of fruits per shoot	*Cirsium rivulare* Number of individuals /25 m²	Height of shoot (cm) $\bar{x} \pm$ SD	Number of fruits per shoot
I	*Cirsietum rivularis*	92	101.8 ± 11.2	36.3 ± 4.2	8	109.9 ± 11.6	4.5 ± 1.1
II	*Lysimachio-vulgaris Filipenduletum*	46	120.2 ± 13.9	40.7 ± 6.8	36	104.4 ± 8.4	4.0 ± 0.7
III	brushwood of *Salix cinerea*						
	within	44	128.6 ± 4.5	48.4 ± 5.7	105	114.2 ± 4.0	3.9 ± 0.3
	edge	9	139.55 ± 20.9	69.2 ± 15.9	15	132.5 ± 13.7	6.1 ± 2.29
IV	*Salicetum pentandro-cinereae*	30	117.3 ± 8.85	31.2 ± 2.3	10	111.4 ± 5.34	5.1 ± 1.0

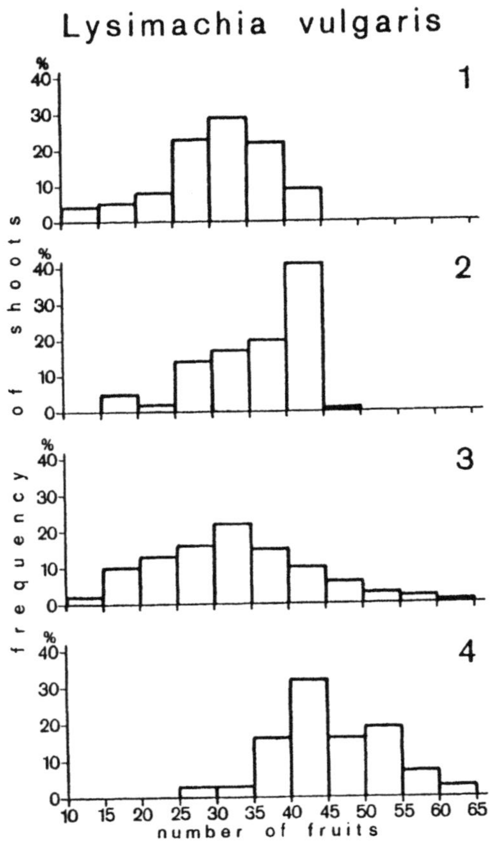

Fig. 4.17. Variation in number of fruits per shoot in meadow (1), sedge swamps (2), on the edge of brushwood of *Alnus glutinosa* (3), and on the edge of brushwood of *Salix cinerea* (4).

number of seeds around willow and alder scrub, and the smallest when they occur in sedge marshes (Figures 4.17; 4.18).

Seedling recruitment. Seedling bank

In regularly mown meadows and after the cessation of mowing (Phase 0 and Phase I) only a small number of seedlings develop, representing a small number of species, i.e, 5 to 12 per 1 m². Seedling density varies between 100 and 150 per 1 m², with a total density of plants of 600 to 800 per 1 m². In the macroforb-community formation phase (Phase II) seedling density decreases almost by a half, whereas after the appearance of willows and gaps in the vegetation cover (Phase III) it grows to reach a level similar to that recorded in the first phase of succession (Phase I) (Figure 4.19). When the willow scrub community begins to form (Phase IV), the density of seedlings is almost three times lower than in Phases III and I. It increases again, almost two-fold, in the forest initial phase (Phase V) when there appear free spaces around trees (Figure 4.19). On the basis of a comparison with other investigations into seedling abundance and density in the forest communities of the Białowieża Forest it may be presumed that the density observed in succession phase V will still persist

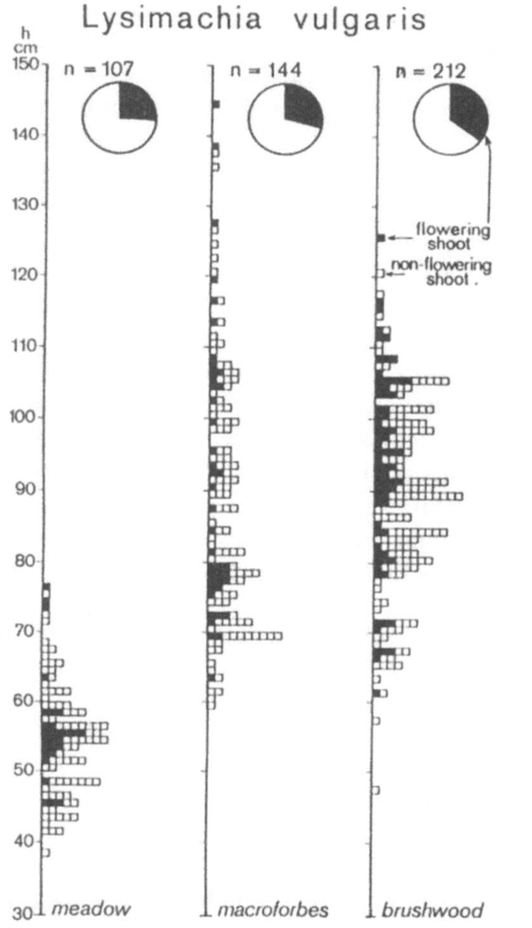

Fig. 4.18. Differences in shoot size in the course of 15 year succession in abandoned meadows, observed in the same 5 m × 5 m plot in species — rich patch.

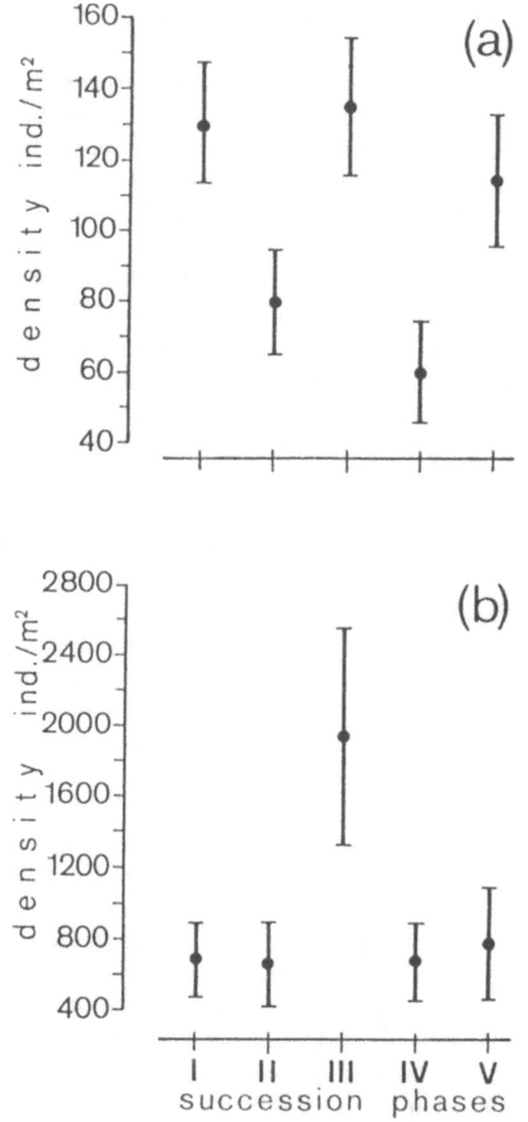

Fig. 4.19. Changes in the density of the seedlings ($x \pm$ SD) in consecutive successional phases: (I) meadow, (II) macroforbes community, (III) willow brushwood, (IV) brushwood community, (V) *Saliceturm pentandro-cinereae.* a) natural vegetation cover; b) after removal of plant cover.

after the formation of a forest phytocoenosis. For similar data have been obtained from studies of seed and seedling banks in a *Tilio-Carpinetum* phytocoenosis (Pirożnikow 1983), as well as in carr and alder-swamp (Falińska 1979, 1985) phytocoenoses.

Seedling density shows a considerable association with the spatial structure of a phytocoenosis. The limiting role of the vegetation cover, especially during a succession, has been noted many times (Holt 1972, Platt 1975, Gross and Werner 1982), Faliński 1986b, Peart 1989a, Falińska 1989a, b).

To understand the effect of the vegetation cover on seedling growth, in areas representing different succession phases (I to V) 1 m^2 plots were set up, 12 plots in each area. In the autumn all the aerial parts of plants were removed from the plots, and in the following growing season the emergence of seedlings was checked every 10 days, using a 1 m^2 grid divided into squares 10 cm in side length.

The investigations have shown a much higher seedling density in all the succession phases following the removal of plants than in the case where the plant cover remained intact. The phytocoenoses with an undisturbed structure and those with a disturbed structure have also been found to differ in the type of similarity in seedling density between the particular succession phases. In the case of undisturbed vegetation the lowest seedling density was recorded in the macroforb communities (Phase II) and willow scrub (Phase IV), this, no doubt, being the effect of the high density and shading. At that time macroforbs and willows prevailed in the phytocoenoses, but in the situation where plants had been removed, the highest seedling density was found in Phase III, that is, when there formed a spatial complex of meadow macroforb and scrub communities, characteristic of a succession series of this type. At that stage the seedling bank is the highest and most varied (Table 4.4, Figure 4.19).

The density of seedlings of the same species varies between the succession phases, but certain regularities can be noticed. Distinct among the species present throughout the whole succession process are those whose abundance in different succession phases is similar, and those which attain a high density only in one phase, being rare or absent in others (Figure 4.20). Generally, species that determine the floristic nature of a meadow play a greater role in the abundance dynamics of seedlings in Phases I and II than in Phase III. In the willow-scrub formation phase (Phase III) a considerable decrease occurs of the seedlings of all meadow species present in the seedling bank before, while the woodland species considerably enrich the seedling bank, the latter being then the largest (Table 4.4, Figure 4.21).

The above differences become even more conspicuous if the seasonal abundance dynamics is compared of seedlings in three succession phases. The phases differ in the number of seedlings and course of its variation, especially the time of reaching the peak numerical level. The earliest, recorded in June, was the peak seedling abundance in Phase I, and that in the remaining two phases, only in August. The peak abundance of seedlings stayed the longest time in Phase III, that is, until September (Figure 4.22). In autumn the number of seedlings is similar in all the succession phases. There are few seedlings, i.e., 2000 to 3000 per 12 m^2, and in the maximum number period this quantity varied between 6000 and 15000 per 12 m^2. A comparison of the survivorship of the seedlings of the particular species making up the phytocoenoses has shown that it varied considerably, between 0 and 100% (Table 4.5). Moreover, populations of the same species vary considerably in this respect, depending on the succession phase (Table 4.5).

In general, during a succession in progress, seedling density is subject to fluctuations. But the specific composition of a seedling bank shows a clear tendency towards seral changes: a reduction in the number of meadow species, and an increase of the number of woodland species, with a general reduction of the species richness (numbers) in the seedling bank (Figure 4.22). A similar nature of seed and seedling-bank has also been found in other studies of the dynamics of vegetation (cf. Leck, Parker and Simpson 1989).

Vegetative propagation

In the case of species with iterative (multiple) growth the proportion of individuals propagating vegetatively as a rule increases in abandoned meadows. In some species the process begins immediately after the cessation of mowing, and in others as late as several years after it. This depends on the structure of the underground organs of the plants. In species with delicate rhizomes and a high growth rate, e.g. in *Lysimachia vulgaris* and *Myosotis scorpioides*, one-year-old and two-year-old modules continually detach from the parent organism, especially if their neighbours are highly competitive plants, e.g. *Carex acutiformis*. In the consecutive succession phases the proportion of individuals producing vegetative offspring is, in the populations of these species, very high, ranging from 70 to 90%, and in some cases even 100%.

In species with durable and strong rhizomes the proportion of the individuals that multiply vegetatively does not increase until the meadows have not been used for several, e.g. 6 or 7, years, e.g. in

Table 4.4. Number and survival of seedlings during succession (Szydłowska, unpublished data)

Species	Succession phases: I spring 1989	I autumn 1989	I survival %	II spring 1989	II autumn 1989	II survival %	III spring 1989	III autumn 1989	III survival %
Lythrum salicaria	9.000	95	1.1	2.042	10	0.5	10.490	28	0.3
Galium palustre	1.865	628	33.7	1.160	273	23.5	48	35	72.9
Cirsium palustre	1.307	721	55.2	1.842	774	42.0	581	506	87.1
Filipendula ulmaria	681	173	25.4	1.033	189	18.3	93	66	71.0
Ranunculus acris	627	431	68.7	1.967	1.416	72.0	82	60	73.2
Lychnis flos-cuculi	338	338	100.0	333	284	85.3	99	38	38.4
Viola epipsila	254	88	34.6	147	75	51.0	37	25	67.6
Lysimachia vulgaris	159	31	19.5	196	6	3.1	17	—	0.0
Geum rivale	136	6	4.4	303	48	15.8	368	186	50.5
Rumex acetosa	121	33	27.3	69	69	100.0	—	—	—
Trifolium repens	91	65	71.4	53	11	20.7	2	—	0.0
Epilobium palustre	89	24	27.0	78	13	16.7	—	—	—
Potentilla erecta	61	17	27.9	24	10	41.7	—	—	—
Salix cinerea	60	12	20.0	84	9	10.7	1.072	3	0.3
Epilobium roseum	51	18	35.3	113	40	35.4	129	129	100.0
Myosotis scorpioides	50	12	24.0	65	40	61.5	10	6	60.0
Mentha arvensis	50	15	30.0	81	55	67.9	56	42	75.0
Ranunculus lanuginosus	43	11	25.6	14	6	42.9	29	11	37.9
Lathyrus pratense	39	23	59.0	481	33	6.9	38	1	2.6
Urtica dioica	32	14	43.8	15	7	46.7	147	107	72.8
Vicia cracca	29	5	17.2	15	—	0.0	2	—	0.0
Veronica beccabunga	21	21	100.0	2	1	50.0	—	—	—
Caltha palustris	12	12	100.0	—	—	—	7	4	57.1
Cerastium holosteoides	11	2	28.2	36	36	100.0	—	—	—
Potentilla anserina	9	7	77.8	19	19	100.0	—	—	—
Geranium palustre	9	4	44.4	1	1	100.0	—	—	—
Prunella vulgaris	9	3	33.3	8	3	37.5	—	—	—
Crepis paludosa	9	2	22.2	6	—	0.0	19	2	10.5
Cardamine pratensis	6	2	33.3	25	2	8.0	—	—	—
Hypericum perforatum	4	2	50.0	3	1	33.3	—	—	—
Senecio sp.	3	2	66.6	—	—	—	—	—	—
Cirsium rivulare	2	2	100.0	—	—	—	—	—	—
Frangula alnus	2	1	50.0	—	—	—	—	—	—
Campanula patula	2	—	0.0	21	21	100.0	—	—	—
Polygonum bistorta	2	—	0.0	1	1	100.0	11	—	0.0
Peucedanum palustre	1	—	0.0	34	34	100.0	52	2	3.8
Scutellaria galericulata	—	—	—	65	31	43.7	25	25	100.0
Aegopodium podagraria	—	—	—	35	17	48.6	2	—	0.0
Plantago major	—	—	—	28	2	7.1	8	—	0.0
Achillea millefolium	—	—	—	8	6	75.0	—	—	—
Potentilla palustris	—	—	—	5	5	100.0	10	9	90.0
Veronica longifolia	—	—	—	2	1	50.0	80	25	31.2
Galeopsis tertrahit	—	—	—	6	—	0.0	3	—	0.0
Lycopus europaeus	—	—	—	5	—	0.0	2.819	683	24.2
Anemone nemorosa	—	—	—	5	—	0.0	1	—	0.0
Symphytum officinale	—	—	—	—	—	—	13	11	84.6
Total	15185	2820	18.0	10430	3549	34.0	16350	2004	12.0

Filipendula ulmaria and *Iris pseudacorus* (Figures 4.6; 4.10). Part III contains a detailed description of the relationship between the vegetative reproduction and the age of polycormonal plant

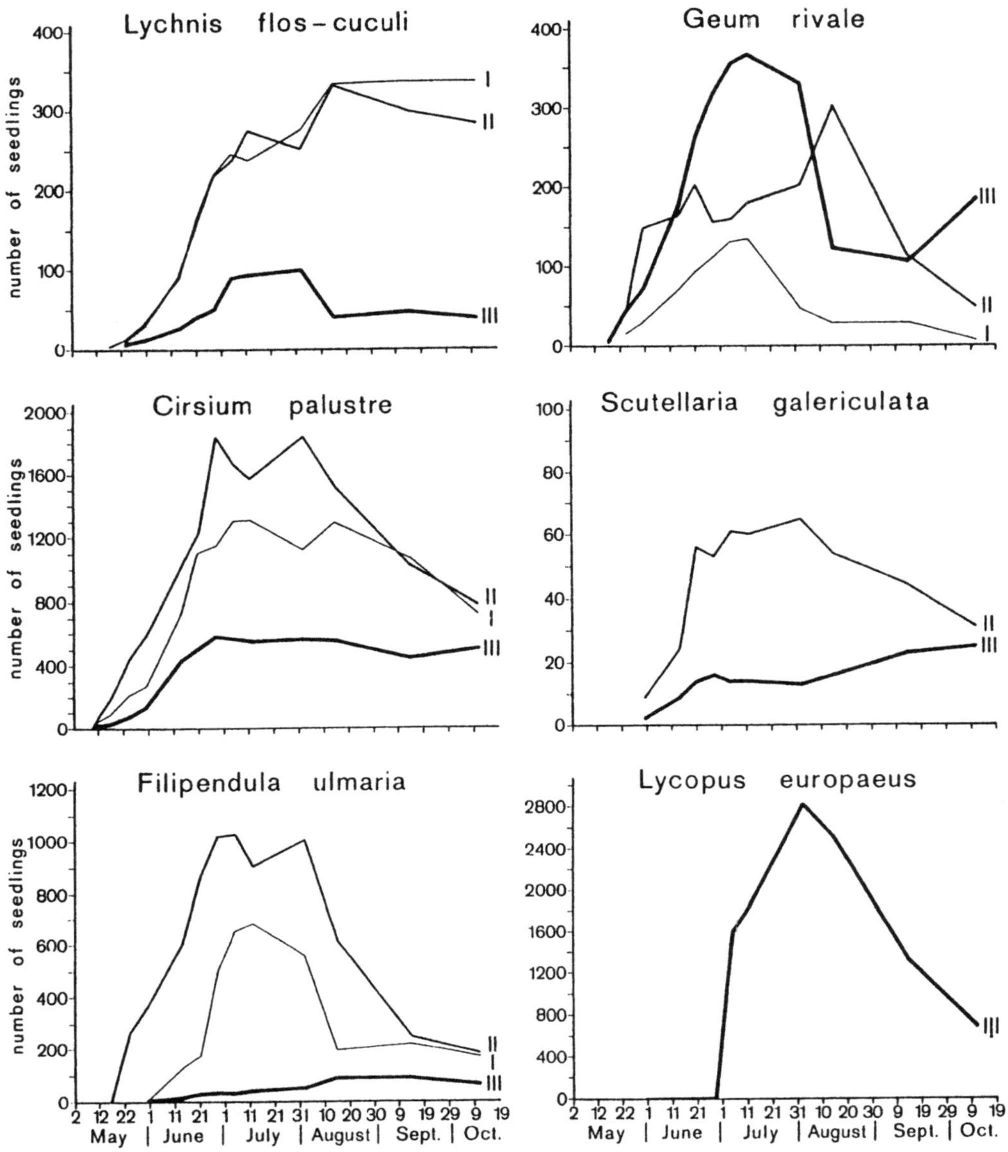

Fig. 4.20. Seasonal variability of number of seedlings in consecutive successional phases: (I) meadow, (II) macroforbes community, (III) willow brushwood (Szydłowska, unpublished data).

organisms. They usually undergo fragmentation when senescence begins to progress in their centre (Figure 3.1).

The proportion of vegetatively propagating individuals grows as the proportion increases of senescent individuals (Figures 3.8; 4.10). It has been found that in many species about 50% of the individuals that have attained this phase divide into several independent entities each, e.g. *Filipendula ulmaria, Iris pseudacorus, Geum rivale, Cirsium rivulare* as well as, though less often, *Caltha palustris* and *Lythrum salicaria.* Hence the growth in number of the vegets in populations was seen as late as several (6—7) years after the actual colonization, that is, in succession phases II and III.

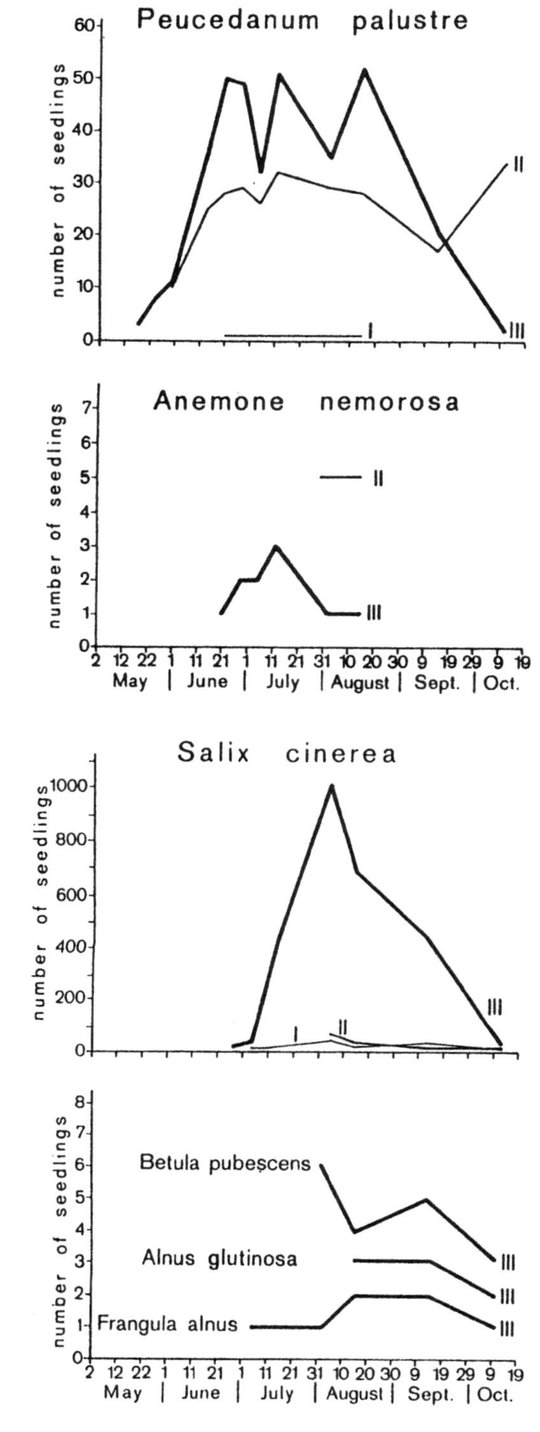

Fig. 4.21. Seasonal variability of number of seedlings in consecutive successional phases: (I) meadow, (II) macroforbes community, (III) willow brushwood (Szydłowska, unpublished data).

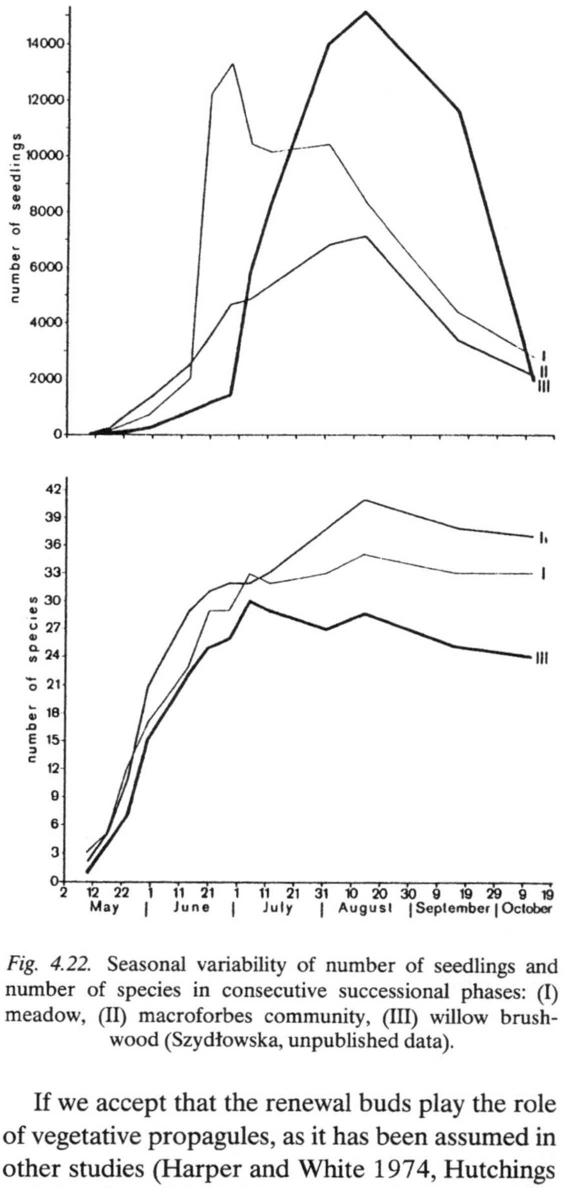

Fig. 4.22. Seasonal variability of number of seedlings and number of species in consecutive successional phases: (I) meadow, (II) macroforbes community, (III) willow brushwood (Szydłowska, unpublished data).

If we accept that the renewal buds play the role of vegetative propagules, as it has been assumed in other studies (Harper and White 1974, Hutchings and Barkham 1976, Maun 1984, Szmeja 1987a, b), the efficiency of vegetative reproduction of many species during a succession will be slightly different. The highest vegetative reproduction efficiency will then occur in succession phases I and II, that is, during the first years following the cessation of mowing. At that stage 1, or up to 10 renewal buds develop at each of the shoot bases in the individuals of many species (Table 4.6). In the first period following the cessation of meadow mowing as many as 60—70% of them attain the

Table 4.5. Number and survival of seedlings during succession (elaboration from material of Szydłowska, unpublished data)

Species	Number of seedlings												Survival %											
	1000—10000			100—1000			10—100			1—10			70—100			40—70			10—40			1—10		
Succession phases	I	II	III	I	II	III	I	II	III	I	II	III	I	II	III	I	II	III	I	II	III	I	II	III
Lythrum salicaria	*	*	*																			*	*	*
Galium palustre	*	*							*									*		*				
Cirsium palustre	*	*				*												*	*	*				
Filipendula ulmaria		*		*					*										*	*				
Ranunculus acris		*		*					*						*	*	*							
Salix cinerea			*				*	*														*	*	*
Lycopus europaeus			*								*													*
Lychnis flos-cuculi				*	*				*				*				*	*						
Viola epipsila				*	*				*											*	*	*		
Lysimachia vulgaris				*	*				*										*				*	
Geum rivale				*	*	*																	*	*
Rumex acetosa				*				*											*	*				
Epilobium roseum					*	*	*								*				*	*				
Lathyrus pratensis					*		*		*										*				*	
Urtica dioica						*	*	*													*	*	*	
Trifolium repens							*	*				*							*				*	
Epilobium palustre							*	*														*		
Potentilla erecta							*	*														*	*	
Myosotis scorpioides							*	*				*					*	*	*					
Mentha arvensis							*	*	*									*					*	
Ranunculus lanuginosus							*	*	*									*	*	*				
Vicia cracca							*	*				*												
Veronica beccabunga							*				*		*	*										
Caltha palustris							*				*								*	*				
Cerastium holosteoides							*	*									*							
Cardamine pratensis								*		*						*				*				
Campanula patula								*		*						*								
Peucedanum palustre								*	*	*							*							*
Scutellaria galericulata								*	*												*		*	
Aegopodium podagraria								*				*								*				
Plantago major								*				*									*			
Potentilla anserina								*		*							*							
Crepis paludosa									*	*	*					*	*				*			
Polygonum bistorta									*	*	*					*								
Symphytum officinale									*												*			
Geranium palustre										*	*					*								
Prunella vulgaris										*	*								*					
Hypericum perforatum										*	*													
Senecio sp.										*									*					
Cirsium rivulare										*														
Frangula alnus										*														
Achillea millefolium											*													
Potentilla palustris											*	*												
Anemone nemorosa											*	*		*										
Galeopsis tetrahit											*	*												
Veronica longifolia									*		*				*									

shoot phase, and only 50% in the macroforb-community formation phase, whereafter the number of shoot buds drops year by year down to 22%, and even 12%, particularly in the willow-scrub formation period. In species like *Lysimachia vulgaris, Myosotis scorpioides, Crisium rivulare* and *Filipendula ulmaria* the vegetative renewal is limited by the increasing plant density following

Table 4.6. Growth index of *Filipendula ulmaria* population during succession

Phases of succession	Number of individuals mature	Number of generative offspring	Number of vegetative offspring	Growth index of population		Ratio of mature (m) shoots to renewal buds
	M	JG	JV	M:JG	M:JV	m:j
I	35	116	–	1:3.31	–	1:4.8
II	58	6	31	1:0.10	1:0.53	1:9.2
III	31	43	18	1:1.38	1:0.58	1:3.1
IV	17	9	6	1:0.58	1:0.35	1:1.1

the cessation of mowing, and due to their being shaded by the growing willow trees. At that stage, i.e., in Phase IV these species are present, but their individuals consist of small numbers of shoots (2–3), but they possess well-developed underground organs, which enable them to survive under the changing conditions during the succession in progress.

Factors affecting the population abundance

It is not an easy task to determine the factors that

Photo 4.2. The patch dominated by *Iris pseudacorus* 15 years after abandonment of the meadow.
Photo: J. B. Faliński

control the abundance of populations growing in multispecific and dynamic systems, such as meadow communities, after the cessation of mowing. To become members of systems that are already established, individuals of a new species no doubt need free spaces for their seedlings to develop and survive until the subsequent developmental phases (Kershaw 1973). This dependence is most conspicuous in sedge marshes made up primarily of *Carex acutiformis*. Seedlings and the mature individuals grow in the gaps that appear from time to time in the dense sward. The occurrence of gaps in meadow phytocoenoses has been emphasized by Rabotnov (1985) who thinks they play a considerable role in determining the abundance of the species making up those phytocoenoses. In the first period of meadow disuse the number of individuals of several species and their spatial distribution depend to a large extent on the occurrence of a few gaps in the compact sedge marsh. (Photos 4.2; 4.3). In populations found in mid-meadow troughs, abundance changes follow a different rule, because under the conditions there the presence of other species does not, as a rule, limit the population growth. The sites are sparsely grown up with vegetation. Hence some species colonize and fill the area within a short time, e.g. *Iris pseudacorus*. In many studies density is believed to have a considerable effect on numbers by acting directly or indirectly, i.e., by limiting the growth and fecundity of, for instance, perennials (White and Harper 1970, Harper 1977) and

Photo 4.3. Aggregation of *Myosotis scorpioides* within a patch dominated by *Carex acutiformis* 8 years after abandonment of the meadow.

Photo: J. J. Faliński

annuals (Wilkoń-Michalska 1976, 1985, Watkinson and Harper 1978, Falencka 1983, Symonides 1983/1984). Observations of the dynamics of abandoned-meadow populations indicate that in the case of perennials, population abundance is determined mainly by the number of free sites to colonize, the competitive power of the species that are already present, or of those entering the area at the same time, the above-ground water level, as well as the morphological features of the plants.

From demographical studies of perennial species, it follows that a significant population-abundance regulating mechanism lies in their morphological and biological characters. Particularly important is the ability of individuals to grow in size, due to which some species can colonize new areas within a relatively short time (5—6 years), even if the number of individuals is small. The ability of individuals to renew senescent fragments, considerably prolongs their life and persistence in the area they have colonized, while the appearance of empty spaces within several-year-old individuals prevents population stagnation, creating conditions for the growth of seedlings and vegetative individuals that arise as the result of polycormone fragmentation, or detachment of polycormone fragments. The formation of polycormones and clones makes it possible for a species to keep an area in a multispecific phytocoenosis, or at high water levels in the spring and autumn, and at the same time it does not prevent successive renewals of a population or its growth in size. Habitat conditions, which modify the rate of individual growth, indirectly affect the dynamics of a population, and especially its numbers and age structure, which can be studied by following the fate of the populations of the same species growing in different conditions, e.g. in a depression-trough, in sedge marches and in willow scrub (Figures 4.5; 4.9).

CHAPTER 13

POPULATION SPATIAL STRUCTURE DYNAMICS DURING SUCCESSION

Introduction

In many papers population dynamics has been described on the basis of numbers or density, and at the same time the role has been ignored of distribution and translocation. It has often been stressed that in order to understand the main mechanisms of the control of numbers, it is necessary to determine the relationship between the numbers and distribution of individuals, that is, to understand the rules that govern the temporal and spatial population-structure (Begon and Mortimer 1981).

In the course of the succession process the relationship between the populations of many species and space is subject to significant changes. The features particularly affected are the frequency of occurrence and manner of space filling. The nature of the changes may vary. Some species tend to considerably increase their frequency, and after several years they will be found all over the Reski Range, while others appear only at one or several sites.

The following colonization-mode diversity has been seen: a) one species almost completely fills its area by forming large aggregations; b) two species form large patches, with other species represented by a small number of individuals; c) a species forms small assemblages dispersed amidst other components of the phytocoenoses (Figures 2.2; 2.3).

In many plant species the cessation of mowing causes new individuals to appear every year, which leads on to their spreading throughout the Reski Range, and then filling the area taken possession of by them. In polycormonal species the process if faster because the individuals are capable of propagating shoots over their long-term life-histories. Depending on the intensiveness of the demographic processes and polycormone growth in size, some species within a short time form large patches in different parts of the Reski Range, and others, uneven-sized aggregations dispersed over a large area (Figure 2.3).

Spatial structure types

Under natural conditions the diversity of individuals differs considerably from the theoretical spatial sturcture type models, i.e., random, uniform and aggregative. Population spatial structure is in reality very complex, and the particular distribution models are only realized in certain simple ecological situations. Chessel (1977) has proposed 6 population spatial structure types: random, regular, clumped, aggregated, patched, and gradient.

A population of the same species often appears to have a random distribution in one part of the biochore, and aggregated or even uniform in another. In such a case Chessel suggests structure types that might be termed as "mixed", e.g.: gradient-aggregated, regular with aggregations, patch-aggregated, gradient-aggregate-clumped.

Gradient-aggregate-clumped spatial sturcture

During the long-lasting succession it could be seen that individuals of the same species in one meadow biochore portion concentrated in large numbers, and in another they were dispersed or formed small assemblages. There are several causes of such a distribution of individuals, viz.: a) abiotic agents, b) inter-population action, c) different meadow abandonment times.

A concentration of individuals of a species in one part of a biochore and their dispersal in another is often the result of gradient variation of abiotic agents. For instance due to variations in the meadow water level, individuals of *Caltha palustris, Iris pseudacorus* and *Myosotis scorpioides* tend to concentrate in biochore portions with a water level of 20—30 cm throughout the growing season. Their density drops, however, gradually as the water level falls to zero.

A similar spatial structure type may also form in a situation where meadows located side by side were abandoned in different times. For example, the gradient-aggregate-clumped structure type represented by *Carex cespitosa* in the 1 ha area resulted from differences in abandonment time. The largest aggregation of this species formed in the meadow biochore part in which mowing had been stopped the earliest, i.e., in the period 1970—1974. In that part of the biochore (right side of the map in Figure 4.23). *C. cespitosa* not only shows the highest density, but also consists of the largest tussocks.

Close to the highest *C. cespitosa* density spot, the tussocks form numerous aggregations, whereas at distances greater than 20—30 m from that spot there, as a rule, occur small, dispersed tussocks, the tussock area coming up to 200 cm^2. In that part of the biochore the meadows were abandoned 4—5 years later, that is in the period 1974—1978. Differences in *C. cespitosa* tussocks in the 1 ha area result from the uneven periods of colonization of the particular biochore parts by the species. The rate of area colonizing, and especially of its filling, is closely related to the growth period of *C. cespitosa* tussocks which lasts about 9—10 years, whereas the formation of a tussock in a 1/4 m^2

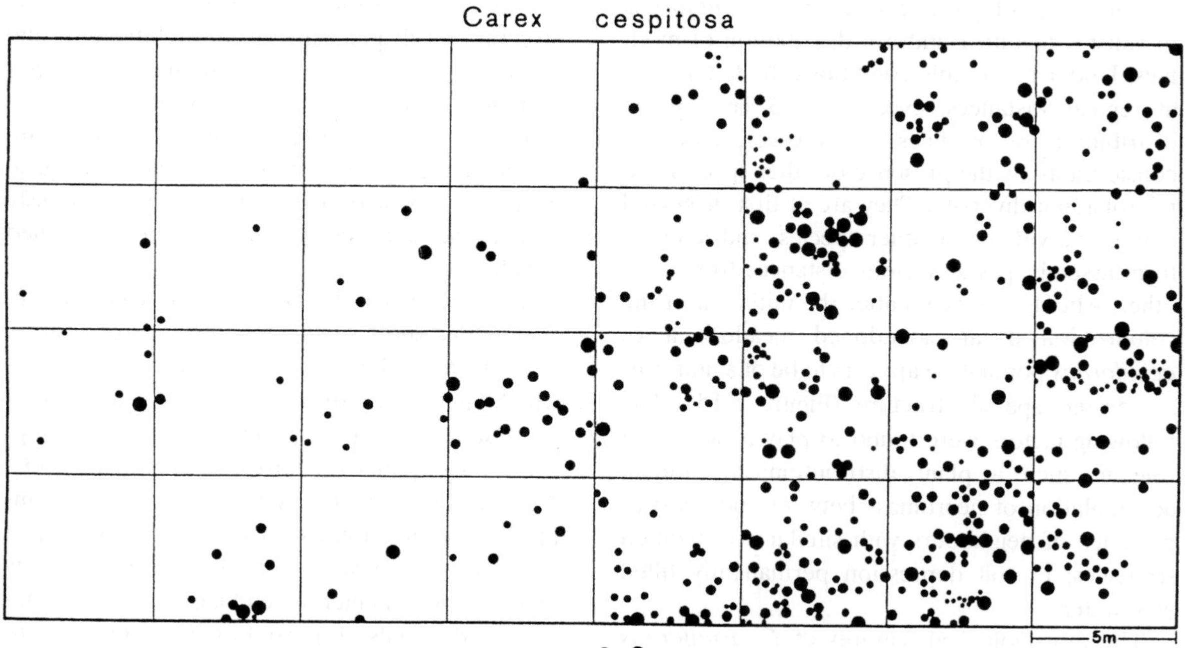

Fig. 4.23. Gradient, aggregate, and clumped spatial structure of population 15 years (on the right of map) and 10 years (on the left of map) after abandonment of the meadow (Brzosko, unpublished data).

area takes at least 5—6 years. Hence the proportion of this species in the seral community sequence becomes more noticeable as late as 5 years after the cessation of meadow mowing, and very high after 9—12 years when the number and size of the tussocks are at their highest (Figure 3.13).

Uniform-aggregated population spatial structure

Due to the yearly rhizome growths and formation of new modules at various distances from the sister modules, species possessing these capabilities can colonize and fill large areas within a comparatively short time. In abandoned meadows, *Carex acutiformis* forms large patches within 8—10 years. In such cases the population spatial sturcture is most often defined as patched or uniform. As a matter of fact, modules form large or small assemblages, therefore in a 100 cm^2 plot there can grow 1 to 20 of them and in a 1 m^2 plot, 30 to 500 (Figure 3.15).

The morphological features of *C. acutiformis* organisms favour a uniform distribution of modules. Under favourable conditions the latter form at regular distances, every 10—15 cm. Spatial distribution of modules is, however, modified considerably by the presence of other species and microhabitat diversity. They are at first dispersed among individuals of other species, and later on form assemblages at various distances from each other. About 5—6 years after the initiation of the colonization of an abandoned meadow, a *C. acutiformis* population appears to be of a uniform-aggregated spatial structure (Figure 3.15). The following factors were found to play a significant role in such a plant distribution: a) gradual accumulation of necromass between modules; b) presence of neighbours with similar colonization strategies; c) soil depression permanently filled with water.

The morphological features of *C. acutiformis* and the manner and rate of meadow colonization by it have been described in detail in the chapter dealing with the inhibitors of succession (Part III, Figures 3.11; 3.15).

Hierarchical population spatial structure

The hierarchical organization type is most frequently encountered in populations of herbaceous plants which reproduce generatively and vegetatively and whose individuals are capable of many times increasing the number of rooted shoots during their life-histories. A considerable association is then seen between the spatial organization of a population and its age structure and reproduction type. The mechanism of this phenomenon is simple: with their progressing age, year by year individuals increase the number of shoots. As a result, after several years polycormones represent shoot assemblages varying in size and permanence. If polycormones attain the senescent stage or a size endangering their physical existence, then they disintegrate to form clones, each of several individuals, which begin their independent life. There may simultaneously develop individuals from seeds, in which case there will form, in the vicinity of the clones, groups similar in their habitus and size to several-year-old polycormones. The relationship between the spatial structure and plant-organism architecture and reproduction type can be exemplified by species that form in the course of succession large aggregations, or even large-area stands. They are species of different morphological and biological characters whose populations grew simultaneously and in the same conditions, i.e., an abandoned meadow.

Cirsium rivulare is a perennial of two reproduction modes and an individual life span of about 5 years. Its individuals have fragile, spongy rhizomes which favour their disintegration into clones. Due to these features, their populations have a group-aggregative spatial structure. In different areas of a *Cirsium* meadow (0.6 ha) there formed assemblages covering areas of 4 to 25 m^2. The largest aggregation to have ever formed in the meadow included 80 smaller assemblages and 55 one-shoot individuals, dispersed or in groups close to several-shoot individuals, usually several-year-old ones. Most of them consisted of 1—5 shoots each, representing 48% of the population, the remainder were polycormones made up of 6 to 25 shoots.

They jointly produced 561 rooted units, including 525 rosettes and 36 generative shoots.

The group-aggregative population spatial structure of *Cirsium rivulare* first of all reflects the morphological and developmental features of the individuals, that is their potential to produce several to about a dozen aerial shoots over their several years life-history. Areas occupied by small assemblages as a rule represent polycormonal, less often clonal areas. The formation of higher-order aggregations, occupying areas of several to about a dozen m^2, is determined primarily by the dissemination biology — a large number of seeds in light infructescences. Vegetative reproduction has a less important influence on the spatial structure of *Cirsium rivulare* populations. Of a total of 135 individuals that grew in an area of 25 m^2 only 14 disintegrated, giving rise to clones.

Lysimachia vulgaris is a perennial that rapidly spreads in abandoned meadows, and, owing to its specific morphological characters, almost totally fills the areas colonized by it. *Lysimachia vulgaris* seeds germinate in spring, and at the end of the growing season each of the survivors is made up of 2—3 shoots. In the next growing season each of them will send off several rhizomes 10 to 150 cm long. Only sporadically, were rhizomes 3—5 cm long recorded. Rhizomes root at varying distances from the parent organism, and give rise to new shoot assemblages. This results in the crossing of the rhizomes of many individuals, and the daughter organisms do not form distinct clones. For instance, a fairly distinct shoot aggregation in 1 m^2 was found to contain 3 individuals derived from different parent organisms, 80, 125 and 168 cm apart. Though *L. vulgaris* individuals are able to grow into polycentric polycormones, under natural conditions it is usually impossible for dispersed fragments of an individual to continue to form an integral entity. The preventing factors are the structure of the rhizomes, which are slender and long, and the action of many environmental agents accelerating their fragmentation. In a population that appeared to be relatively stable after its 6 years' growth a stand-aggregation-tussock type of spatial structure was found. In a 25 m^2 area there grew 270 one-shoot individuals: 58 developed from seeds, and 212 due to the rooting of the rhizomes extending from 191 individuals. The individuals produced jointly a total of 838 rooted shoots, at a density of 7 to 83 per 1 m^2, representing a lower-order assemblage, as illustrated by Figure 3.12. A *Lysimachia vulgaris* population inhabiting a 0.6 ha wet and waterlogged meadow formed large aggregations (100—200 m^2) made up of smaller (5—25 m^2) assemblages including lower-order assemblages (100—400 cm), due to whose irregular distribution there were high-density and low-density zones in the population area (Figure 4.24).

Filipendula ulmaria individuals' characters play a decisive role in the formation of a hierarchical population spatial structure. At the end of an 8 years' population growth period in a 25 m^2 area there were 5 large aggregations, occupying 1 to 2 m^2 each. They were clones jointly including a total of 27 vegetative-origin individuals with 6 generative-origin individuals interspersed among them.

The area was inhabited by a total of 113 individuals: 44 single-shooted and 69 multi-shoot individuals, and polycormones made up of 2 or up to 10 shoots represented the largest proportion, i.e., 80% of the population, while the remainder included polycormones made up of a very variable number of shoots (from 11 and up to 32). The aggregation-group type of population spatial structure of *Filipendula ulmaria* is the result of both the morphological and developmental properties, that is, individual ability to multiply shoots over the long-term life-history, and vegetative propagation. High-order aggregations are clones consisting of several smaller groups of 2 to 7 vegets that arise due to the disintegration of polycormones (Figure 3.8). Moreover, within the particular clones there develop generative individuals which form small groups.

Under favourable conditions, at an intensive polycormone growth and a high-rate of generative and vegetative reproduction, a colonized area becomes filled completely. This occurs when the inter-polycormone and inter-clone spaces become filled due to the widening of individual areas, or

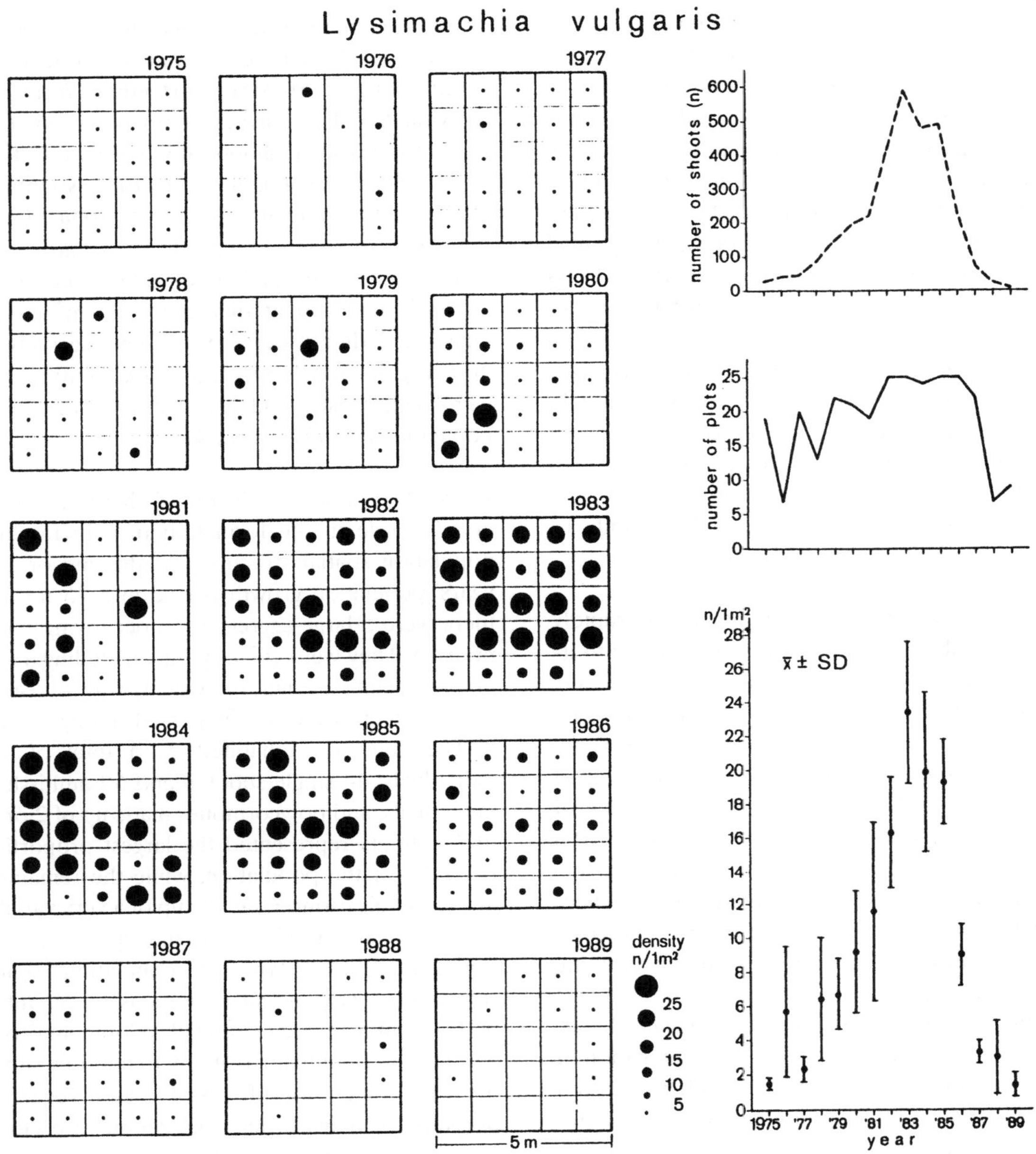

Fig. 4.24. Spatial dynamics of population for 15 years in a 5 m × 5 m plot in abandoned meadows.

appearance and fast growth of the generative and vegetative offspring. Both processes cause, within a relatively short time, the inter-individual and inter-clone boundaries to disappear. As a result, there forms a monospecific stand with distinct high-density and low-density zones (Figures 4.25; 4.26).

Iris pseudacorus. The formation of a hierarchical spatial organization of *Iris pseudacorus* populations will be illustrated by presenting the distribution of individuals in a 25 m^2 area, a fragment representative of the conditions required by the species, i.e., in a fertile and moist meadow, in a vegetation-free depression, with water staying

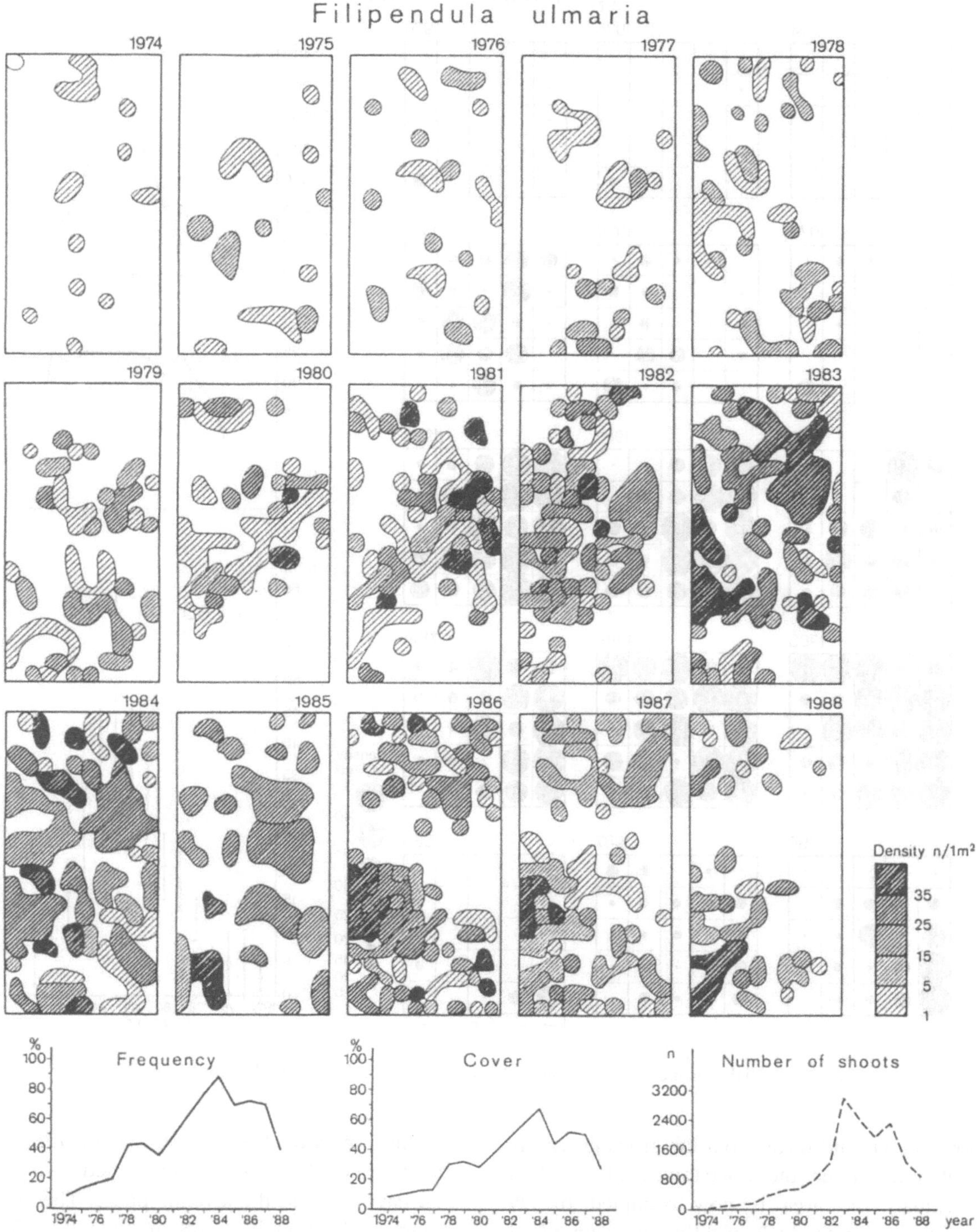

Fig. 4.25. Spatial dynamics of population for 15 years in a 200 m^2 plot in abandoned meadows.

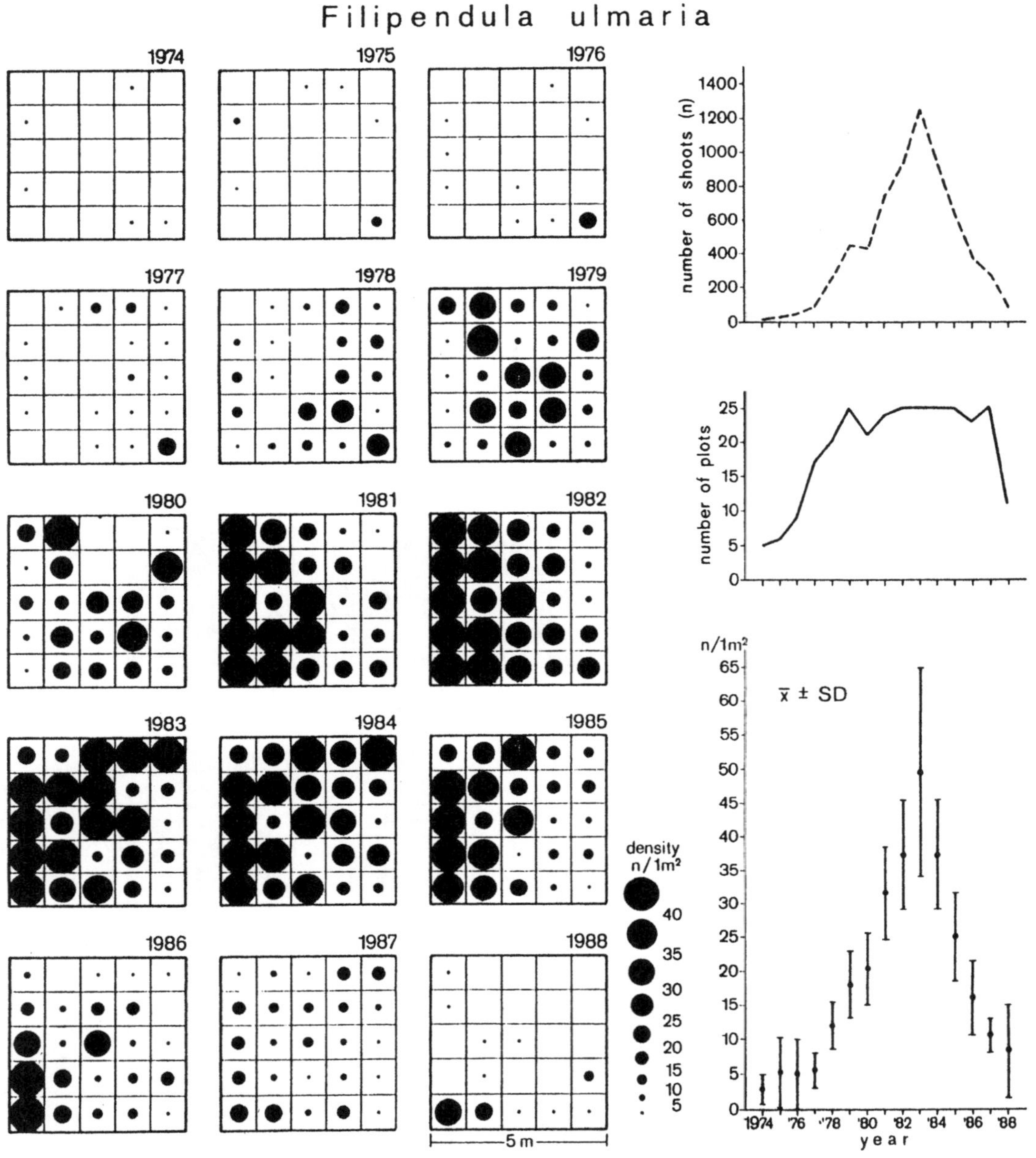

Fig. 4.26. Spatial dynamics of population for 15 years in a 5 m × 5 m plot in abandoned meadows.

above the ground throughout the growing season. Eight years after the appearance of the first individuals the area is now occupied by 99 uneven-sized individuals of different origin: 60 of generative and 39 of vegetative origin, grouped in 7 clones (Figure 4.10). They jointly produced 522 rooted rosettes and 65 flower shoots. As regards to their individual structure, 31 of them are monocormonal (one-shoot) units, aged 1—3 years, that have attained the juvenile phase. The largest group of the remaining 68 individuals includes polycormones with 3—15 rooted shoots, and only 3 of them consist of several dozen (37, 45, 58) shoots each, and include 120 rosettes and 8 flower shoots. Only 3 in 25 plots are devoid of *Iris pseudacorus* and all other species. If the relationship between the population area and total area is estimated by comparing the spaces occupied by rooted rosettes and flower shoots, then it is found that they jointly occupy less than 5 m^2, that is, 1/5 of the total area. The average number of rooted units per 1 m^2 is equal to 24.86 ± 12.24. If, however, polycormone architecture is taken into account, then it is found that the area is filled almost completely by *Iris pseudacorus*. This is because the rooted rosettes marked on the population spatial structure map as the only points are in reality made up of about a dozen leaves each: after attaining a height of 50—70 cm they lie flat on the ground and cover it tightly. The same is true of the generative shoots. Individuals of other species seldom grow under the polycormone canopy. They will only appear in an *Iris pseudacorus* polycormone area when there have formed empty spaces released by necrotic polycormone fragments, usually in the polycormone centre (Figure 3.5).

The formation of the hierarchical population spatial structure of *Iris pseudacorus* is favoured by the heterogeneous environment, e.g., the varying groundwater table, and by individual morphological features and the ability to propagate vegetatively. As a result, there arise (within a population) aggregations occupying 1—3 m^2, and being as a rule clones each consisting of several or a dozen or so individuals made up of a varying number of rooted units (Figure 4.10). Smaller assemblages are usually polycormones, less often groups of generative individuals. In this case the intensiveness of aggregating to a large extent depends on the population age structure according to the following rule: the greater the number of several-year-old individuals, the more the aggregation size resembles the size of individuals; the more there are individuals aged 6 up to 9 years, the more the large-size aggregation areas resemble the clone structure. The cause of this lies in the fact that most polycormones aged 5 years disintegrate due to the senescence of their underground organs, or because of having attained such large sizes that their physical existence becomes endangered.

Spatial dynamics types

It will be worth while to consider the durability and reproducibility of the population spatial structure for at least three reasons: 1) if and how stable is the spatial structure after the abundance equilibrium has been reached; 2) if a particular spatial structure type is peculiar to the species or phytocoenosis type; 3) if the spatial structure type varies with the growth phases (abundance increase, decrease and equilibrium).

Previous researches into the population spatial structure of meadow and woodland phytocoenoses have revealed that: a) conspecific populations under different conditions differ by their spatial structure; b) within the same phytocoenosis, species differ considerably from one another in their spatial structure; c) the density and distribution of individuals in a population vary with the growing seasons and many-year periods (Falińska 1989a, b).

The range and rate of changes in the density and distribution of individuals may vary, depending on the habitat conditions and biological properties of the species. They can be of a considerable magnitude, and can, after some time, result in considerable population spatial structure changes, e.g. from a random to an aggregated type, or they can be manifested by density fluctuations of little significance. According to the range and duration of the changes, three spatial structure dynamic types have been distinguished: (I) successive progressing density changes related to the population growth stages; in such cases each population growth stage can represent a different spatial structure type; (II) fluctuating at a certain point of time, the density of individuals in a population changes, but it returns to the original level after some time; changes of this type occur every growing season or every several years; (III) pulsating — small density variations, peculiar to almost every population, the magnitude of which is low and different in each part of the population area; such changes do not alter the spatial structure, nor cause a deviation from the model tended towards in a particular situation.

Gradual population spatial structure changes

The above spatial dynamics type is represented by species whose populations twice or three times over the 15 years changed their spatial structure type from a random to an aggregated, and then to a patch-aggregated type to return to the random structure after about a dozen years (Figure 4.26). Population spatial structure changes of that sort were seen in species the presence and proportion of which had been significant before the cessation of meadow mowing, and in those whose frequency had never been above 10%, and cover 1—5%.

Species whose proportion in the cover in regularly mown meadows is high, significantly increase their areas within several (5—7) years following meadow abandonment. Over the Reski Range they appear at many sites, and within a short time fill all the newly occupied areas, forming more and more new aggregations. As time passes, there occurs a reverse process, that is, a reduction in density and frequency (Figures 4.27; 4.28). Among the species which respond in that way to the cessation of mowing are: *Lythrum salicaria, Lysimachia vulgaris* and *Filipendula ulmaria* (Figure 4.25).

If density and distribution changes start at a low frequency and at small proportions of species in meadow communities, then within several years their population spatial structure changes from a random to an aggregated and then to a random one. After 6—7 years these species often retreat, partly or completely, form areas occupied by them to return there again when there have formed, for instance, willow scrub. Sparse individuals of those species will then occur on the margins of the scrub. The following species have been included in the group: *Cirsium palustre, C. rivulare* and *Geum rivale* (Figure 4.29).

A successive spatial dynamics is peculiar to species with a highly effective generative reproduction, that is, a high seed production and self-sowing rate, and a high seedling survivorship (Table 4.4), or with two reproduction modes functioning simultaneously. The successive, gradual structure changes will be illustrated with a number of selected examples.

Lythrum salicaria is a typical representative of species that quickly invade abandoned meadows and become dominant at a certain stage of seral vegetation changes.

Over the 15 year period of meadow disuse, the *Lythrum salicaria* population size to area relationship changed significantly, as indicated by the frequency, cover and density (Figure 4.28). In periods of regular mowing the species had formed uneven-sized assemblages of individuals, with an even density of 1—5 per 1 m^2 (Figure 4.28).

A seral population spatial structure transition to a patch-aggregated one is caused by the increase of: a) frequency from 10—20% to 30—70%, b) cover from 10% to 80%, and c) density from 1 to 30/m^2.

A comparison of the population spatial structure in 200 m^2 and 25 m^2 plots shows that during the first 6 years of meadow disuse the number of aggregations increased, and so did their size and, above all, plant density. The maps and charts illustrate variation in the aggregation-occupied area size, and in the micro- and macro-scale density. Over the first years of meadow disuse the density was uniform, ranging from 1 to 5 per m^2. In the period between population growth years 4 and 6 the density increased 10-fold, and frequency came up to 100% (Figure 4.28).

After 7 years there began a gradual retreat of this species from the sites occupied by it. During that time (9—15 years) the cover dropped from 80—90% to 10—30%, while the density returned to its previous low and little diversified level. A similar density had been seen immediately after the cessation of mowing (Figure 4.28).

Over the 15 years' *Lythrum salicaria* population growth period, the spatial structure changed from a random to an aggregated and patch-aggregated type, and then returned to the aggregated form. The significant density and distribution change observed between population growth years 8 and 15 was caused by the growth of the willows (*Salix cinerea*) that had encroached upon the area simultaneously with *Lythrum salicaria.* They did not at first have any significant effect on the density and distribution of *L. salicaria* individuals. Their influence was not noticeable until

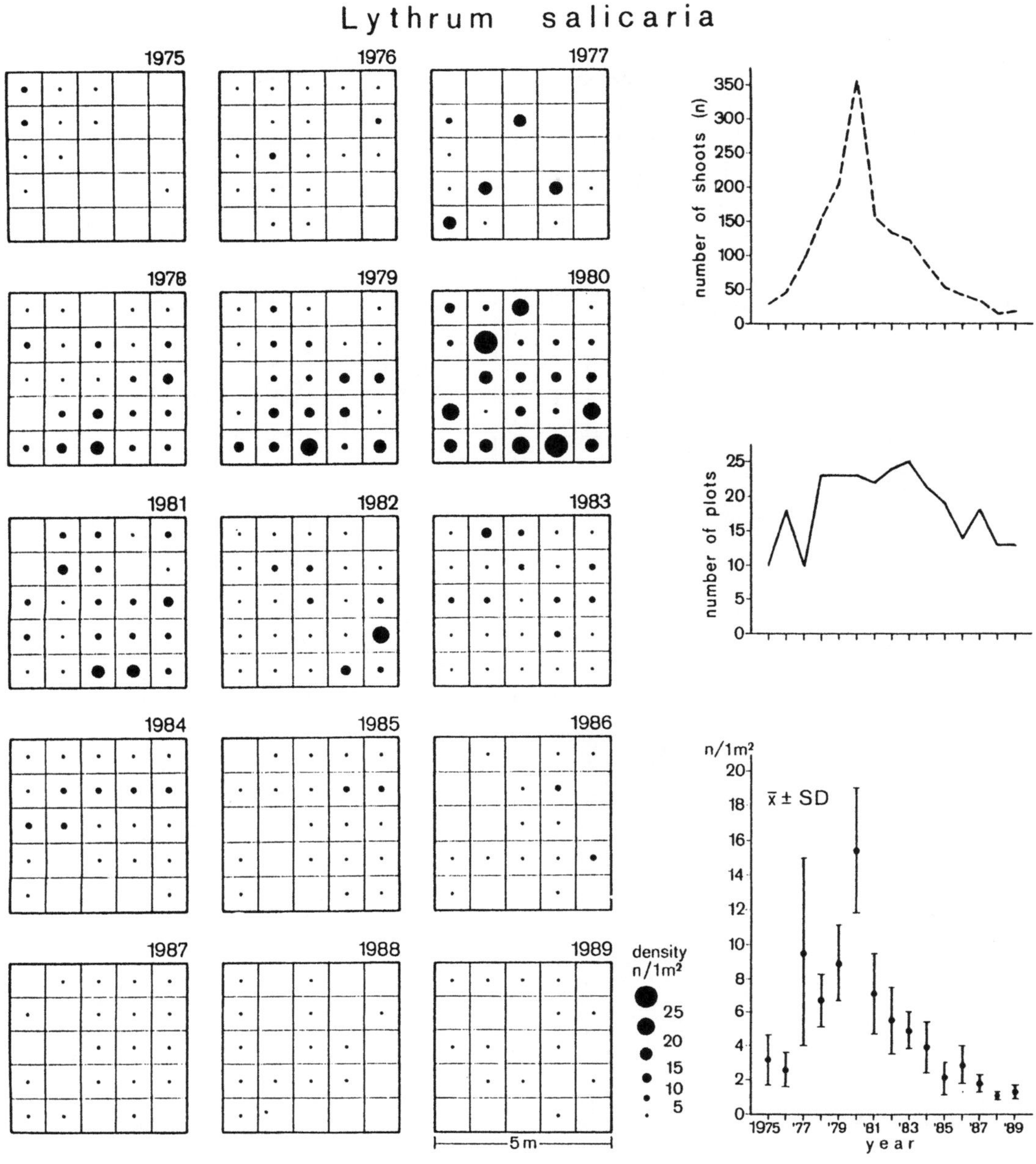

Fig. 4.27. Spatial dynamics of population for 15 years in a 5 m × 5 m plot in abandoned meadows.

several years later, after there had formed tall willow scrub.

In those meadow parts where *Lythrum salicaria* had become dominant, together with *Carex acutiformis*, the population spatial structure of both species was over the 15 year period of a patch-aggregated type. In such cases, changes in plant density and distribution in a *Lythrum salicaria* patch were to a lesser extent the result of demographic processes, and to a greater extent of variations in the number of shoots and size of the areas of the long-lived polycormones.

For instance, in a 400 m² area, in 1975, a total of 220 *Lythrum salicaria* assemblages were found,

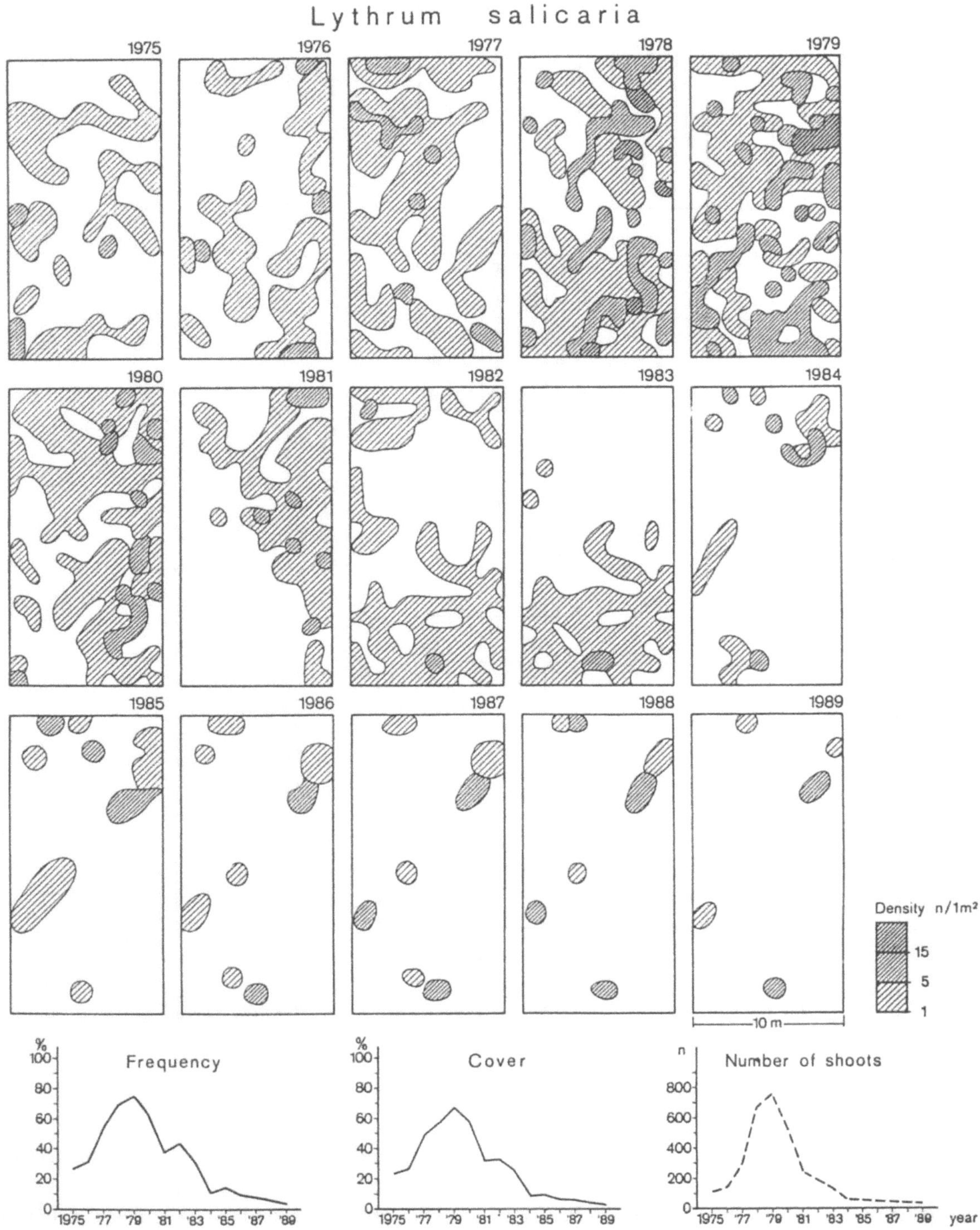

Fig. 4.28. Spatial dynamics of population for 15 years in a 200 m² plot in abandoned meadows.

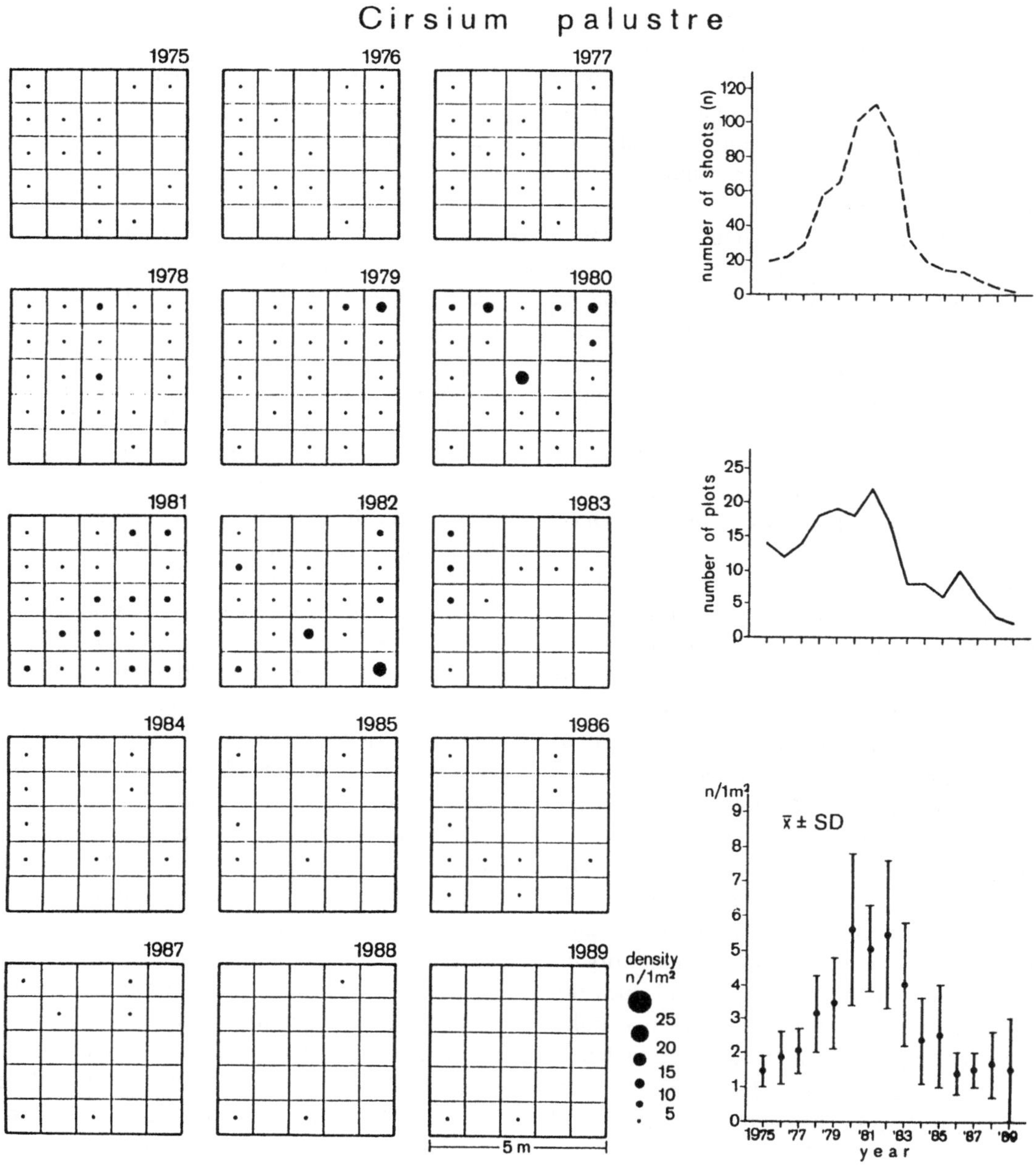

Fig. 4.29. Spatial dynamics of population for 15 years in a 5 m × 5 m plot in abandoned meadows.

and in 1989, that is, after 15 years, 109. Over that period the percent cover changed only slightly (from 32 to 36%), whereas subject to significant variations were the shape and size of the aggregations, and the variation increased of the density of plants, that is, it varied between particular parts of a patch, ranging from 1 to 35 plants per m^2, while in the initial period a density of 1 to 10 per m^2 prevailed.

The population spatial structure dynamics of *Lythrum salicaria* in abandoned meadows is primarily determined by a high efficiency of the generative reproduction (a high seeding rate, germinating capacity and seedling survivorship)

(Table 4.4). Over the long-term history of a population there appear many individuals of generative origin (Figure 3.23), and the aggregated type of population spatial structure is moulded by seedlings growing in large assemblages, and by individuals increasing the number of shoots during their long-term history.

A similar spatial-structure dynamics type during a succession in progress is represented by *Filipendula ulmaria*. Over the 15 year period of meadow disuse the species occurred throughout the Reski Range area. In some parts of the range it formed large patches varying in their durability (Figure 2.1). As in the case of *Lythrum salicaria*, the population spatial structure was subject to gradual changes: from random or aggregative to patch-aggregated and then back to the random or aggregative type (Figure 4.25).

By contrast to *L. salicaria*, the population spatial structure variation of *F. ulmaria* is to a larger extent caused by a fast rate of growth in size of the polycormones. Several year-old *F. ulmaria* polycormones are made up of about a dozen or several dozen shoots (Figure 3.2), while those of *L. salicaria* usually of several shoots only (2—5). Apart from this, large *F. ulmaria* aggregations are in the period of becoming dominant, i.e., after 8—10 years, clones that have formed due to polycormone disintegration (Figure 3.8). In *L. salicaria* polycormone fragmentation was seldom observed. The type of its underground organs does not favour organism division and does not contribute to the formation of large-area multi-shoot polycormones. An increased aggregateness of the populations of these two species results from different factors. The aggregated structure of *Lythrum salicaria* populations is primarily the result of the generative reproduction and longevity (due to the durable underground organs) of the organisms. In *F. ulmaria* aggregateness is the result of two modes of reproduction and the ability of organisms to form multi-shoot polycormones. At a certain time during the growth of a population this capability plays a greater role in the determination of the spatial structure (Figure 3.8). High densities at some points in a patch arise due to the presence of many-year-old polycormones, each made up of over a dozen or several dozen shoots. Density variations between years in 1 m^2 plots result primarily from varying numbers of shoots in polycormones. The situation is illustrated by the charts worked out for a 25 m^2 plot that represented a part of a large *F. ulmaria* aggregation (Figure 4.26).

Filipendula ulmaria differs from other polycormonal species by the highest micro- and macro-scale density. The species often forms large and long-lasting aggregations in abandoned meadows, with low proportions of other species (Figure 3.22). For instance, in areas differing in habitat conditions, the maximum density ranged from 500 to 5000 shoots per 200 m^2, 1200 to 2600 per 25 m^2, and 60 to 82 per 1 m^2.

Low-density *F. ulmaria* populations are found in abandoned meadows if the meadows are colonized also by *Carex cespitosa*. *F. ulmaria* individuals then form small aggregations dispersed within *Carex cespitosa* patches. In a 25 m^2 plot their number ranges from 2 to 6, and each consists of several to over a dozen shoots. The individuals sometimes grow on sedge tussocks, in which case they are monocormonal forms and their density is low (Figure 3.13). This type of distribution and density may continue to stay unchanged until there forms willow scrub in the area. The population density may then increase, because numerous individuals again appear on the willow scrub margins.

If *Filipendula ulmaria* colonizes an area jointly with *Carex acutiformis*, then its populations are for many years of an aggregated spatial structure, and their density fairly quickly rises and drops (Figure 4.5). In such a situation dominance belongs to the sedge which covers 70—80% of the area. *Filipendula ulmaria* and some other species then form small assemblages in rather compact sedge patches (Figure 3.11).

Lysimachia vulgaris populations vary widely in their response to the cessation of meadow mowing. When the meadows were mown every year, the species occurred throughout the Reski Range, and formed aggregations, varying in size, dispersed in the meadow sward and sedge swamp. Immediately after the cessation of mowing the species was found to quickly retreat from the different sites, particularly those in which species like *Filipendula ulmaria, Lythrum salicaria* and *Carex cespitosa* had become dominant within 5—6 years. The proportion of *Lysimachia vulgaris* at that time dropped from 30—40% to 5—10%, but the species appeared again at new sites, e.g. on the margins of willow scrub, or in gaps in *Carex acutiformis* patches. The *L. vulgaris* populations of that period were characterized by a fast density growth, followed (2—3 years later) by a decline and stabilization at a low level that lasted until the return of forestation. At the border between different community types, e.g. woodland, meadow, scrub and the sedge marsh, *L. vulgaris* populations showed a tendency towards higher densities (Figure 4.13), this being similar to what was recorded for *Lythrum salicaria.*

There also occurred ephemeral, lasting several (5 or 6) years, aggregations. Their growth usually started at a high density level that increased for 2—3 years, and then fell rapidly, and the populations retreated from the areas they had been occupying.

The spatial dynamics of *Lysimachia vulgaris* populations is, in general, successive in nature. The species differs from other species of this group in that it rarely forms monospecific aggregations and seldom becomes dominant in those phytocoenoses that arise in the course of a succession. More often it is a co-dominant species among such species as *Carex acutiformis, Filipendula ulmaris* and *Salix cinerea.* Apart from this, *L. vulgaris* aggregations that form after the cessation of meadow mowing are short-lived. *L. vulgaris* aggregations disappear from some areas of the Reski Range and new ones form in others, as a rule in the gaps or on the margins of the aggregations of other species (Figures 3.11; 3.13). *L. vulgaris* individuals easily colonize new areas because they produce long rootstocks (up to 5 m, extending in several directions) which readily root among individuals of other species. If, however, plants with strong underground organs appear in *L. vulgaris*-occupied areas, the species will be eliminated very soon, since the rhizomes of its individuals are very delicate. This can account for the rather quick rotation of this species in the course of succession and its fast relocation over the Reski Range. Permanent *L. vulgaris* aggregations were, however, seen there persisting for 15 years (Figure 4.24).

In their nature *Lysimachia vulgaris* population spatial structure variations are similar to the spatial dynamics of species like *Filipendula ulmaria* and *Lythrum salicaria,* but only in situations where in abandoned meadows polydominant communities form within several years, followed by macroforb communities. That is when none of the tall herbaceous species forms large aggregations. Only in such a situation will the spatial structure of a *Lysimachia vulgaris* population change within a relatively short time from a random to an aggregated type, and then in the period of willow scrub formation it will (locally) assume a patch-aggregated form, especially around *Salix cinerea* aggregations (Figure 4.13).

Fluctuant spatial dynamics

Some of the species present in the meadow-utilization period do not show a high expansion rate following the cessation of mowing. They, rather, change their density in previously occupied places, the changes occurring within a relatively short time (3—6 years). Several years later the species retreat from many sites, and where they stay over the Reski Range their individuals tend to be randomly distributed in the abandoned-meadow willow scrub or sedge marshes that have formed there within 8—12 years.Very often these species considerably change the sites of their occurrence, although they do not increase their contribution to the plant community structure of the Reski Range. It means that the species entirely retreat from some places and appear in others. This applies, particularly, to species of a delicate structure and a comparatively short life-history (2—4 years). This type of population response to the cessation of mowing was found, for instance, in *Cirsium palustre, Ranunculus acris, Lychnis flos-cuculi, Polygonum bistorta.* These species continue to be present until the formation of willow scrub.

Sometimes their individuals present in willow scrub are randomly distributed there, or form small assemblages on the margins of large-sized willow aggregations.

Frequency and density

Short-lived species present in regularly mown meadows and in abandoned meadows change their population density locally, that is, in areas previously occupied by them but their frequency on the Reski Range seldom changes; the area they occupy does not increase significantly. They are species associated with the early succession stages and play some role at the stage of polydominant meadow and macroforbes community formation. Typical representatives of this type of spatial structure changes are: *Cirsium palustre* and *C. rivulare.*

In meadows that are mown every year individuals of the above species appear to be randomly distributed or occur in assemblages, on macro- and microscale. Their distribution over the Reski Range area can be described as "patched", with a tendency towards a random distribution of individuals. For example, over a 15 ha meadow stretch there occurred in the first year following the cessation of mowing 12 large *C. palustre* assemblages, occupying areas from 20 to 400 m^2. The density of individuals within the assemblages varied widely — between 1 and 11 per 1 m^2. Dispersed among the assemblages were solitary individuals. After 10 years disuse of the meadows there remained 3 assemblages, occupying 25, 40 and 100 m^2, and after 15 years there persisted 2 assemblages, occupying 20 and 25 m^2. An example has been presented (Figure 4.29) of density and frequency changes in a 25 m^2 plot located in an area where a *C. palustre* aggregation was present for 15 years.

In the first years following the cessation of mowing the frequency of this species over the Reski Range, that is on a macroscale, dropped, but in some parts of the area its density increased considerably, particularly in the period when there formed macroforbes communities, i.e., between meadow-disuse year 5 and 8 (Figure 4.29). After the formation of scrub communities the density and frequency of *Cirsium palustre* returned to the original situation. Single individuals or small assemblages of *Cirsium palustre* are found growing around willow aggregations. In a 25 m^2 plot several individuals are found (Figure 3.20).

Ten years after the meadow abandonment the species was found to have completely retreated from some parts of the Reski Range, and several years later it appeared again in the same areas. Individuals then grow singly on sedge tussocks or on willow scrub margins. During the first succession phases the population spatial structure of *Cirsium palustre* is aggregative (Figure 3.20), with very wide spatial and temporal density variation. In meadows that are mown yearly, and after the formation of scrub communities it shows a high stability.

A similar spatial dynamics type is represented by *Cirsium rivulare.* The species was present in yearly-mown meadows and during the succession in progress in the Reski Range. Its population spatial structure is aggregative. With progressing succession, variation in its density increased gradually — in some plots several individuals and in others several score individuals were found (Figure 4.30). This type of population response to the cessation of meadow mowing could be observed in meadow-disuse years 5 and 7. The highest *C. rivulare* density and frequency were recorded during the formation of macroforbes communities, and much lower values of these two parameters were found during the formation of sedge marshes and willow scrub (Figure 4.12).

Sometimes the above two species jointly form large aggregations, in which case their highest population density occurs 5—6 years after the cessation of mowing, and can stay for several years (Figure 3.21). In such a situation one of the species is dominant. Their aggregations usually occur among other species, particularly amidst tall herbs, sedge (*Carex acutiformis*) tussocks, and on the margins of *Filipendula ulmaria* and *Salix cinerea* aggregations, and on *Carex cespitosa* tussocks (Photos 3.7; 4.4).

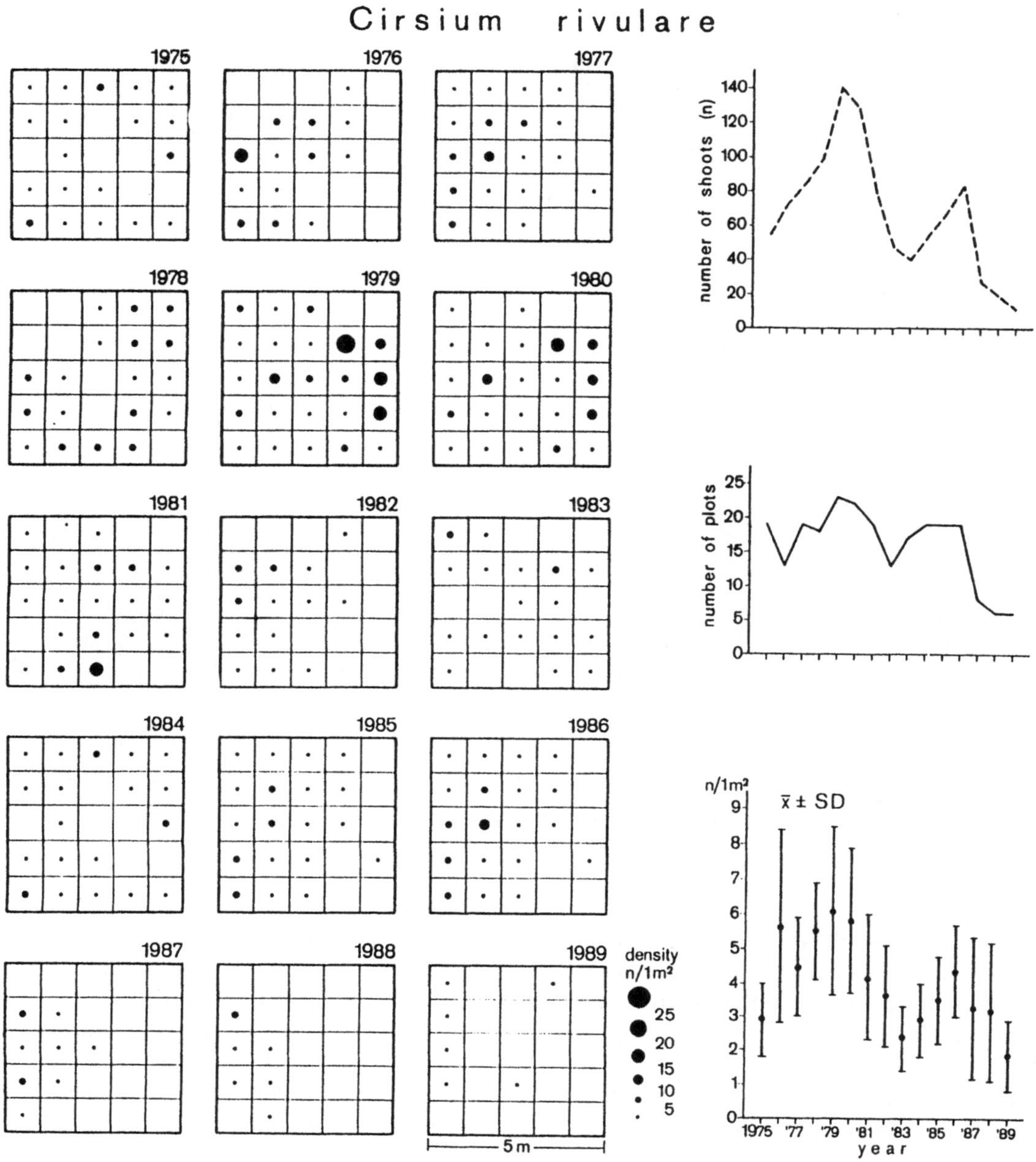

Fig. 4.30. Spatial dynamics of population for 15 years in a 5 m × 5 m plot in abandoned meadows.

Plant relocation and rotation

During the vegetation changes in abandoned meadows a number of populations of different species showed no significant spatial structure changes. They have preserved for 15 years their aggregated structure and even the same number of assemblages and their distribution in an area. In the same period variation could, however, be seen in the distribution of particular species over the Reski Range, defined as a "relocation" of individuals". Variations of this type occur particularly in

Photo 4.4 The meadow with few dominant species 8 years after abandonment of the meadow.
Photo: J. J. Faliński

those species whose distribution of individuals is closely associated with the spatial structure of the species that are dominant in the progressing succession, e.g. *Caltha palustris, Myosotis scorpioides, Galium palustre* and *Polygonum bistorta*. Small assemblages of these species are found in patches dominated by *Carex acutiformis* (Figure 4.31).

The distribution of individuals of these species clearly depends on the number and size of the gaps formed within *C. acutiformis* aggregations. Gaps appear in different parts of an aggregation every several (3—5) years, and are within a space of time again grown up by expanding sedge rootstocks. Temporary gaps in *C. acutiformis* aggregations are colonized by various species, which will some time later be eliminated by the sedge returning to its earlier sites. For this reason, species like *Myosotis scorpioides, Galium palustre* and *Caltha palustris* disappear from some places of an aggregation and appear in others. The, as it were, relocation of their individuals is the result primarily of the growth, necrosis and renewal of *Carex acutiformis* rhizomes, and depends also on the amount of accumulated necromass and rate of its decomposition within a *C. acutiformis* aggregation.

Some species appear to possess considerable capabilities of colonizing necromass-covered areas and persisting in them for 4—5 years, that is, until the necromass is totally decomposed. They are, however, ousted by stronger competitors, e.g. *Carex acutiformis*, after the completion of necromass decomposition (Figure 3.11). A quick species rotation in the gaps arising within sedge aggregations results from the fact that the gaps are colonized by species of a delicate body structure,

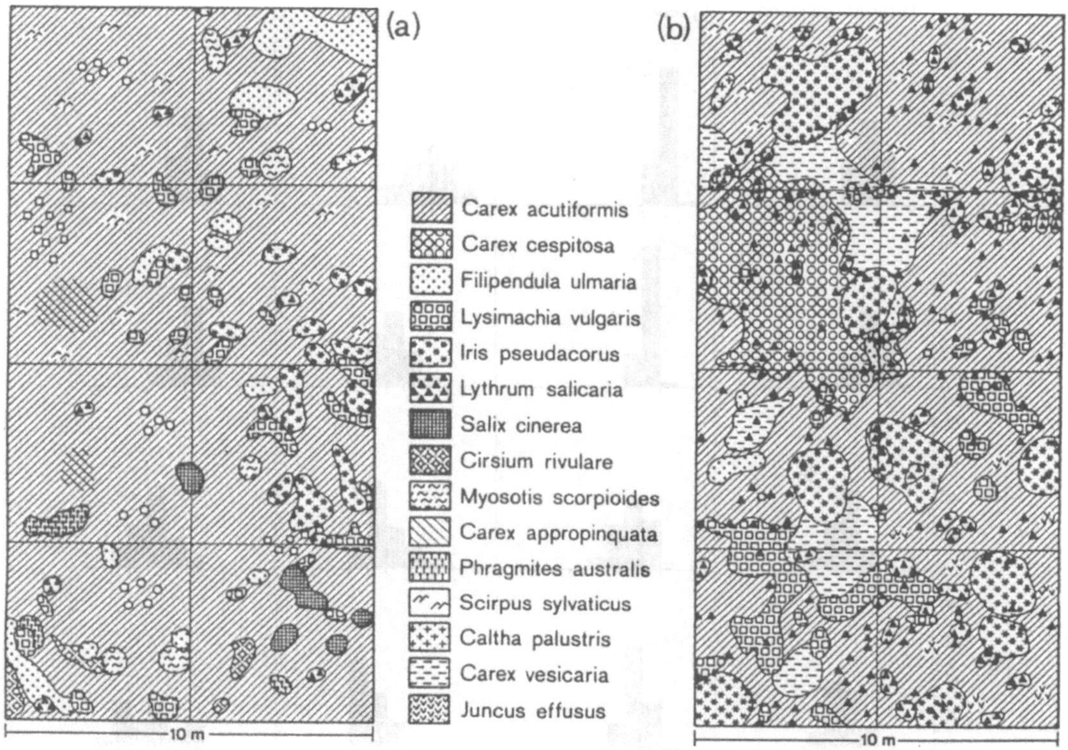

Fig. 4.31. Population mosaic of species inhabiting in-meadow mucks: a) muck dominated by *Carex acutiformis*; b) mosaic of populations of few species.

especially those with weak underground organs. Hence their presence in gaps is short-lived. They fill the gaps for only a few years, that is, before there appear in them species of a fast growth rate and with strong and durable underground organs. The mechanism of such changes has been described in the chapter dealing with succession inhibitors.

Pulsations

Minor variations in plant density only in some places over a population area, represent a variety of the fluctuant population spatial dynamics. They have been named "pulsations". This type of population spatial dynamics is peculiar to species that colonize specific-type habitats, like mires and various hollows and depressions temporarily or permanently filled with water. Plant populations making up the vegetation of the mires do not conspicuously change their spatial structure in the course of a succession, even when there form willow or alder scrub around them. They are populations of species like *Iris pseudacorus, Caltha palustris* and *Carex vesicaria.* Having colonized and completely filled a mire, they stay in it for many years, maintaining a fairly stable density and distribution of individuals. As a rule, one of such species is dominant, and within a patch there occur aggregations including individuals of 2—4 species, e.g. in an *Iris pseudacorus* stand, *Caltha palustris* and *Myosotis scorpioides* assemblages are often found, while in a *Carex vesicaria* patch, *Iris pseudacorus* assemblages will develop. In 6 mires located in the Reski Range area, there occurred during the 15 year period, various population spatial structure modifications of the species that inhabit them.

The following mire-size-related regularities

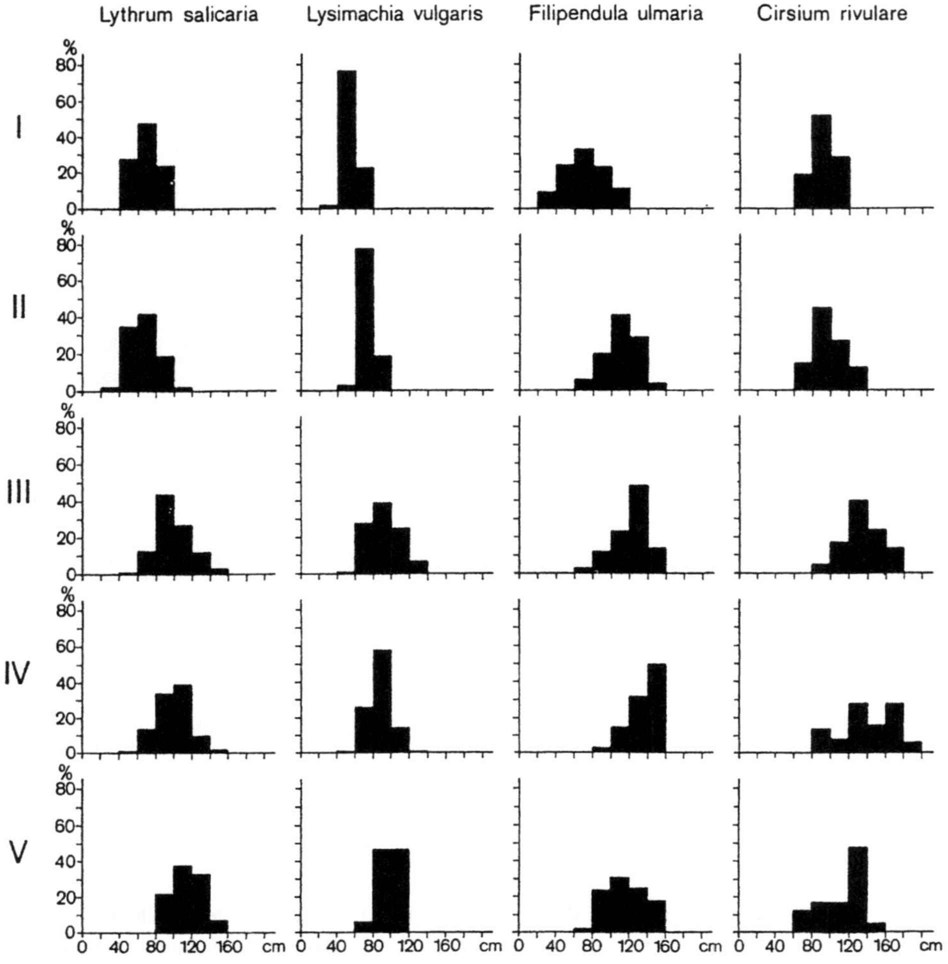

Fig. 4.32. Size structure of the population of long-term promoters during succession: (I) meadow, (II) macroforbes community, (III) willow brushwood, (IV) brushwood community, (V) *Salicetum pentandro-cinereae.*

have been found. In two mires, 800-1000 m^2 in surface area, the spatial structure of the dominant (*Iris pseudacorus*) and co-dominant (*Caltha palustris* and *C. vesicaria*) species was subject to minor variations during the first 10 years of meadow disuse in the Reski Range area. More conspicuous changes were observed in the later period, i.e., between meadow-disuse years 10 and 15. At that stage the willow-alder scrub growing around the mires attained a height of 2—3 m, and considerably shaded the vegetation of the mires. At that time the proportion decreased of flowering shoots, and significant changes could be seen in the individual-size structure of the populations overgrowing the mires. A predominance could be noted of individuals below the average height. The changes coincided with the appearance of a new dominant species that significantly limited the proportions of species like *Iris pseudacorus, Caltha palustris* and *Myosotis scorpioides.* Among the species that played such a role in the vegetation changes of the mires were *Carex vesicaria* and *C. acutiformis.*

In mires 200—400 m^2 in surface area, population spatial structure changes could be seen as early as the first 5—6 years after the cessation of mowing. Among species that had been present in a particular mire since an earlier stage there appeared species known to have been present in adjacent areas when yearly mowing was still continued (e.g. *Lythrum salicaria, Polygonum bistorta, Filipendula ulmaria, Carex cespitosa* and *C. acutiformis*). Several (5—6) years later the contribution of the immigrants from adjacent habitats to the mire vegetation was similar to that of the species that had been present there before. Ten years after the abandonment of the meadows, the population spatial structure of the species inhabiting small mires attained a relative stability (Figure 4.32). In subsequent years there could only be seen variations that could be referred to as "pulsations", i.e. density changes only at some points over the area of a mire.

CHAPTER 14

POPULATION INDIVIDUAL-SIZE STRUCTURE IN SUCCESSION PROCESSES

Introduction

Individual characters' differentiation in the course of growth plays a significant role in the moulding of many population characteristics. Traits then established, determine the intra-population and the individual-environment interface relations. The strength of the interactions between members of the same plant population, and of different species, as well as their action on the habitat depends on their habitus and growth. The role of individuals in their population and habitat can thus be expressed by their size (height, breadth) and biomass, since these characteristics are usually the expression of their chance to survive and reproduce. Due to the presence of various growth-modifying agents, body size differences occur even between even-aged individuals in the same growth phase. These differences become much more conspicuous if a population includes uneven-aged individuals, because individuals size and biomass change with the age of the individuals (cf. Part 3).

The transition of an individual from the vegetative to the generative stage is associated with a specified body size and biomass. There are many factors that significantly influence the individual body size, e.g. competition (Platt 1975, Werner 1977, Grubb 1977), habitat resources (Harper 1986, Silvertown 1983), degree of vegetation-cover of canopy-closure (Falińska 1979, 1989a, b, Werner 1979, Barkham 1980a, b). The latter two factors in particular, play a significant role in the life-history and growth of individuals, and in the moulding of the population dynamics of the plant species taking part in a succession. It has been found that at the stage of succession initiation, the most important is the role of the increasing population density of the meadow species, and the inter-species competition that is just beginning among them (Figures 2.2; 2.3). Several (8—10) years later the effect becomes conspicuous of the diversifying vertical structure of the plant communities. Due to the dominance, at first of tall herbs and then of shrubs and trees, individuals of the meadow species become dwarfed, and there occur changes in their density, leading on to their total elimination (Figure 3.20).

Types of individual-size structure changes

The relationship between the individual-size structure of a population and density and variations in habitat conditions is clearly visible during the vegetation changes that go on in abandoned meadows. Over the consecutive plant community succession phases there occur significant individual-size structure changes in the populations of the species that have occupied the same habitats for many years (Figure 4.32).

Two types of size-structure changes have been identified:

1. The individual-size structure of a population undergoes significant changes: a size-structure distribution that was similar to the normal in yearly-mown meadows becomes asymmetric (negatively-or positively-skew) within several years following the cessation of mowing whereafter it returns to the normal distribution.

2. Within a population individual-size structure,

similar to the normal distribution the following changes occur: a) the proportions are changed of individuals in particular classes; b) the individual-size diversity increases, as evidenced by a higher number of size classes. Throughout the time, however, there persists a size distribution similar to the normal.

Density and individual-size structure

Density seems to be related to the population individual-size structure, as evidenced by a considerable correlation of the two characteristics over the population growth-history. It means that in the successive population growth phases, the density is not only subject to change, but also the degree and range of individual variation, as indicated by the presence of different size-structure types (Figure 4.33). There is a widening range of variation from hardly noticeable between individual differences in size, biomass, habitatus, e.g. number of ramifications, or number of flowering shoots (Figure 4.18). This is to a large extent due to the fact that with the advancing growth of a population there also occur age-structure changes, and the diversity of individuals in perennial plant populations is further increased by the inner variation of individuals within the same age class.

The following relationship has been found between population density variations and individual size. Following the cessation of mowing, increasing population density is accompanied by a high-rate plant growth, and most individuals attain a body size above the average (Figure 4.18). Individual size is diversified yet more by the density that increases year by year. Relationships of this type have been presented by the histograms of the individual-size structure of the populations of *Filipendula ulmaria* and *Cirsium rivulare* (Figure 4.32). At a certain population-growth stage (6—8 years) the largest-sized individuals prevail and the size distribution is similar to the asymmetric type, e.g. in macroforbes communities

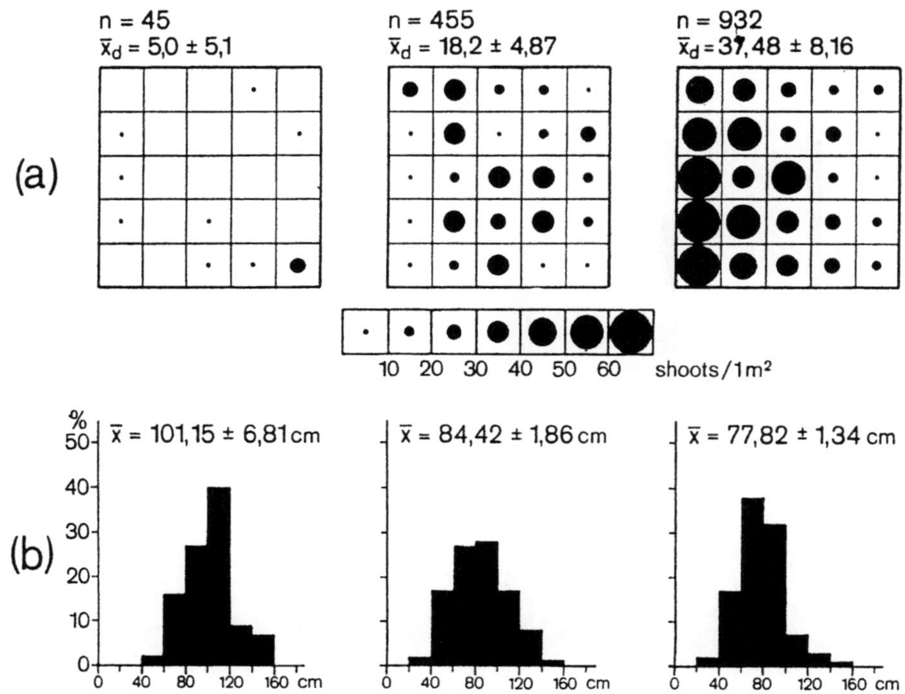

Fig. 4.33. Shoot-size variation (b) in relationship to density in a 5 m × 5 m plots (a) in an abandoned meadow. n — number of shoots in a plot 5 m × 5 m $\bar{x}_d$ — mean (± SD) density of shoots in a plot 1 m²; $\bar{x}$ — mean (± SD) height of shoots.

(Phase II, III). Sometimes there occurs an opposite response, where small-sized individuals dominate for a long time. The former situation was usually observed in monospecific aggregations, and the latter in cases where there appeared highly competitive species. For *Cirsium palustre*, for instance, a competitor of this kind is *Carex acutiformis*, and for *Filipendula ulmaria, Salix cinerea* and *Carex acutiformis* (Figure 4.34). Several (8—10) years later, the asymmetric size distribution gradually reverts to the normal type.

A direct relationship between density and individual-size diversity was usually observed at the stage when there formed monospecific aggregations. In the areas occupied by them there soon occurred a fast growth and decline, for example, in *Filipendula ulmaria, Lythrum salicaria* and *Iris pseudacorus* populations. However, even in such cases, the relationship between individual size and density is not unequivocal (Figure 4.34), for with the advancing age of the individuals of these species, subject to changes is not only the number of shoots, but also the size. Young (1—3 years of age) individuals as a rule produce short shoots, and mature shoots are twice as tall. Shoots produced by older individuals are similar in size to those produced by young individuals, (Table 3.1), but their size varies much more. In perennial species, variation in individual size often coincides with a population density increase (Figure 3.8). It is, therefore, difficult to separate the range of action of either of these factors (density and age structure) in the determination of individual size structure in populations, particularly those of polycormonal species, e.g., *Filipendula ulmaria, Lysimachia vulgaris, Iris pseudacorus* and *Lythrum salicaria*. A considerable density growth in these species is often accompanied by an increase in individual height and biomass, since the dominant proportion of a population includes juvenile individuals whose individual size is on the increase. For example at high densities during the colonization period, *Filipendula ulmaria* individuals can attain heights of 100—120 cm, while at low densities in the retreating period when senescent individuals prevail, their height varies between

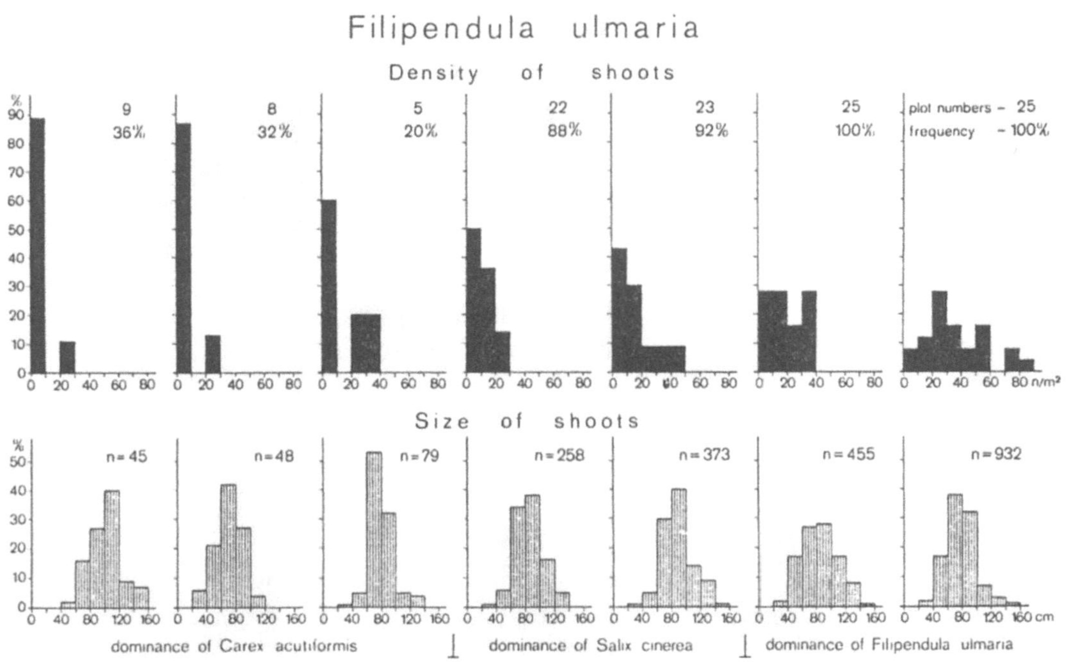

Fig. 4.34. Changes in density and size of shoots in populations colonizing three patches of meadow: patch dominated by *Carex acutiformis* and patch dominated by *Salix cinerea* and patch dominated by *Filipendula ulmaria.*

90—100 cm (Table 4.7). Similar differences were seen between populations of the same species, living under different conditions. Tall-herb populations, for instance, attain the highest density levels in ecotones between woodland and meadows. In spite of a population density five times as high as in the meadow, individuals living in the ecotone are much taller, their size distribution being similar to the negatively-skew asymmetric type (dominance of tall individuals). Dominant in the meadow population are average-sized individuals, whereas their size distribution is similar to the normal type (Figure 4.32). For example, most of the ecotone *Filipendula ulmaria* individuals reach heights of about 140 cm, whereas those of the meadow, 90 cm (Table 4.6).

In populations of the same species individual-size structure type diversity can occur even if the populations have attained a relative-density equilibrium, but are living in different plant communities (Figure 4.34).

During its seral changes the abandoned-meadow vegetation is highly mosaic-like in its structure. Some elements of the mosaic pattern appear to be highly durable, that is they persist for many years, even if forest communities are returning to adjacent areas. These characteristics are peculiar to macroforbes communities at the forest-meadow boundary, where the populations of species like *Filipendula ulmaria, Lysimachia vulgaris* and *Lythrum salicaria* appear to be highly stable as regards to their density and individual-size structure. Considerable differences are, however, found between conspecific populations that are members of different communities, e.g. of sedge marsh and macroforbes communities (Table 4.7).

In the populations of some components of the phytocoenoses that form during a succession, coexisting individuals of other species can exert an influence similar to that of the individuals of the same population. Mutual growth-limitation of heterospecific plants may assume the form of direct individual-individual relations leading on to a mechanical limitation of their development, and thereby to individual-size structure variations. In relation to meadow species such a role is played at first by macroforbes and sedges, and then by willows which limit the growth of plants in abandoned meadows as late as 10 years after the abandonment. The growth of herbs is at first impeded by shade, as a result of the differentiation of the vertical structure of herbs and scrub, and

Table 4.7. Differentiation of shoot size of few species in various communities and their ecotone

Communities	*Lysimachia vulgaris*		*Lythrum salicaria*		*Filipendula ulmaria*		*Myosotis scorpioides*	
	Number of individuals /25 m²	Height of shoot (cm)	Number of individuals /25 m²	Height of shoot (cm)	Number of individuas /25 m²	Height of shoot (cm)	Number of individuals /25 m²	Height of shoot (cm)
Cirsietum rivularis	36	74.92 ± 2.44	57	67.63 ± 4.11	94	93.8 ± 2.4	255	53.6 ± 4.33
Caricetum acutiformis	212	87.68 ± 1.71	231	96.62 ± 3.18	112	103.8 ± 5.0	2158	38.32 ± 3.0
Lysimachio vulgaris-Filipenduletum	108	92.6 ± 3.4	111	99.17 ± 3.7	243	117.8 ± 4.67	17	17.79 ± 1.53
brushwood of *Salix cinereae*	107	54.13 ± 1.58	105	69.73 ± 2.57	330	91.2 ± 4.3	205	49.46 ± 1.3
Salicetum pentandro-cinereae	114	101.8 ± 3.9	210	113.87 ± 5.0	364	97.2 ± 5.1	120	27.9 ± 2.4
ecotone between meadow-forest	739	105.9 ± 11.3	111	120.5 ± 3.5	804	139.9 ± 2.6	12	40.0 ± 6.2

then due to mechanical actions. Numerous willow ramifications tightly cover the meadow vegetation and thereby destroy and eventually eliminate it (Figure 4.4). The retreating of individuals from their area is usually preceded by individual-size structure changes in a population, expressed by individual-size equalization, or increase in the proportions of very tall and very short individuals. In such cases the population vertical structure consists of two layers, that is, small, vegetative-stage individuals are found under the canopy of tall individuals.

The above type of response to changing neighbours during a succession is found in short-lived species like *Cirsium palustre* (Figure 4.4), and in long-lived ones, but of a delicate structure, e.g. *Lysimachia vulgaris, Cirsium rivulare* and *Myosotis scorpioides* (Tables 4.3; 4.7).

In some herbs a high density within the same population and in neighbour populations causes a rapid plant growth to heights above the average. This phenomenon was observed, for example, in *Filipendula ulmaria* and *Lythrum salicaria* populations.

Population density stability as the result of individual variation

In studies of plant population stabilization, it is assumed that the stabilized population is an ecological system where density is subject to no directional changes, but only to seasonal and long-term fluctuations. It is assumed that a stable density is the result of a stabilized relationship between reproductiveness and mortality, density itself modifying those individual characters which significantly affect population processes. The latter character group includes primarily individual fecundity and survivorship, highly correlated with individual habitus and biomass (Figures 4.35; 4.36). For some species a negative correlation has been found between individual biomass and density increase, the finding having become the source of the population self-regulation concept (Yoda *et al.* 1963). Later studies of the relationship between population processes and individual characters have, however, disproved the common applicability of this direct relationship (White and Harper 1970). Watkinson and Harper (1978) suggest that such a rectilinear relationship can apply to monospecific systems with no competition from other species.

The above regularity has been confirmed by data from researches into individual character variation and population dynamics during a succession in progress. It mainly concerns those species which are dominant in the particular meadow patches, and at the same time belong to the floristically rich phytocoenoses that arise during the succession. In *Lysimachia vulgaris* and *Filipendula ulmaria*, for example, density has been found to be associated with individual size (height, biomass) and fecundity only in those patches where these two species were dominant. Each population of these species has a different threshold density value above which individuals never attain the average height (size). For *F. ulmaria* the value is of the range of 40—80 plants per 1 m^2, (Figure 4.37) and for *Lysimachia vulgaris*, 30—40. If, however, either of the species is one of many components of a phytocoenosis, then individual size is to a larger extent determined by the neighbours and their density. That is why conspecific populations occurring in different plant communities differ considerably in their individual-size structure, even at similar densities (Tables 4.3; 4.7).

The view that individual-size variation is a population density regulating mechanism is supported by the high plant plasticity. Owing to this feature, individuals of many species can survive under deteriorating conditions during a succession by only limiting their growth (Figure 4.38).

During a succession, increasing densities or worsening habitat conditions do not necessarily lead to an increased death rate. They only intensify individual variation, and thereby determine the stability of population density. This is possible most often due to considerable individual size and architecture variation owing to which individuals can form several layers. At high densities a multi-layered system, resulting from individual size and habitus diversity, enables a particular population to better utilize and fill a hibitat. It has also been

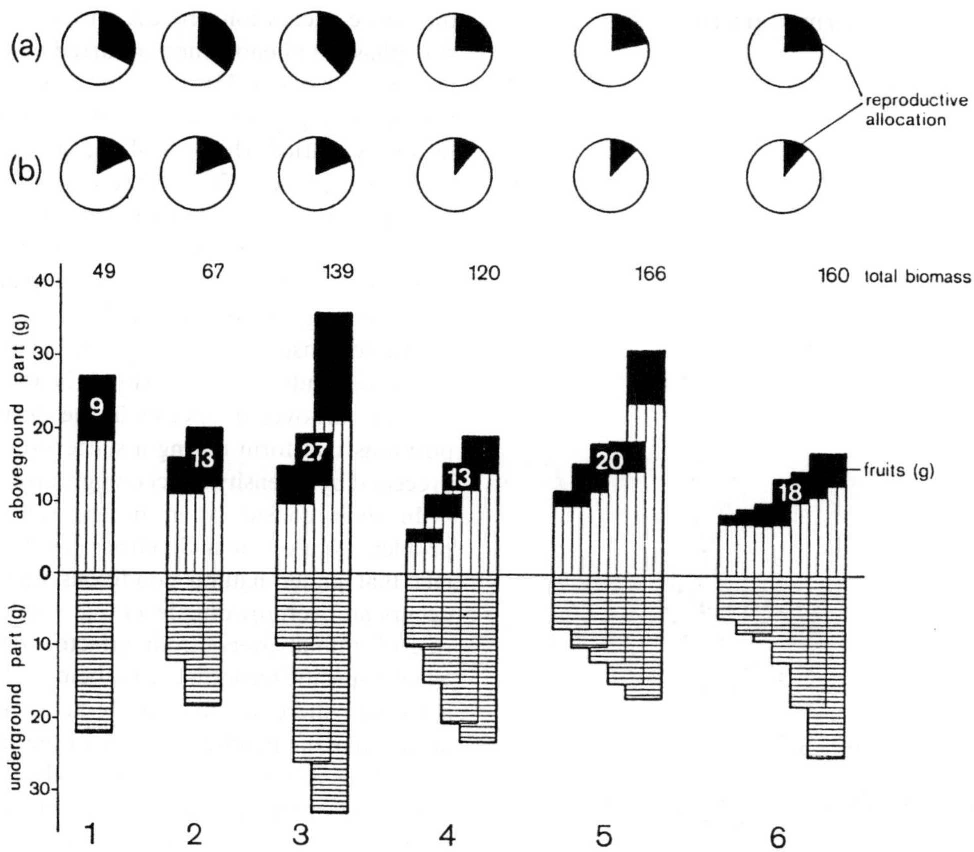

Fig. 4.35. Biomass allocation for reproduction (fruit) in individuals (a) varying in weight and number of generative shoots (1—6), and in evaluation of the reproductive effort; (b) proportion of fruit biomass in relation to the biomass of whole individual; (c) proportion of fruit biomass in relation to shoot biomass.

found that under conditions fully meeting the needs of a species, even at very high densities nearly all individuals attain the maximum height. This kind of response has been found in a *Filipendula ulmaria* population in succession Phase IV. Mature individuals in such phytocoenoses are much taller than those found in meadows and woods. The majority, almost 80%, of individuals attain heights of 120—160 m while those in other phytocoenoses hardly ever attain such heights (Figure 4.32). *Filipendula ulmaria* individuals attain this height at the expense of their architecture which must be reduced, since in this situation the leaves and inflorescences concentrate only in the top 1/4 of a shoot. Studies of individual modifications in many plant species during a succession in progress permit the presumption that this type of plant response is common when population density and competition are on the increase (Falińska 1979).

Due to the occurrence of diverse individual responses to density and a variety of relations between demographic processes and habitat conditions, views supporting the self-regulation concept arouse considerable criticism (Harper 1977, Watkinson and Harper 1978, Falińska 1978, 1986a, Smith 1983a, b). Doubts are aroused particularly by the rectilinear relationship sometimes found between density and individual characters. The reason for the doubts is the fact that for every population there is a certain density range over which density does not yet have a vegetative effect

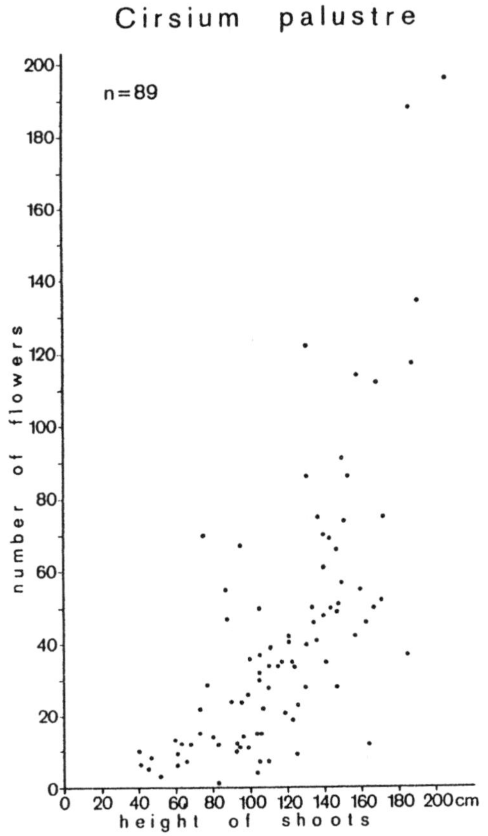

Fig. 4.36. The relationship between the size of the shoots and number of the flowers in a 5 m × 5 m plot in an abandoned meadow.

on individual survivorship or reproduction, that is, when density is optimal for the growth of the population.

Optimum density is understood as the number of individuals per unit area at which the highest survivorship and fecundity are attained. Titov (1978) is of the opinion that differentiation must be made between group effects and density effects. According to him, the former are individual reactions to density which do not lead to a higher mortality, but increase the individual character diversity in a population, and the latter are those situations in which density causes increased mortality rates, often also referred to as the overcrowding stress.

It has been demonstrated that in the populations of the species that were present throughout the entire succession process, in the first succession phases a phenomenon occurred that might be termed, following Titov's view, a group effect, because with growing densities, individual size and fecundity varied (Figures 4.17; 4.18). In later succession phases (III—IV) there occurred, in the same populations, a density effect followed by the recession of some species. It may be well to emphasize that in a particular population, an overcrowding stress can be brought about by increased densities of the neighbours, and not only by its own individuals' density. It is a known fact that the turnover of species in the floristic compositions that form during a succession is always preceded by a density effect or a group effect.

In their consideration of the self-regulation problem, Watkinson and Harper (1978) stress the fact that under natural conditions there seldom occurs an overcrowding level that would play the role of a numbers-controlling factor. At the time when a specific meadow area is being colonized by a species, there simultaneously act a number of agents, like the number and size of the sites where individuals can grow, presence of other species or their concurrent encroachment, all of which affect the population size. Due to the effect of the above factors, there rarely occurs a population overcrowding level that could function as a controlling mechanism. The approach to population density as a population dynamics controlling mechanism also stems from the fact that the phenomenon is often studied on a microscale, i.e., at a selected site occupied by a particular population. It is a known fact that as a rule populations occupy heterogeneous habitats, as evidenced by density variation. For this reason, direct relations found in a selected area cannot always be referred to a whole population. This is confirmed by population studies accompanied by macroscale (1 ha) and microscale (25 m^2, 200 m^2) density assessments (Figures 4.25—4.28).

It has been found that within a population area, the frequency is low of points where a negative effect of density on fecundity and survivorship is noticeable, and they cannot play any significant role in the determination of the population size. Everything indicates that aggregation-size diver-

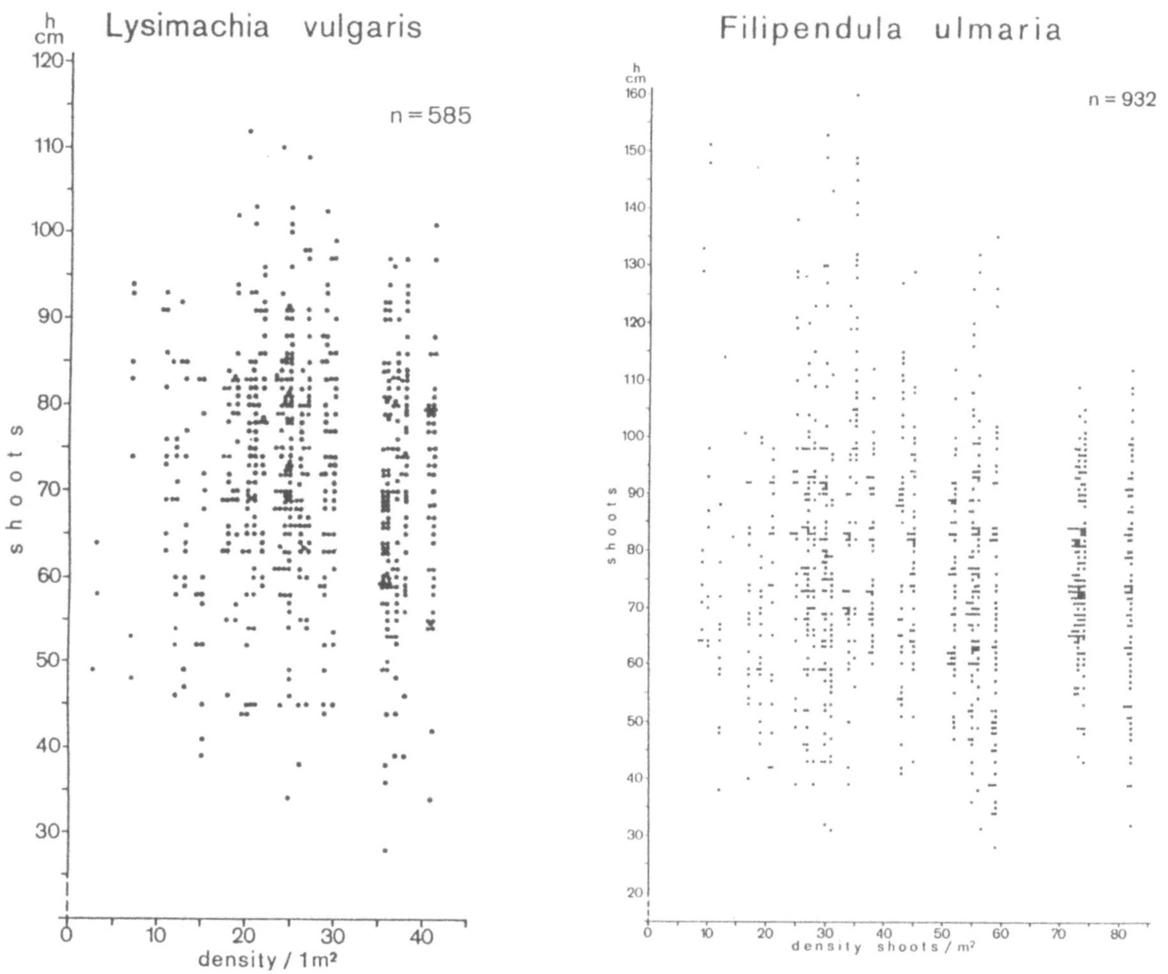

Fig. 4.37. Changes in density and size of shoots in macroforbes community in a 5 m × 5 m plot.

sity and considerable density variation within aggregations, enable species to fully utilize the meadow community biochore heterogeneity. To sum up studies of the plant population self regulation in the course of a succession, it can be stated that the phenomenon was more conspicuous: a) in short-lived rather than in long-lived species; b) in monospecific rather than in polyspecific patches.

The occurrence of directional density changes during the growth of a population, and the simultaneous change of the individual-size structure permits the assumption that intra-population variation is one of the factors responsible for population size. It is not only a proper space occupation by individuals and its filling that matter, but also a considerable effect of the structure on the population processes, since individual size and habitus are correlated with the generative and vegetative reproduction of individuals, as well as with their survival rate. Individual variation within a population is to a large extent maintained by the environment heterogeneity, significant in the vegetation turnover in abandoned meadows (Figure 4.38). This relationship and the density-stabilizing influence of individual variation have been the subject of many model studies (White and Harper 1970, Harper 1977, Stenseth 1980). The opinion that individual varia-

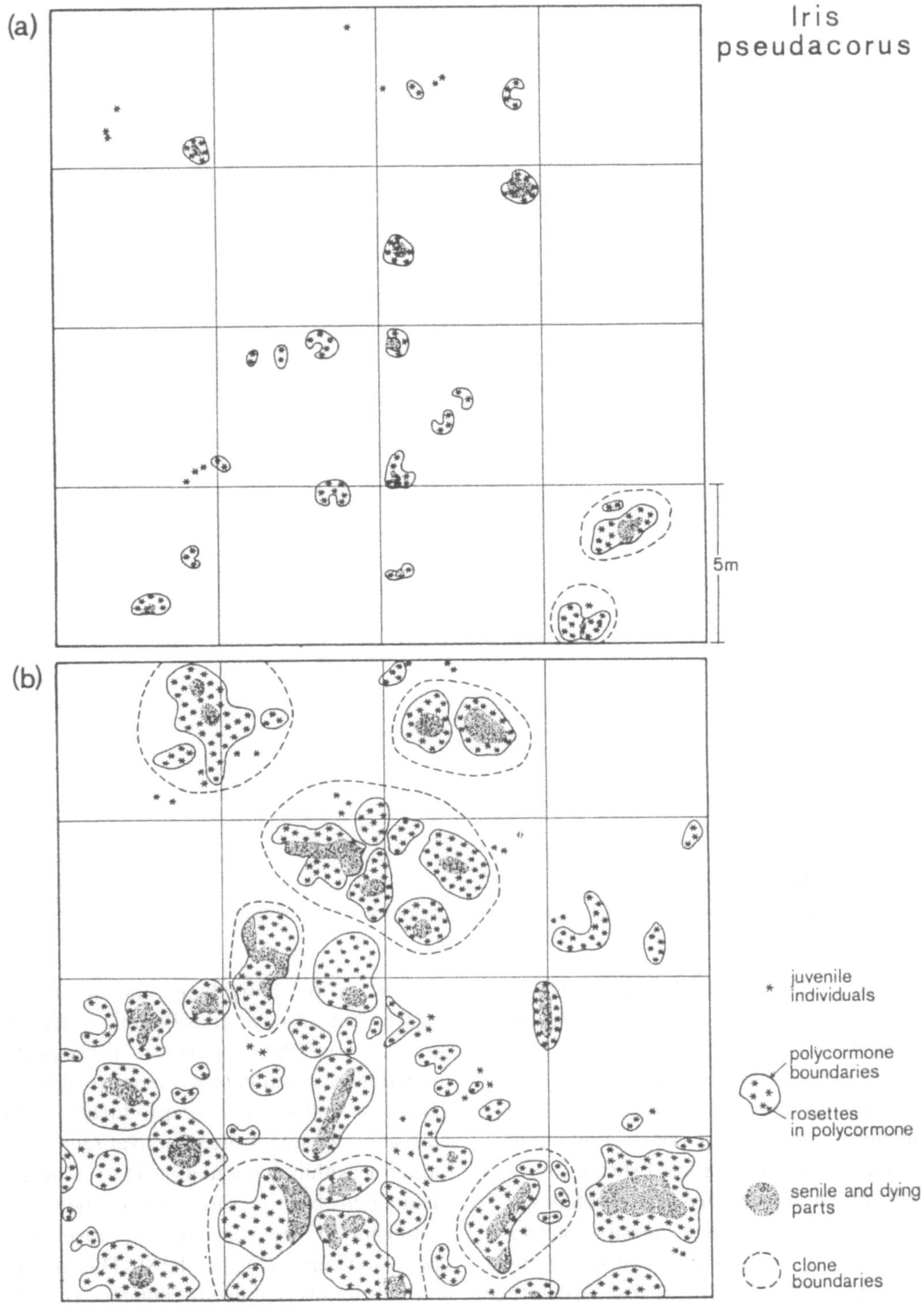

Fig. 4.38. Size and shape of polycormones and clones, in the sedge communities (a), and mucks (b), 8 years after abandonment of the meadow.

tion favours the persistence of a relative population density stability is fairly common, and has been dealt with in many theoretical and methodological analyses (White and Harper 1970, Harper and White 1974, White 1979, Andrzejewska and Falińska 1983).

The view that the individual-size structure of a population plays the role of a density-controlling mechanism is substantiated with the following statements:

1. Plant response to variable habitat conditions is manifested by individual habitus, size and biomass variation (Figure 4.16). The latter directly or indirectly affect the population processes: reproductiveness and mortality.
2. Correlated with individual size and biomass are individual viability, fecundity and survivorship, that is, these characters determine the role of the individual in the population (Figure 4.18).
3. Dependent on the biomass of the individual is its reproductive potential, that is, when and what proportion of its biomass it will allocate to propagation. The larger its size, the more biomass can an individual allocate to propagation in conformity to the "reproductive effort" rule (Figure 4.35).
4. Dependent on individual size and habitus is the population density. These two traits directly determine the individual area (space occupied by an individual's underground and aboveground organs), and thereby affect the "individual-individual" relationship between conspecific or heterospecific individuals (Figure 4.38).

Role of individual-size variation

The size of individuals determines their relationship to space and neighbours. Harper (1986) is of the opinion that the size of modular plants, as a method for occupying space is an evolutionally-developed form of adjustment of multibodied plant organisms enabling them to find and use habitat resources. Individual body size is one of the very plastic characters, and represents a recognized measure of the habitat conditions of a population. Individual-size variation within a population is usually related to the heterogeneity of the abiotic and biotic environment in which individuals develop. This relationship, and the effect of individual-size variation on the population dynamics have been dealt with in many studies (White and Harper 1970, Harper 1977, Begon and Mortimer 1981).

Large individuals, especially polycormonal forms, usually play a more important role in the population and environment, and their effect on neighbours is stronger. In a population with a varied individual size there as a rule occur: 1) a better area-utilization, as uneven-sized individuals occupy different niches; 2) a variable fecundity, since correlated with the individual size is the number of fruits and seeds; 3) a varied action on neighbours.

The relations between size structure, spatial structure and individual fecundity become very conspicuous during a succession. In the initial succession phases small individuals are often dominant and occupy various gaps that form in meadow communities following the cessation of mowing. Individual size increases year by year, and individuals totally fill available space, and then eliminate the neighbours (Figure 2.2). The process is particularly visible in polycormonal species whose individuals gradually, year by year, increase their size by multiplying shoots, and thus growing in width. Varying with age is also the shoot height. In meadows unmown for several years the shoot height and shoot biomass, and thereby the weight of the whole plant organism, can be almost twice as high as before the cessation of mowing.

It has been found that with the passing of the successive population-growth stages in many species, the proportion changes considerably of large individuals, and so do the proportions of fruiting individuals and fruit abundance (Figure 4.17).

Individual-size structure variation can be defined as the following response types:

1) a particular population includes even-sized (small or large) individuals;
2) the range of individual-size variation is wide, as expressed by the presence of a large number of classes;
3) individual-size distribution can be asymmetric if the number of individuals increases only in some size-classes (negatively-skew distribution) or decreases only in some classes (positively-skew distribution);
4) the frequency of individuals varies with size classes, while the size-distribution of the population is normal.

The realization of the particular structure types of plants depends on the biological properties of the species, habitat conditions and population growth stages. The individual-size distribution of a population may, therefore, vary with its growth phases, and with habitat conditions (Figure 4.33).

In the course of population growth, changes in the above structure tend toward a higher individual diversity. It is often manifested by changed numbers of individuals in particular size classes. At a certain stage the proportions of individuals in particular size classes vary in such a way that the distribution becomes normal. The picture may be disturbed by changes of environmental agencies and intra-population factors. If the changes are short-lived, or the intensity of the population-modifying agents is not very high, then after a certain time the population will revert to the original size structure, this often being correlated with the return to the original density.

Such a size-structure type variation has been observed in the populations of several species which took part in the forest succession in abandoned meadows, e.g. in *Filipendula ulmaria, Lythrum salicaria, Lysimachia vulgaris* and *Cirsium rivulare* (Figure 4.33).

Population-growth phases and the size structure

There is a strong relationship between the individual-size structure and the population-growth phases (Figure 4.32). In many species taking part in a vegetation succession, a relative dynamic equilibrium is attained by populations found in mown meadows and woodland communities. This is also reflected in the individual-size structure of a population, as indicated by an individual-size distribution similar to the normal. Such an individual-size variation picture was seen primarily in species that are important components of forest and meadow communities. In such cases the cessation of mowing not only causes changes in the density of individuals, but also significantly affects their size. The following regularities have been found:

1. Following the cessation of meadow mowing in the populations of some species a growth in numbers occurred, followed by an increased frequency of the tallest individuals. The process continued, depending on the species and habitat, for 6—8 years, and the individual-size variation was expressed by an asymmetric, negatively-skew distribution.
2. At the relative numeric equilibrium stage there occurred a further individual-size diversification, accompanied by growth of the number of average-sized individuals — their distribution approached the normal type;
3. At the population-size decline stage, individual-size in a population was subject to rapid changes, as indicated by the asymmetric size distribution, that is, dominant were either small or large individuals;
4. If the numeric decline did not end with a complete recession of a particular species from the abandoned meadows, then there occurred a gradual increase of average-sized individuals, and several years later their distribution was again similar to the normal. This applies particularly to those species which took part in the formation of scrub and woodland communities.

CHAPTER 15

RECAPITULATION: POPULATION SIZE AND POPULATION CHANGES

Introduction

The size of a population is defined by its abundance and spatial range. The two characteristics are included among the fundamental population characters that make it possible to assess the current and future fate of a population and its role in vegetation structure and dynamics. Population abundance determines the potentials of a species as regards to the use of the environmental resources, and establishes its relations with other species.

In the case of many meadow species the cessation of mowing triggers off a rapid growth in numbers. Significant population size changes are characteristic of those species whose individuals are characterized by a high growth rate and a high fecundity. They also possess morphological properties which lead to the elimination of neighbours, that is, to the releasing of space by competitors, e.g. by growing in width. Space as a factor plays a significant role in population-size changes, especially in a succession (Werner 1977, Gross 1980). After the cessation of mowing some meadow species attain, within a short time, an absolute dominance over other phytocoenotic components, by increasing (2—10-fold) their numbers and the area they occupy (Figure 3.22).

The life, reproduction and death of individuals, and thereby the population dynamics of plants are effected by many biotic, abiotic and intra-population factors (Harper 1977). However, often only one of them is of key importance in the successive stages of the life-history of individuals (Grime 1979), and in particular phases of population growth (Tamm 1972a, b, Inghe and Tamm 1985, Rabotnov 1985, Falińska 1985, 1986a, b). For instance the key factors may be: plant competition (Sarukhan and Harper 1973, Gross 1980, Turkington and Harper 1979), closed vegetation cover (Platt 1975, Grubb 1977, Barkham 1980a, b, Gross and Werner 1982, Schenkeveld and Verkaar 1984), habitat patchiness (Connell and Slatyer 1977), and sometimes weather conditions, especially in the case of annual plant species (Wilkoń-Michalska 1985, Watkinson and Harper 1978).

Spatial dynamics

During succession the diversity in the distribution of individuals differs widely from the classical models of spatial structure, that is, random, uniform and in assemblages.

Since there is a tendency to form single or two-species patches in the process of succession, investigations into the spatial structure of populations are actually limited to analyses of the aggregateness, its intensity and scale. The origin of the assemblages of individuals of each of the different plant species is different.

As we have seen, plant assemblages were: (a) groups of individuals developed from seeds; (b) collections of individuals of vegetative origin, i.e., clones; (c) collections of rooted shoots belonging to single individuals — a polycormone; (d) collections of individuals of generative and vegetative origin, differing at the same time in habitus, size, growth stage and age (Figure 3.8).

Aggregateness as a general type of spatial

structure of plant populations inclines us to consider its causes and consequences, especially the consequences of this type of distribution of the individuals for their survival and reproduction. Among the many abiotic factors enhancing aggregation of plants of prime importance, is the unequal distribution of nutrients and water, non-uniform light access and the heterogeneity of the habitat. Among the phytocoenotic factors, an important role is played by the interaction of the component populations occurring according to the principle: aggregation of the species evokes that of individuals of other species (Figure 3.22). From among the biological properties of plants, a strong aggregative influence on individuals is exerted by: group spread of seeds and thereby clumped development of seedlings (Figure 4.10), as well as the morphological properties of perennial individuals, that is, the capacity of multiplying the above-ground shoots and underground organs (Figures 3.22; 4.10). In extreme cases one individual can overgrow an area of several score or several hundred square metres.

The dynamic nature of the spatial structure depends on the morphological properties of individuals and the type of reproduction and also on the structure of the phytocoenosis to which the particular species belongs. From the aspect of the range and duration of density and distribution changes of individuals, three types of transformations of this structure are distinguished: (1) successive directional changes in density, connected with the successive phases of population development; (2) fluctuating seasonal and long-term density changes; (3) pulsating — small changes in density different in the particular parts of a population.

Density is the expression of the numerical relations of the populations and the simplest expression of the spatial structure. Thus density is an attribute of the spatial organization of the population and an index of its size. Therefore, the same parameter serves as a basis for the evaluation of different properties of the population — its structure and processes. This brings about many theoretical and methodical implications in the demography of plants. There are few populations for whose true size and the occupied area would be known.

In considerations on the stabilization of plant populations, it is assumed that a stabilized population is such an ecological system in which density does not undergo directional changes, but only seasonal or multiple fluctuations. It was observed that the individual-size structure diversity and heterogeneous meadow habitat favour a stable density (Figure 4.34). Many other features are correlated with the individual size, such as viability, fecundity, survival rate and durability (Figures 4.15—4.18).

Density has a many-sided influence on the growth and development of individuals, and may be a factor affecting the size of the population. This has found its expression in the theory of population size regulation, where the role of the regulating mechanism is attributed to density. This view has found both approval and contradiction, and to this day is the subject of theoretical and metodical considerations. There are at least two points of view in phytoecology on the mechanisms controlling population size. One of them tends to attach more importance to population size regulation by intrapopulation factors, and the other, by environmental factors. These two different concepts have arisen because demographic investigations are conducted either in simple systems and on species with a short life span, or in complex systems and on long-lived species. It is difficult at present to decide whether these are different strategies of population size control or whether suitable methods are lacking which would allow us to understand the demographic processes in long-lived species.

Change in population age structure during succession

The high effectiveness of generative reproduction, that is high seed production and germination capacity (Tables 4.4; 4.5) of all herbs and the high survival of their individuals to the reproduction phase allows these species to rapidly settle on unmown meadows (Figure 3.22). For complete occupation and filling of the settlement site, however, the ability of forming polycormonal forms is most important. The increase in the number of shoots in geometric progression and the durability of the underground organs are decisive for the competitive strength of the species and their role in building the phytocoenosis. Only gradual dying of the rhizomes and disintegration of the polycormones increases the chances of other species participating in the formation of new phytocoenoses. The places relinquished by senile polycormones are mostly occupied by willows which initiate the development of brush-wood communities (Figure 3.22). These data show the strong dependence between the age of the individuals and their capability of competing with other species, as demonstrated also by Kershaw (1962) for *Carex bigelowii* individuals. It results from investigations on promoter species populations that the age structure is the property responsible for the course of population phenomena, for instance for the spatial structure, and thus the participation of this species in the successively developing phytocoenoses. Most frequently the mosaic pattern of vegetation is interpreted as a reaction to the heterogeneity of the habitat or as the effect of non-uniform seedlings, whereas the increased clumping in the herb population of the successive phases of their development is above all the effect of the spread of the individuals into multi-shoot polycormones frequently of extensive dimensions (Figure 4.10). Disintegration of senile polycormones or dying back of their oldest parts frequently leads to the formation, or rather return, to the random-clumped spatial structure (Figure 4.28). Therefore, in herbaceous species with a life span of many years the age pyramid in the form of an "urn" is not always evidence that the population is attaining the regression phase. Very often it is the type of age structure which ensures maintenance of the population in the phase of equilibrium. Division and gradual death of senile polycormones makes, namely, possible the development of new individuals of generative origin, this favouring permanent rejuvenation of the population after a state which might be referred to as "ageing of population" (Figure 4.39). Criteria are,

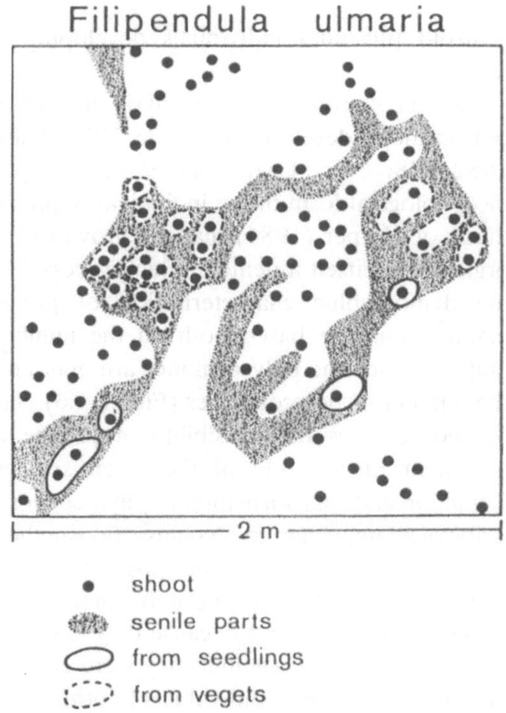

Fig. 4.39. The distribution of necrotic parts in a 10-year-old polycormone of *Filipendula ulmaria* and encroachment of new individuals of generative and vegetative origin on the vacated places.

however, lacking for precise evaluation of the age of individuals in perennial herbaceous plants and this greatly limits the recognition of the relations between population processes and the age structure. This concerns particularly the determination of the age of individuals of vegetative origin. Kershaw (1973) for instance says that gemmulae of *Carex bigelowii*, derived from individuals 25 years old, are aged eight to twelve years as indicated by their viability. The development of individuals of for instance *Filipendula ulmaria* from vegets (parts of the polycormone) indicates that they are of the same age as the maternal plant. New individuals arising by the division of the polycormone usually continue the processes of ageing of the maternal organism. On the other hand, units formed by the separation of peripheral fragments of the polycormone may be considered as juvenile. They are units with a similar degree of advancement of growth and development as one-year-old (juvenile) individuals developing from seeds.

Results of up-to-date research on the biology of polycormone development (Fiala 1978, Falińska 1982, 1986a, b, c, Czarnecka 1986), as a specific morphological form of the individual, support the thesis of Harper (1986) that the growth of such organisms is itself a demographic process. Hence the demographic characterisation of perennial plants should be based both on the number of units of which the polycormones are built and on the number of polycormones (Figure 4.6). Such a procedure gives the possibility of precise establishment of the role of the given species in formation and transformation of phytocoenoses. It is above all important to recognise: (a) whether the declining role of the given population in the structure of the phytocoenosis is connected with dying out of individuals because they reached the end of their life, or whether they were eliminated form the given site by individuals of other species; (b) whether the increasing role of the population in the phytocoenosis is connected with enhanced reproductiveness or whether it results from the congenital predispositions of the individuals to multiply the above-ground and underground parts in the course of their life, enabling them to occupy even wider spaces; (c) how does the role of a given species change in the building of phytocoenoses in dependence on the change of the relation between the number of generative and vegetative progeny. Observations of labelled individuals allowed the elaboration of the age structure of the population and its changes in the course of succession for several species.

The investigations demonstrated that, in populations of certain species, the age structure does not undergo significant changes in the succession process, although the changes in population size are very wide (Part V). There also are species in which significant changes have occurred in the age structure of the population, as illustrated by different types of pyramids — flat, slim and urn-shaped (Figure 4.40). Such a type of age structure variability in the development of the population was noted in such species as *Iris pseudacorus*, *Lythrum salicaria* and *Filipendula ulmaria*, thus in

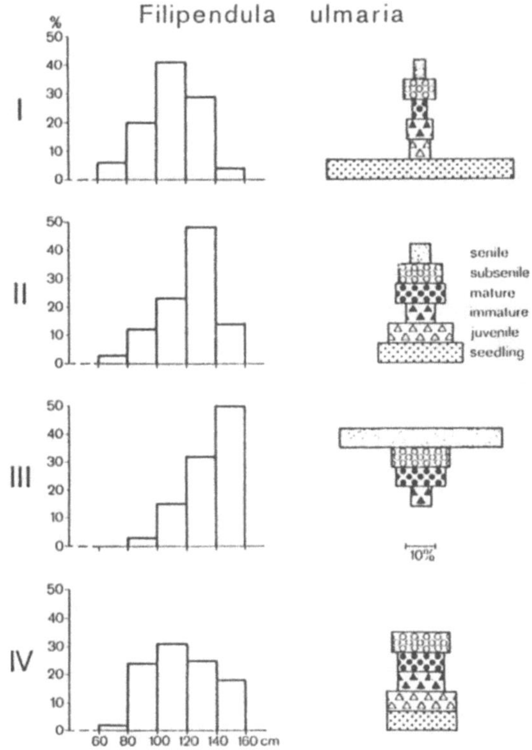

Fig. 4.40. Changes in the shoot size and population age structure of population in the period 1974—1988 in consecutive successional phases: (I) meadow, (II) macroforbes community, (III) willow brushwood, (IV) brushwood community.

long-living species. The process of exchange of individuals in the populations of these species is slow, because the perennial individuals live long and occupy wide areas completely filled with numerous shoots. In this situation there is little chance for the appearance of seedlings. Thus, the participation of old individuals increases in the population, and the age pyramid takes the form of an urn (Figure 4.40). The process of "populaton ageing", may, however, be arrested or the population may become "rejuvenated", when the senile individuals will reach the phase of disintegration. Then a chance opens for the development of seedlings (Figure 4.39). Moreover, the peripheral fragments of the old individuals are usually young, and after detaching themselves from the maternal organism, they start to grow intensively as is characteristic for juvenile individuals. The structure of the population age then changes and the population is characterised by a bell-like or slim pyramid. Many data indicate that such a type of pyramid represents an age differentiation of the population which is most advantageous for its development and stabilization.

In dependence on the life strategy of the species, the age structure of the population and its changes may take different forms. As a rule, polycormonal species, after reaching an age distribution close to urn-like, return to the bell-shaped pyramid. Other perennial species, especially those multiplying for the most generatively, may for a long time be represented by the bell type pyramid and never reach the age distribution type of the "urn". The diversity of types of age structure and their changes are shown in the diagram (Part 5).

It was found that the subsistence of withdrawal of the given species during succession depends on whether the changes produced by individuals of this species in the environment will be adopted by the progeny of the same or of other species.

PART 5

(SYNTHESIS)

POPULATION DYNAMICS AND VEGETATION DYNAMICS

CHAPTER 16

SUCCESSION: A POPULATION PROCESS

Introduction

The dynamics of vegetation are, above all, dynamics of populations and phytocoenoses. The aim of a study of vegetation dynamics is to gain a knowledge of the fate of a given phytocoenosis from its beginning to the achievement of dynamic equilibrium, and of the fate of the populations of species which built the phytocoenosis. In investigations on succession it is endeavoured to find out the fates of the successive phytocoenoses and to become acquainted with the mechanisms of species turnover (Table 5.1). It is suggested recently, that for gaining a better knowledge of the vegetation dynamics it is necessary to understand the population structure, its variability in time and space (Noble and Slatyer 1980, Houssard *et al.* 1980, Peet and Christensen 1980, Faliński 1980, 1986, White 1985, Falińska 1989a, b, Peart 1989a, b, c).

White (1985) stresses that the character of the vegetation depends on the features of the individuals which compose it. There are many indications that population dynamics is determined by the particular phases of the life cycle of the individuals, whereas abiotic and biotic factors affect the course of these phases, and in this way influence the structure of the populations constituting the given community. Population dynamics depends not only on the biological properties of the species, but also on the character of the phytocoenoses of which it is a component. Sometimes the influence of neighbours is much more efficient in modifying the individual and population traits than the abiotic factors (MacArthur 1972, Mack and Harper 1977, Silander and Antonovics 1982, Turkington *et al.*, 1985). Thus, every phytocoenosis is in a certain sense an unrepeatable phenomenon, because it arises by the interaction of the components under the given conditions. This is not in contradiction with the earlier finding of the repeatability of certain phytocoenoses under repeatable ecological, bioclimatic and historical conditions (Matuszkiewicz 1981). Certain differences between phytocoenoses of the same type, therefore, with a similar species composition, result from somewhat different properties of the populations. The reaction of populations to different life conditions is manifested in respect to many traits. This depends largely on the life strategy of the species and the type of the phytocoenose in which they participate. Such individual traits as size and fertility (Figures 4.15—4.18) and population features such as the number of individuals and units of which they consist (Figure 5.1), are most readily modified. It may be that the number of individuals of polycormonal species remains unchanged for long years in the given population, whereas the number of rooted shoots differs (Figure 5.1). The variability of the density of individuals and of the number of rooted shoots, decides the different contribution of the same species to the building of different types of phytocoenoses. In perennial species, the variability of individuals is connected with age, abiotic factors, and the influence of neighbours. It is a basic mechanism of differentiation of the spatial population structure, and thus it determines their cenotic role in the particular types of phytocoenoses (Figures 3.22; 4.28).

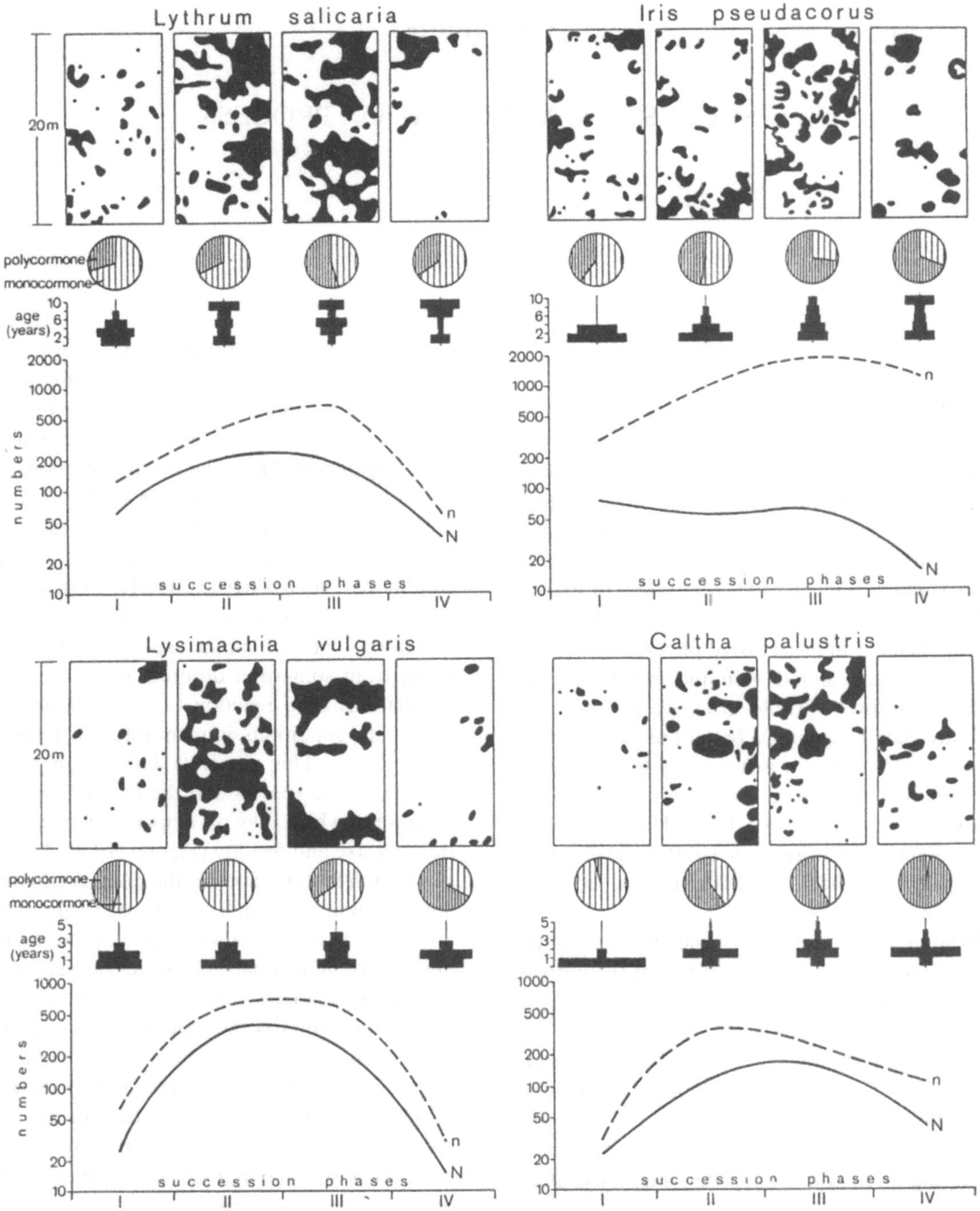

Fig. 5.1. The behaviour of populations of few species in consecutive successional phases: (I) meadow, (II) macroforbes community, (III) willow brushwood, (IV) brushwood community. Change in some traits of populations: coverage in a 200 m^2 plot, proportion of monocormonal and polycormonal forms, age structure and abundance dynamics: N — number of individuals (monocormones and polycormones), n — number of shoots.

Succession phases and developmental phases of the populations

The simultaneously conducted investigations on the dynamics of populations and the dynamics of plant communities demonstrated that the process of transformation of one species' composition into another may be caused by only certain species during succession (Table 5.2). It was found that the long-lasting role of a species in the succession process on abandoned meadows finds its reflection in the following relation: consecutive succession phase — consecutive development phase of the population (Figure 5.1). It was noted that if the phase of maximum population size persists for a longer time, then the given species plays a significant role in limiting the growth of populations of other species. A decrease in the population numbers, on the contrary, makes possible an increase in the population sizes of the neighbours or species arriving onto the relinquished space (Figures 3.15; 3.22).

The opinion is expressed that an increase in population density under extreme conditions is chiefly under control of abiotic factors, and under optimal conditions, of many ecological factors, while moderately favourable conditions of life enhance the inter-species competition.

The good agreement between the developmental phases of the populations of many species, and the succession phases indicates that these populations play an important role in the meadow vegetation transformations when mowing is discontinued. It also indicates that the demographic processes in these species influence the changing conditions from optimal to more or less favourable in the course of succession.

As illustrated, the relation between population dynamics and plant community dynamics may serve species which participated in the successive phases of long-lasting succession on the abandoned meadows (Figure 5.1). These species are present in flood-plain forests (*Circaeo-Alnetum*) and are a permanent, though not always most important, component of secondary meadow communities (Faliński 1966). They are species which greatly increase their participation on unmown meadows, form vigorous macroforbes, persist in the phase of willow brushwood, and are present in spontaneously formed secondary flood-plain forests.

Demographic studies of species partcipating in the succession process indicate that several of them reach the developmental phases in the consecutive succession phases (Figure 5.1). Namely:

(I) The phase of settlement and growth in size falls to the initial phase of succession (I), in

Table 5.1. General trends in various ecological characteristics changing during succession

Succession stages	initial	temporary	final
Type of communities	meadow macroforbes	macroforbes and brushwood of *Salix*	forest
Number of species	decreasing	increasing	stability, fluctuating
Concentration of dominance	increasing	maximal	decreasing
Life strategies (sensu Grime 1979)	C – R – S	C – CR – CS	C – CR
Mechanisms of succession (after Connell and Slatyer 1977)	tolerance facilitation	inhibition tolerance	tolerance
Role of species in succession (after Falińska 1989a, b)	promoters neutral	promoters inhibitors	neutral
Coverage	increasing	decreasing	stability, fluctuating

Table 5.2. Characteristic of principal species estimated by role in succession

	PROMOTERS										INHIBITORS			
	long-term						short-term							
	Filipendula ulmaria	*Lythrum salicaria*	*Lysimachia vulgaris*	*Geum rivale*	*Potentilla palustris*	*Iris pseudacorus*	*Salix cinerea*	*Cirsium rivulare*	*Cirsium palustre*	*Caltha palustris*	*Carex acutiformis*	*Carex cespitosa*	*Scirpus sylvaticus*	*Phragmites australis*
Growth types														
unitary									*					
iterative	*	*	*	*	*	*	*	*		*	*	*	*	*
Architecture types														
monocormone									*					
polycormone														
size														
large	*			*	*	*	*				*	*	*	*
small		*	*					*		*				
integrity														
short			*		*			*		*				
long	*	*		*		*	*				*	*	*	*
Clone types of different origin														
gemmaes										*				
vegets	*	*		*		*	*	*		*		*	*	*
ramets			*								*			
Colonization strategy														
phalanx												*	*	
guerilla			*		*						*		*	
mixed	*	*		*		*	*			*				
Colonization type														
monopolist	*					*	*				*	*	*	*
dominant	*	*	*			*	*				*	*	*	*
co-dominant	*	*	*	*	*	*	*	*	*	*		*	*	*
Manner of space filling														
clumps	*	*	*	*	*	*		*	*	*				
aggregations	*	*	*			*	*		*	*	*	*	*	*
patches	*					*	*				*	*	*	*
Regenerative strategies														
vegetative expansion	*		*	*	*	*	*				*			*
seasonal regeneration								*	*	*				
persistent														
seeds				*				*		*		*	*	*
seedling	*	*				*	*							
numerous wind dispersed seeds	*	*					*	*	*	*	*			
Reproductive types generative (G); vegetative (V)	G ≈ V	G > V	G < V	G ≈ V	G ≈ V	G < V	G < V	G > V	G	G ≈ V	G < V	G > V	G < V	G < V
Life strategies (sensu Grime)	C	C R	C S	C	C	C	C R	C R		C S	C S	C R	C S	C R

which quantitative and spatial changes start among the populations of species, building the meadow phytocoenosis;

(II) the phase of maximal numbers in the population is reached in phase (II) of succession, that is of development of a new species composition — the macroforbes phytocoenosis;

(III) the phase of decrease of population size starts in phase (III) of succession, that is in the period of initiation of willow brushwood and may last over the period of brushwood phytocoenose formation (phase IV of succession);

(IV) the phase of dynamic equilibrium of population size is reached in forest communities. In this period, the changes in population size are of the character of fluctuation. These species attain a relative stability of population size at a very low level (Figure 1.2).

A different behaviour of populations was also observed in the succession process, namely, withdrawal of the species completely or partially from the occupied site for several years, and then its return in the period of willow brushwood formation. This type of reaction was observed in such species as *Lychnis flos-cuculi, Ranunculus acris* and *Cirsium palustre* which after an absence of several (5—9) years often return to the edge of the brushwood, but as a rule grow on large sedge tussocks. Another type of reaction was observed for instance in *Caltha palustris* populations. This species, at the time of macroforbes and brushwood formation, withdraws almost completely, and when the forest communities begin to form, appears on some special sites — niches with periodical or permanent water (Figure 5.2).

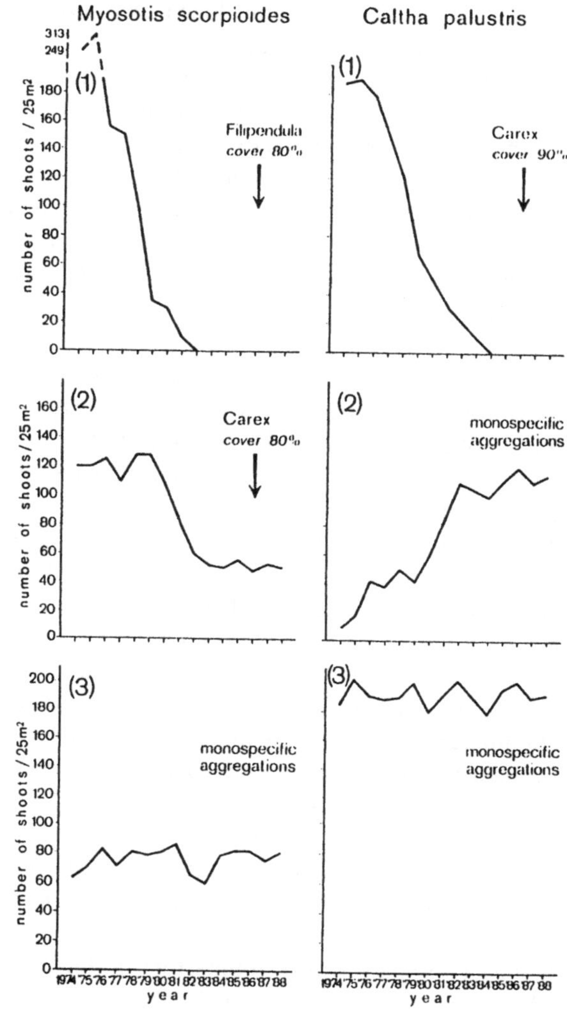

Fig. 5.2. Population dynamics of two species during succession, in the presence and absence of competitive species (Falińska 1989b). *Myosotis*: (1) meadow dominated by *Filipendula ulmaria*; (2) meadow dominated by *Carex acutiformis*; (3) mucks dominated by *Myosotis scorpioides. Caltha*: (1) meadow dominated by *Carex acutiformis*; (2) gap in meadow, (3) mucks dominated by *Caltha palustris*.

Long-term population dynamics

In demographic investigations of species taking part in a succession, a few population growth types have been distinguished according to the abundance variations of individuals and shoots:

1. If individuals produce a small number (3—10) of aerial shoots, and their growth becomes stabilized in year 2 or 3 of their life, then there is a considerable concordance of variations in the number of shoots and individuals, e.g. in *Lythrum salicaria* and *Caltha palustris* populations (Figure 5.1).

2. If the growth of individuals proceeds at a high rate, and in life year 5—6, the individuals become multi-shoot polycormones, then with a continuous increase in the number of aerial shoots, for several years, the number of individuals in a population become stabilized within a relatively short time; in such populations the number of individuals stay at a low level at a simultaneously high and varying number of above-ground shoots, e.g. in *Filipendula ulmaria* (Figure 1.2).

3. If individuals assume the form of large-sized

(over a dozen or several dozen m^2) polycormones within a comparatively short time (5—6 years) then in the initial colonization period, the growth in number of the aerial and that of the individuals are to an extent concordant. But several years later the number of individuals drops considerably, this reduction being as a rule accompanied by a growth in number of the above-ground shoots, e.g. *Iris pseudacorus* (Figure 5.1).

Variations in the number of individuals are the basis for identifying and describing the growth phases of a population and forecasting its fate. The data presented in this book indicate that for the same population, different estimates are arrived at if variations in the value of N are taken into account than when the assessment is based on variations in the value of n. The n curve may, for instance, indicate a population growth phase, and the N curve, an equilibrium or even decline phase (Figure 4.9). It must be emphasized that the shoot abundance curve is usually of the shape of a curve peculiar to populations of annual species colonizing new areas. This is natural as in most macroforbes perennials — the above-ground parts live one or two growing seasons, and during the first years of individual growth the increase in number is of the geometric progression type (Fiala 1978, Falińska 1986a). This accounts for the fact that when the number of individuals of a population becomes stabilized or reduced, the number of shoots usually grows rapidly (Figure 1.2), because in *Filipendula ulmaria* and *Iris pseudacorus* populations attaining the abundance-equilibrium phase, several-year-old individuals constitute a large group; they produce the largest number of shoots (Figure 5.1).

Studies of the populations of several species differing in their growth type, e.g. *Caltha palustris, Lythrum salicaria* and *Filipendula ulmaria*, indicate that the numeric ratios of the individuals and shoots differ very much. If over their long-term life-history individuals produce few (2—4) shoots, then the difference between the N and n curves is small. But if individuals grow in size to form multi-shoot polycormones with the particular entities interconnected with durable rhizomes, then variations in the values of N and n differ very significantly, as has been illustrated with the curves of population abundance dynamics of several species (Figure 5.1).

Earlier population dynamics studies which took into account variations in the number of individuals (N) and shoots (n) suggest the following conclusions: 1) The understanding of variations in the abundance of individuals makes it possible to identify the population growth phase and forecast the fate of a population; 2) The understanding of changes in shoot abundance can be used for assessing the rate of area filling by a species, and thereby makes it possible to understand the relationship between a population and the environment.

Dynamic value of species

The importance of the particular species for the stabilization or transformation of a phytocoenosis will depend on the population dynamics, thus its size, the age- and size-structure of its individuals, as well as on its spatial structure. There are many species which participate in the whole succession series, even if it lasts as long as the forest returns to meadows. Among these species three groups were distinguished in respect to their dynamic value. Group one includes species which are components of several interchangeable phytocoenoses. They constitute in some phytocoenoses the basic “core”, whereas in others they are only one element among many. Change in the role of these species in the successive phytocoenoses are connected with the population size dynamics (Figure 5.1) determined by the effectiveness of the vegetative and generative reproduction in the period of settlement, the varying individual life span dependent on the habitat conditions and the ability of expanding into multi-shoot polycormones. Included in this group were species such as *Filipendula ulmaria, Lysimachia vulgaris, Lythrum salicaria, Geum rivale, Carex acutiformis* and *Carex cespitosa*. These plants usually are dominants or subdominants in the phytocoenoses of the subsequent succession phases. Group two comprises species which play an important role in the structure of phytocoenoses only at certain succession stages, but do not much influence their transformation. This means that they have no significant effect on the variability of the populations of other species. They are, as a rule, subdominants only in the first period following the cessation of mowing. To this group belong such species as *Cirsium rivulare* and *Cirsium palustre*. Group three comprises species occupying specific niches temporarily of permanently filled with water. They play a rather important role in the structure of mid-meadow mires, included are: *Myosotis scorpioides, Galium palustre* and *Carex vesicaria* (Figure 5.2).

Life history

It has been repeatedly demonstrated in the succession series described, that annual and biennial species are replaced by perennial ones, and these in turn by shrubs and trees.

This type of sequence is true for primary or secondary succession beginning in abandoned fields. However, a succession starting in floristically rich meadows presents a different type of sequence of species (Figure 5.3, Table 2.5). In such a situation, annual and biennial species play, above all, a negligible role in the succession process. Moreover, for a certain time after the abandonment of the meadows new species do not arrive, and the decrease in meadow plants is gradual. They are eliminated by species which become dominant several years after the abandonment of the meadows. Only phases II and III of succession are characterized by a decline of floristic richness and an increase of dominance. The retreat of one floristic composition and the appearance of a new one is as a rule preceded by the dominance of one species, different in each meadow patch (Figure 2.3).

The appearance of the first forest species was observed after eight to ten years. Only sporadically do they arrive earlier, and then perish rapidly (Figure 5.3).

Ten years after the abandonment, the number of forest species, both herbs and trees (Figure 5.3), increases from year to year in the abandoned meadows. The meadow species still persist, although in small numbers, up to the initial phase of forest, that is, formation of willow-alder brushwood. Only then does the proportion of forest species significantly increase in the structure of the phytocoenosis, and is expressed by the percent cover and frequency (Figure 3.22).

Forms of life and growth

In succession, especially in its first phases, an essential role is played by perennials. However, in this group of plants there are wide differences as regards to the life span of individuals, their life cycle lasting several or several score of years. Since with age of the individual its size fertility, architecture and area occupied change, therefore, its relation to space and neighbours also changes (Figures 3.11; 3.22). It has been demonstrated that the role of species in the process of succession depends on the length of life of the individuals. In view of the wide differences in the life history among the perennials, the following groups of species were distinguished.

(1) *Facultative biennials*. These are species which with progress of succession prolong their lives to four or five years (Caswell and Werner 1978). With every year there are less individuals in the populations which attain the phase of flowering in the second year of life, and more of them remain for several years, lagging in the juvenile phase, that is, in rosette form as, for instance, *Cirsium palustre* (Figure 4.4).

These species cause elimination of plants with delicate structures, owing to the development of large rosettes, 30—50 cm in diameter, which by occupying the same site for many years physically restrict the development of their neighbours (Figure 3.21). They play an important role in the first and second phase of succession, thus, in the period of decomposition of meadow phytocoenoses.

(2) *Pseudobiennials*. These are species in which

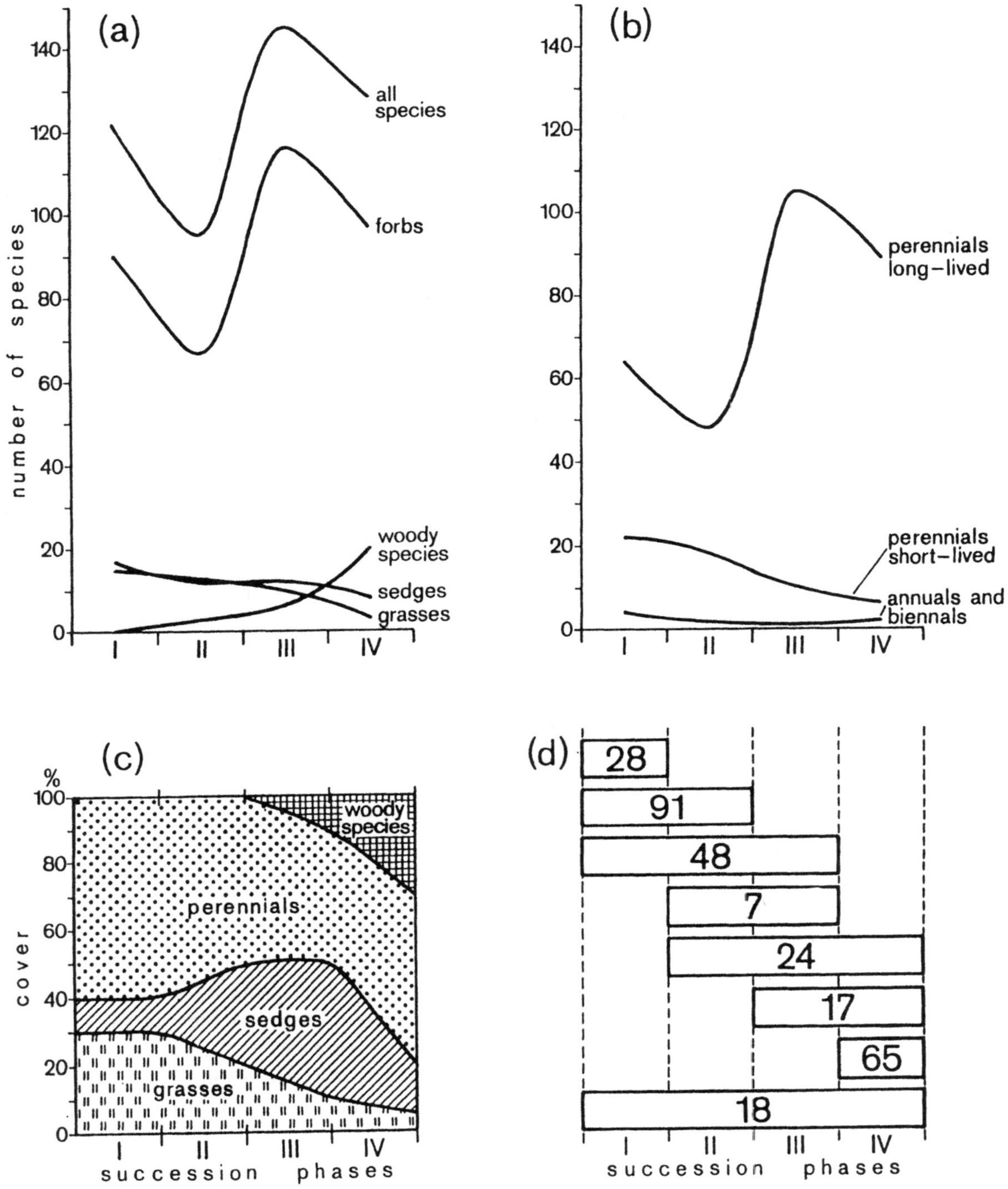

Fig. 5.3. Changes of life-form categories in consecutive successional phases: (I) meadow, (II) macroforbes community, (III) willow brushwood, (IV) brushwood community, (V) *Salicetum pentandro-cinereae.* a) Number of species representing each life form during succession; b) Number of species: annuals and biennials, short-lived perennials; long-lived perennials; c) Participation of life forms in consecutive successional phases; d) Number of species present in the meadow only (I), in the forest only (IV) and the number of species common for the consecutive succession phases.

the modules survive for about two years. The plants produce regeneration buds at the base of shoots in summer or autumn. In the next season the buds develop into rosettes. In summer, part of the rosettes develop flowering shoots, but most attain this phase only in the next season, thus after a lapse of two calendar years. Similarily, as in group (1), the modules of these species may remain in the juvenile phase in the form of rosette for three or four years as, for instance, *Myosotis scorpioides*.

These species usually occupy various kinds of gaps which in the succession process arise frequently and are subject to incessant rotation as they are eliminated by perennial species which after a certain time overgrow these sities.

(3) *Short-lived perennials.* To this group were classified species, the individuals of which seldom grow for more than five years even under control conditions in the experimental garden, as for instance *Caltha palustris* and *Cirsium rivulare.* Individuals of these species have delicate, not very durable, underground organs. Therefore, there is a constant substitution of individuals in the population by numerous competitors (Figure 3.21). Since the new individuals mostly develop from seeds, after a certain time, owing to increasing crowding with other plants and shading by macroforbes and shrubs, their recruitment is restricted. The competitors are mainly species with two ways of reproduction, so that under worse conditions they can multiply vegetatively.

(4) *Long-lived perennials.* Species, the individuals of which can live for a dozen or so or several score of years, owing to the permanent renewal of the senile and dead fragments of the plant. The particular parts of these plants, usually polycormonal are, however, different-aged, for instance in 10-year-old individuals of *Iris pseudacorus*, the oldest modules were aged three to four years, but most of them did not exceed three years. In ten-year-old *Filipendula ulmaria*, the oldest modules were aged four to five years, and the majority were two to three years old. In both cases the modules were joined with senile rhizomes (living but not regenerating) and dead ones which serve as scaffolding to the living parts.

Species classified to this group play an essential role in the transformations of meadow vegetation towards forest communities. Many of them played the role of promoters or inhibitors of succession (cf. chapter 3 and Table 5.2).

CHAPTER 17

PLANT STRATEGIES IN SECONDARY SUCCESSIONS

Life strategies

In general, the participation of life strategies changes markedly with succession, this being in agreement with the theory of plant life strategies (Grime 1979, MacMahon 1980). Grime's (1979) concept was applied in the classification of species. The features of plants which served as a basis for classification of the particular species according to their respective strategies are listed in Table 5.2. In the process of succession, especially in the initial stage, those species dominate the individuals of which exhibit properties that enable them to gain space and access to water, light and nutrient resources rapidly (strategy C). These species are classified as succession promoters and inhibitors (Table 5.2) such as, for instance, *Filipendula ulmaria, Lythrum salicaria, Lysimachia vulgaris, Geum rivale, Potentilla palustris, Carex cespitosa, C. acutiformis* and *Salix cinerea.* They remain dominant for 10—15 years, usually to the time of appearance of willows on the meadow.

The presence of species representing strategy R was noted only in early phases of succession. A typical representative here is *Cirsium palustre* and some grass species (Table 5.2). Other annual and biennial meadow plants withdraw immediately when mowing is discontinued. Only *Cirsium palustre* persists on the abandoned meadows until willow brushwood appears, that is to the initial phase of forest. At this time, however, *Cirsium palustre* exhibits a number of traits, indicating that this species is also resistant to stress and other disturbances, for instance when life conditions deteriorate (increased crowding and shading). The individuals remain latent in the form of rosette for several years. As a rule, *C. palustre* settles in various kinds of gaps among macroforbes and rushes, it also overgrows sedge tussocks and the disintegrating necromass (Figure 3.11). In the period of decomposition of the phytocoenoses, meadow species with strategy C are prevailingly dominant. In later phases, more and more species show features of "mixed strategy", especially C-R. Individuals of these species produce a large number of seeds, but they also have strong durable root systems (e.g., *Lythrum salicaria*). They occupy various gaps, but are also components of stable phytocoenoses (Figure 1.2). The classification by Grime (1979) of *Caltha palustris* to strategy S proved controversial. In the process of succession this species reacted visibly to shading depression of the soil water level and the presence of necromass. After three to four years *Caltha* individuals ceased to produce flowering shoots, and after five to six years the population withdrew from the site. During succession, however, this species appeared in many gaps, filled periodically or permanently with water. *Caltha palustris* should, rather, be classified to strategy R-S. It seems that this type of strategy represents species settling in various gaps with water.

Adaptive strategies of species in succession

Owing to the evolutional adaptation of species to a definite and limited range of habitat conditions, in various succession phases and with varying environmental stress, groups of species with various adaptive strategies dominate (Grime 1979, Huston and Smith 1987). It is suggested that in early succession phases the species which are dominant are those of which the life strategy consists of the production of numerous readily and widely spreading seeds deprived of food reserves. Their characteristic feature is that, in general, the plants are small anf fulfill their full development cycle before being superseded by stronger competitors. Such plants utilise only a few assimilation products for building vegetative parts, and use most of their energy to produce diaspores.

In further succession phases, gradually species adapted to a low level of environmental stress begin to dominate. They slowly gain terrain, producing a relatively small number of seeds disseminated to small distances. Owing to their longevity and large dimensions and also to their well developed root systems and accumulation of food reserves in them, they attain dominance in plant communities.

Drury and Nisbet (1973) stress that both these strategies may be useful in the whole succession series. It seems that extremal life strategies in the subsequent succession stages are characteristic for primary or secondary succession, but only on open areas and poor soils. In theoretical considerations, as a rule, short lasting successions are taken into account, characterized by a small number of species. In such a case the existence of simple models of succession mechanisms is confirmed. One does not, however, get such a clear picture of the efficiency of various strategies in the successive phases as when this phenomenon is considered in the succession series of long duration with its characateristic floristic richness.

Much attention has been devoted in recent years to the biology and ecology of the species participating in the process of succession. It was found that there does not exist one characteristic adaptive strategy in the sense of a combination of established evolutional features owing to which plants would have the chance of settling, survival and production of progeny. Investigations on succession on abandoned meadows indicate that species of plants playing a major role in succession may be characterized by opposite features, for instance:

— production of a great number of small seeds (e.g. *Cirsium palustre, Lythrum salicaria, Filipendula ulmaria*) or a large number of big seeds (e.g. *Iris pseudacorus*);

— the presence of structures facilitating distant transport of seeds (e.g. *Cirsium*) or lack of any flying apparatus (Iris, *Lysimachia, Filipendula, Lythrum*, etc.).

— a high effectiveness of generative reproduction (*Cirsium palustre, C. rivulare, Lythrum salicaria*) or prevalence of vegetative reproduction (*Lysimachia vulgaris, Geum rivale, Potentilla palustris, Carex acutiformis*) or highly effective reproduction of both types (e.g. *Filipendula ulmaria*).

In the light of the foregoing data, the wide spread concept of r and K selection, according to which organisms undergo the influence of type r selection, were classified to environments disturbed and undurable, in contrast to type K characteristics for relatively durable and stable ones (MacArthur and Wilson 1967, Pianka 1970, Gadgil and Solbrig 1972, Grime 1977). Moreover, species appear in successions which in respect to their type of reproduction (many small seeds, high germination capacity) should be classified to r-strategy, and on account of long-living and strong underground organs to K-strategy as for instance *Lythrum salicaria*.

In the light of the latest results of investigations, the thesis of the high reproductive effort of r-strategists and the small effort of K-strategists do not stand the test of criticism. It has been demonstrated that this effort is highly variable and depends on biotic and abiotic conditions under which populations of the given species live (Raynal and Bazzaz 1973, Pickett 1976, Grubb 1977, 1982, Law 1979, Falińska 1979, Fekete *et al.* 1988).

During succession on abandoned meadows an important adaption of many species consists in a plastic division of resources for building the above ground and underground parts. In the successive phases the participation of underground organs in the total biomass of individuals, increases markedly, this greatly enhancing their chances of survival in the growing competition and crowding of the plants (Figure 4.16). Changes in the proportion of biomass directed to growth and reproduction in the total biomass of the individual resulted in an increase of vegetative reproduction with considerable diminution of generative reproduction (Table 4.1). It appeared that in the succession process not only species reproducing generatively are substituted by vegetatively multiplying ones (Knapp 1974), but in populations of the same species the quantitative relations between propagulae of vegetative and those of generative origin change in the successive stages (Figure 4.7). As a consequence of these changes, the number of individuals of vegetative origin gradually increases in the process of succession in populations of many species (Figure 5.4). The important role in the transformations of meadow vegetation with iterative growth indicates that the spread of polycormones in the succession process, frequently even under deteriorating life conditions (increased crowding and thickness of necromass of the soil surface, appearance of new competitors) is an adaptation mechanism, preventing the exhaustion of the environmental resources and giving the plant the possibility of translocation in the search for more favourable conditions of growth and development (Gross 1980, Noble *et al.* 1979, Falińska 1982, 1986c, 1989a, b).

On the other hand, disintegration of the polycormones and clone formation prevents ageing of the genets and the offspring individuals have a chance or further growth and expansion. Grime (1979) describes this process as "vegetative expansion" which is characteristic for numerous species participating in the succession process.

All species classified to promoters or inhibitors of succession on the abandoned meadows exhibit such a set of traits which seems to indicate rather mixed or intermediate strategies than a decidedly r or K selection.

The investigations demonstrated that the success of species in the process of transformation of the vegetation on the abandoned meadows is a resultant of many traits of individuals of these species, namely: (1) high fertility, frequently increasing when mowing is discontinued; (2) a relatively high and variable germination capacity of the seeds; (3) tolerance of seeds and seedlings to a wide spectrum of variability of abiotic and phytocoenotic factors (crowding, shading, water relations and thickness of necromass; (4) high effectiveness of processes of renewal, spread and vegetative reproduction.

These properties ensure to many species a wide spread on the extensive area of the abandoned meadows and persistence under various variable conditions (Figure 3.22).

In the course of succession the following regularities were observed: (a) successive depression of the reproductive effort indicating the diminished importance of generative reproduction; (b) a decrease of seed production in the subsequent succession phases indicates a considerable modification (adaptation) of the reproductive strategy in populations participating in succession of long duration; (c) increase of the biomass of underground organs in relation to that of the above ground ones, indicating an increased importance of vegetative reproduction.

In time, however, there appeared, in the succession process, factors against which the adaptive strategies of herbaceous plants proved insufficient and which brought about a regression of the populations of many species which took part in succession for 10—15 years.

To the factors distinctly limiting development and growth of these species belong: crowding, and above all, shading. Both of these factors have a negative influence on the size and habit of plants and as result cause regression of the population. It was noted that the increasing density of the plant cover during succession was the cause of the decrease of the number of generative and renewing buds in populations of many species (Figure 4.15, Table 4.6). Inhibition of growth and the renewal process of individuals led in time to their gradual death (Figures 3.9; 3.10). The willow appearing at this time (phase III of succession) greatly speeded up this process by shading the

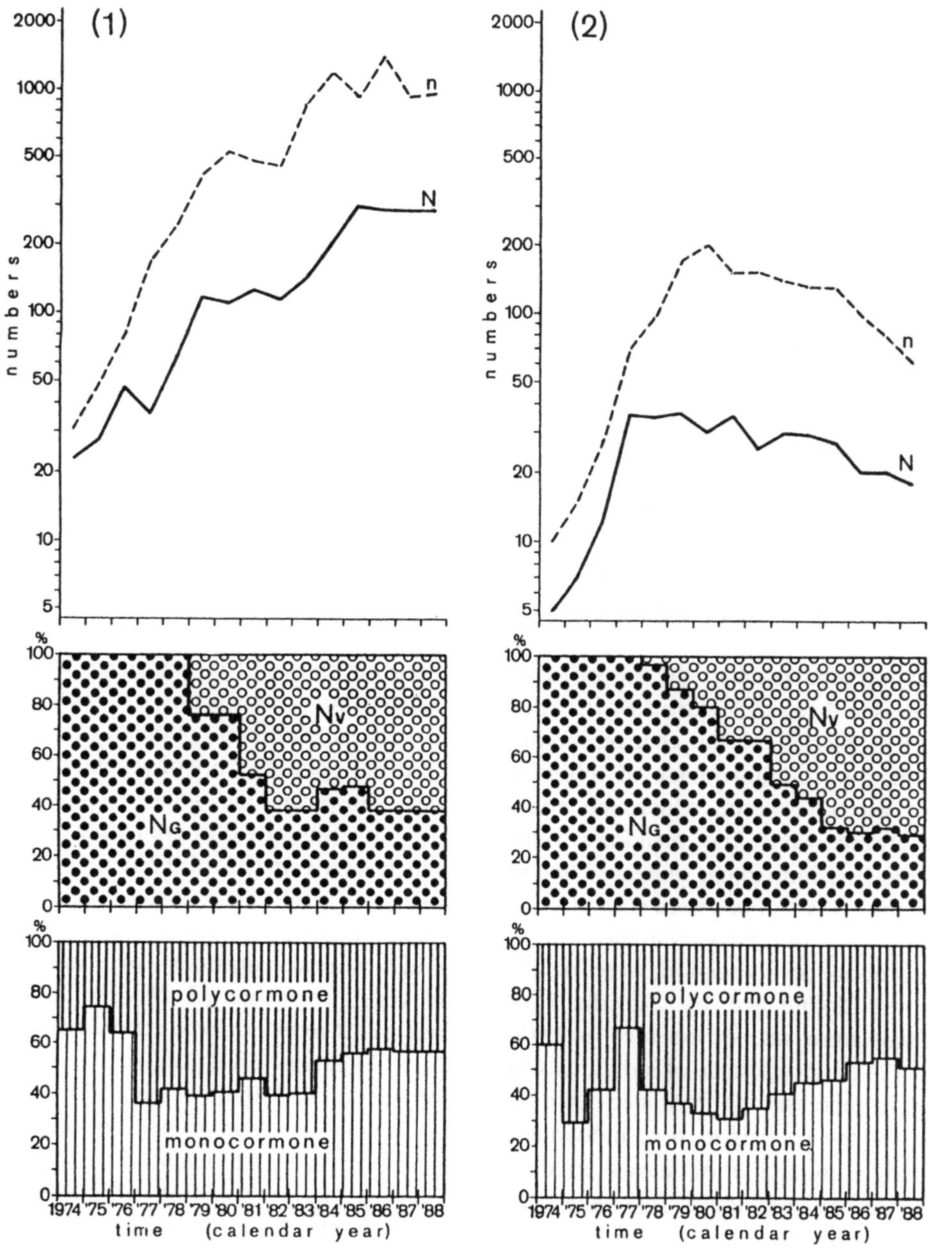

Fig. 5.4. The behaviour of *Filipendula ulmaria* populations during succession in communities of meadow: 1) species-rich patch; 2) meadow dominated by *Carex acutiformis*. N — number of polycormones, n — number of shoots; proportions of polycormones of generative (N_G) and vegetative origin (N_V) and of polycormonal and polycormonal forms in the population.

plants and mechanically eliminating them from the occupied area.

In the particular succession phases, inhibition of growth of the plants and of generative and vegetative reproduction by development of the tree canopy, was the cause of regression of the populations of many species which played an important role in the transformation of the meadow vegetation.

Diversity and genetic monotony in populations during succession

Generally in populations of perennials the genetic variability changes significantly during succession. In its first phases (I, II) the number of individuals developing from seeds — vegets increases — and in phases III and IV individuals of vegetative origin become more and more numerous (Figure 4.6). In phase IV a considerable increase of the participation of individuals of vegetative origin was noted in many species (Figure 3.23). In dependence on the type of growth and of reproduction, represented by these species, this proportion is different.

This evaluation was reached on the basis of labelled seedlings and vegets (Figure 1.8). In order to obtain a clear picture of the changes in the quantitative relations of individuals of different origin, a plot was chosen on which the given species was absent while the meadows were utilized, and the first individuals appeared at the moment when the investigations were started.

The following types of variability in the populations in the process of succession were distinguished: (1) the dominance of individuals of generative origin increases, for instance in *Lythrum salicaria*; (2) the dominance of individuals of vegetative origin increases, for instance *Lysimachia vulgaris*; (3) the participation of individuals of different origin is similar or shows an only small prevalence of individuals of vegetative origin, for instance in *Filipendula ulmaria* and *Iris pseudacorus*.

In all the mentioned cases, the increased role of vegetative reproduction is connected with the type of growth, and especially the structure and durability of the underground organs. For instance, individuals of *Lythrum salicaria* exhibit durable and strong roots which seldom undergo division. The scarce individuals of vegetative origin arise, owing to a mechanical lesion of the polycormone, caused especially by sphagnum or willows. *Lythrum salicaria* shows no natural predisposition to disintegration. They form compact and durable polycormones. Under deteriorating conditions they do not expand, but form layered underground organs. All this contributes to the survival of intact polycormones for many years. Hence the participation of individuals of vegetative origin increases to 20 percent only as late as phase IV.

It was found that the rate of increase of the number of vegets in the popoulation is most dependent on the degree of integrity of the polycormones. The earliest and greatest decline of genetic diversity was observed in trailing species, the shoots of which take root at considerable distances from one another and exhibit a delicate structure, for instance *Lysimachia vulgaris*. When, on a surface area of 25 m^2 there grew, in succession phase III, 270 individuals, only 21 percent of them developed from seedlings and 81 percent from scions. A similar type of variability was noted in *Caltha palustris* populations in which 81 percent in phase IV are of vegetative origin. In this case, however, the great decrease of genetic diversity was not observed. Individuals, namely, which survived (52 of 123) to phase IV represented various genets. In *Lysimachia vulgaris* populations, however, the decrease in diversity was much more conspicuous, for instance individuals of vegetative origin (212) were derived from as few as 12 individuals of generative origin. Therefore, 70 genets (58 + 12) were represented in the population of 270 individuals.

In the *Filipendula ulmaria* population, after 15 years of its development, 128 individuals were present, among which 79 were derived from the division of 12 polycormones and 49 developed from seedlings. At this time the individuals represented 61 various genotypes (49 + 12). Thus, the genetic diversity diminished in the course of 15 years by 48 percent.

It is noteworthy that, in spite of the high effectiveness of vegetative reproduction in the examined species, the genetic diversity in their populations persisted during succession. It appeared that this diversity is greater than expected on the basis of the morphological and biological properties of these species. Other authors (cf. Harper 1977, Silvertown 1987) call attention to a similar regularity, after evaluating the number of genets in populations of species reproducing vegetatively.

CHAPTER 18

MECHANISMS OF SUCCESSION

Successional pathway

According to the succession model of Clements (1936), the species occurring in later phases of ecological succession always have a positive influence on the subsequent arrivals, facilitating their appearance and development.

Connell and Slatyer (1977) critically envisage such a presentation of the succession mechanism, and they state that the influence of species from earlier stages on the later ones may be positive, negative or null. Therefore, I suggest three alternative models of the mechanisms of species exchange in the course of succession: (1) facilitation, (2) tolerance and (3) inhibition.

Lately papers have appeared critically evaluating the usefulness of Connell and Slatyer's models (Pickett *et al.* 1987, Walker and Chapin 1987, Osbornová *et al.* 1990). This is confirmed by investigations on secondary succession on abandoned meadows. It was demonstrated that their usefulness is but limited if this process does not start on a protected area. For instance in the "facilitation" model it was assumed that modification of the environment by the first arrivals (pioneering species) favours the appearance of subsequent species. Such a role may also be ascribed to species present on the meadow at the moment of initiation of succession, and remaining there up to the return of forestation. In the development of populations of these species the properties of the individuals change with age (area, number of shoots, fertility, etc.) and these changes frequently facilitate the appearance of individuals of other species. It was considered, however, that these species cannot be classified to the "facilitating" ones. Their role is limited, rather, to stimulation of various interactions between individuals and populations, resulting in an exchange of species. In this connection, it was assumed that a more suitable term for these species would be "promoters of succession", because they contribute to the transformation of one floristic composition into another. On the other hand, the term "inhibition" was used in the sense proposed by Connell and Slatyer (1977) and as applied in the description of the phenomenon "space capture" (Begon and Mortimer 1981).

The model, "tolerance", is also controversial. It assumes that the environment modification occurring under the influence of the first colonists neither increases not reduces the chances of appearance of the consecutive species. Such a role may be attributed to species characterized by a small population size and small colonization ability. They seldom are dominant and play a "neutral" role in the succession process. It is not clear, however, if they are more tolerant than other species to the diminishing level of food reserves. These species exhibit rather a number of adaptive strategies which ensure to them survival under deteriorating life conditions, and some participation in the successively arising floristic composition. As a rule, all species participating in long lasting succession exhibit a certain tolerance to the changing life conditions, but their role varies and changes in the successive phases.

Success of species

When considering the relations between colonization and "inhibition", Peart (1989a, b, c) notes that colonization ability and the effects of colonization should be treated separately. In the present studies it was considered useful to evaluate the colonizers according to their role in the vegetation dynamics.

Among the species which played a major role in the transformation of meadow vegetation, two types of colonizers were distinguished according to the criterion of the effectiveness of their colonization: colonizer dominants and colonizer monopolists. Included in the former group were species which, after colonizing a new site, tolerate the presence of a co-dominant. As monopolists, those species which, in the first period of colonization, eliminate, almost completely, the meadow components and occupy the conquered space for a long period, were considered.

Type 1 — Dominants. They gradually occupy new areas, but do not completely eliminate the floristic composition of the meadow. As a rule, the dominant, together with the species which have not been eliminated when the meadow was abandoned, initiates the development of a new floristic composition. Dominants are represented by such species as: *Filipendula ulmaria, Lysimachia vulgaris, Lythrum salicaria* and *Iris pseudacorus*. The group of species which can exist among the dominants includes: *Cirsium rivulare, Geum rivale, Potentilla palustris, Polygonum bistorta* and *Cirsium palustre*.

Type 2 — Monopolists. They rapidly take possession of the new site, eliminating nearly all meadow components. They occupy a particular site almost exclusively for many (about 10) years. After several (6—8) years, however, there sporadically appear among the monopolists, individuals of meadow species, that is, those which occupied the site at an earlier stage. For instance, in a patch of *Carex acutiformis* or *Carex cespitosa*, individuals of *Cirsium rivulare, Cirsium palustre, Lysimachia vulgaris, Lythrum salicaria, Filipendula ulmaria* as well as *Ranunculus acris* and *Lychnis flos-cuculi* appear. If forest species do not appear simultaneously, especially willows, then the meadow community is restored for a short time here, but floristically poorer, and with a large proportion of the monopolist (30—40%). This community is, however, short-lasting (5—6 years), since the retreat of monopolists is followed by the appearance of willows which after several years form large aggregations. The meadow species are once more eliminated and remain only at the edges of the brushwood. These willows, especially *Salix cinerea*, become monopolists in various parts of the abandoned meadow for several years, occupying about 70—80% of the space.

Priority of settling and colonization success

The success of a species in colonization depends on the floristic composition of the patch in which it has settled, and on whether it is the only colonizer or is accompanied by other species (Figure 5.5). When, for instance, two species with monopolistic tendencies penetrate into the meadow, such as *Carex cespitosa* and *Carex acutiformis*, for the first five years *Carex acutiformis* is dominant due to the rapid expansion of its stolons. Subsequently, by expansion of tussocks from year to year this position is gained by *Carex cespitosa*. After ten years a system arises in which *C. cespitosa* is the dominant and *C. acutiformis* the co-dominant. The situation is similar when *C. acutiformis* and *Filipendula ulmaria* settle in a meadow patch at the same time. When, however, *Carex acutiformis* occupies a patch of the meadow not accompanied by *F. ulmaria*, then within eight years, sedge becomes the dominant, and after ten years may become a monopolist for a long period. This colonization process is similar when *Filipendula ulmaria* is the first and only colonizer in the abandoned meadow. In this case, however, this species is accompanied, as a rule, by other tall perennials.

Colonization success in relation to reproductive strategy

Species which succeed in colonization are, above all, those characterized by two ways of reproduction. Species with such properties rapidly occupy

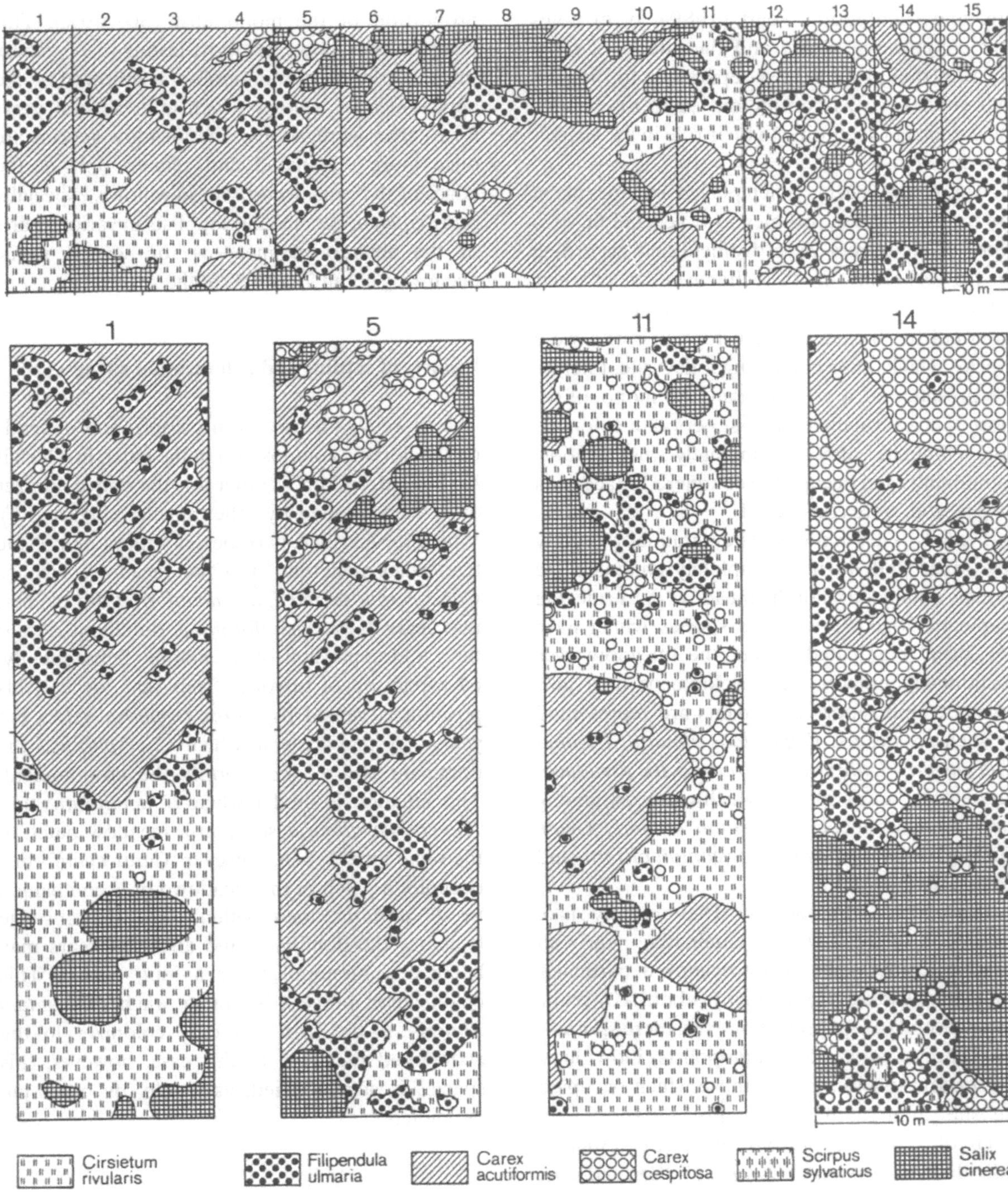

Fig. 5.5. Mosaic of patches dominated by few species 15 years after abandonment of the meadow (top). Differentiation of priority of colonization in abandoned meadow, of chosen plots (bottom).

wide stretches and persist in them for a long time. On the contrary, species with only one type of reproduction, achieve but little success of short duration. The following groups of species were distinguished:

(1) Species which spread in the abandoned meadows by generative reproduction exclusively (e.g., *Cirsium palustre*). They take part in the long-term process of succession thanks to the high production of seeds, survival of seedlings and the ability of prolonging their life, (e.g. biennial species of up to four or five years). The individuals persist in the juvenile phase (rosette) for several years. The rosettes exhibit competitive abilities, that is, they keep other species away from the area occupied by them. They settle, due to high dispersion of seeds, in every newly appearing gap between the dominants, and also grow on sedge tussocks. They are present until willow-alder brushwood is formed, that is, to the initial phase of forestation (Figures 3.20; 3.21).

(2) Species which spread in abandoned meadows, owing to high seed production and survival of the seedlings. They manage to remain on the acquired site because their underground organs are durable (e.g., *Lythrum salicaria* or like the species of the first group, by way of the rosette mechanism and low vegetative reproduction, e.g., *Cirsium rivulare*).

(3) Species which spread in the meadows by two ways of reproduction. In the first period of succession their presence depends on a large number of seeds and a high ability to move to various sites. In later phases, when some of the individuals reach maturity, their spreading depends on the growth of rhizomes, that is, on vegetative reproduction (e.g., *Carex acutiformis* and *Iris pseudacorus*) (Figures 3.11; 4.10).

(4) Species which spread in meadows by simultaneous generative and vegetative reproduction. With the progress of succession, the quantitative relations change in a population between propagules of generative and vegetative origin. After some time (phases III and IV of succession) vegetative reproduction dominates (Figure 3.23). Included in this group have been species which produce special propagation units (e.g., *Caltha palustris*) and those the individuals of which undergo disintegration. The most typical representative of this group is *Filipendula ulmaria* (Figure 5.4).

(5) Species which spread in abandoned meadows mainly by vegetative reproduction. Individuals arising from seeds are rare, and generative reproduction is only of a supplementary nature (e.g. *Lysimachia vulgaris*).

Colonization success in relation to growth type

Stimulation or inhibition of succession is no doubt connected with the type of growth of the colonizers. Species with a unitary type of growth do not show much colonization ability. They have but little chance of remaining for a longer time on one spot in view of the increasing competition in the process of succession. *Cirsium palustre* is a certain exception in this group since it may survive in sedge swamps and between macroforbs thanks to the "rosette mechanism". If the individuals of this biennial species prolong their life span to four or five years, their rosettes and roots greatly expand and in this way they are able to stand the competition with other species for several years (Figure 3.21). *C. palustre*, however, remains in few patches of the meadow. It usually uses small gaps among other plants and can persist there in small numbers up to the initial forest phase (Figure 3.20).

Most successful in colonization are species with iterative growth. They are a highly diversified group as regards to their morphology. Earlier on (cf. Chapter 3), among the polycormonal individuals (Part 3) the following types were distinguished: (1) polycormones with common shoots of long duration, characterized by strong and durable underground organs, (2) polycormones with common shoots of short duration, this resulting from the delicate structure of their rhizomes, (3) polycormones with temporarily common shoots, for one or two seasons, that is, when the particular modules live no longer than two, or up to three, years.

In view of the morphological diversity of the

plants, the rate of colonization of new sites and time of survival on them, are not equal. Species representing the first type of polycormones colonize at a rapid rate and saturate the space, owing to the multiplication of shoots in geometrical progression. Their long persistence is the result of the structure of their underground organs and ability to renew and regenerate senescent parts. It is only disturbed relations between polycormone senescence and renewal that give other species the opportunity to encroach on their territory (Figure 3.5).

Species with polycormones of the second type, owe their expansiveness to the rapid spread of their rhizomes, for instance, in one season their length increases from 30 cm to 1 m. The long and delicate rhizomes penetrate between other plants and root there, detaching themselves from the parent organism and continuing growth in several directions. These species rapidly possess new sites and usually leave them soon, settling on other meadow patches (e.g., *Lysimachia vulgaris*). They are characterized by a rapid translocation in meadows in the process of succession, in spite of the fact that the recruitment of seedlings is not important here (Table 4.4).

Species whose polycormones represent structure type 3, have the smallest chance to colonize new sites and survive at them during a progressing succession. The structure of their shoots is delicate and so is that of their underground organs, so they are readily eliminated by other plants, especially by species representing polycormone structure type 1.

Harper (1986) distinguishes two colonization strategies, namely: "guerilla" comprising plants forming stolons, and "phalanx" characteristic of species growing in tussocks. Many intermediate strategies have been noted in the process of succession. *Lysimachia vulgaris* and *Carex acutiformis*, for instance, doubtlessly represent the "guerilla", while *Carex cespitosa*, the "phalanx" strategy. Such species, however, as, for instance, *Filipendula ulmaria, Iris pseudacorus* and *Lythrum salicaria* play an important role in succession owing to their high colonizing capacity, although their structure is not stolonate or tussock-like. At the stage of settlement, their juvenile individuals elongate the rhizome internodes and can, in this way, penetrate between other species, thus they use the "guerilla" strategy. On the other hand, older individuals shorten their internodes, especially when competition increases in the succession, and build polycormones with a compact structure, thus increasing their competitive ability and making possible their survival in multi-species systems ("phalanx strategy").

To sum up, success is attained in colonization by species with iterative growth, two types of reproduction, and a long life span. A measure of this success is the rate of colonization and the size of the area colonized, as well as the survival in it for a longer time.

CHAPTER 19

SPATIAL VARIATIONS

Mosaic pattern and heterogeneity succession

The mosaic pattern is characteristic of many plant communities, but the cause of its formation are not fully understood (Whittaker and Levin 1977, Belsky 1986). The opinion that investigations into the biology of populations contribute to a better understanding of the factors responsible for the mosaic pattern of plant communities has been confirmed in many papers (Grubb 1977, Grime 1979, Stenseth 1980, Giller, 1984, Faliński 1986).

Simultaneous observations of individuals, populations and phytocoenoses, for many years on the same sites, provided a basis for the determination of the extent to which the mosoaic is dependent on the heterogeneity of the habitat, and on the ability of the plants to form single or several-species patches. This phenomenon was followed on a macroscale in the Reski Range area and on a microscale within the plant communities.

Mosaic pattern of vegetation

The meadows in the Reski Range area occupy what was once the habitat of fertile and damp flood-plain forest (*Circaeo-Alnetum*) (Figure 2.1). The varying water level (10—50 cm), permanently or temporarily above the soil surface found its reflection in diverse floristic compositions, also in meadows mown every year. Two types of phytocoenoses were then dominant: (1) temporarily submerged habitats were occupied by floristically rich phytocoenoses of *Cirsietum rivularis* with a weakly expressed mosaic pattern. When mowing had been discontinued there arose within five years a mosaic of patches dominated by various species (Photo 5.1). (2) habitats submerged for long periods are occupied by floristically poor *Caricetum acutiformis* phytocoenoses, characterized by a mosaic pattern. Over many hectares the dominant species is *Carex acutiformis*. Within the sedge swamp there are patches dominated by *Iris pseudacorus* or *Lythrum salicaria*, or else by *Lysimachia vulgaris* or *Scirpus sylvaticus*.

On a macroscale, a primary mosaic pattern of vegetation is determined by the heterogeneity of the habitat and then by the colonization abilities of the settling species. In the process of succession many species exhibit colonizing abilities which were inhibited in periods when the meadows were mown every year. This applies particularly to species, the individuals of which expand and produce polycormonal forms (see Part 3). That is why the mosaic pattern of vegetation due to the heterogeneity of the hibitat increases with the progress of succession (Figure 5.5), the more so that, in the meadow patches dominated by various species different floristic compositions are formed.

The rather small vegetation complex of the Reski Range became, within a dozen years or so, a markedly mosaic system, as illustrated by the maps of the vegetation of the Reski Range prepared at various times (Figure 2.1).

Mosaic pattern in phytocoenoses

The mosaic pattern within communities is the effect of the occurrence of diverse niches and interspecies competition as well as of various variable morphological and biologicial properties of species (Levin and Paine 1974, Harper 1977,

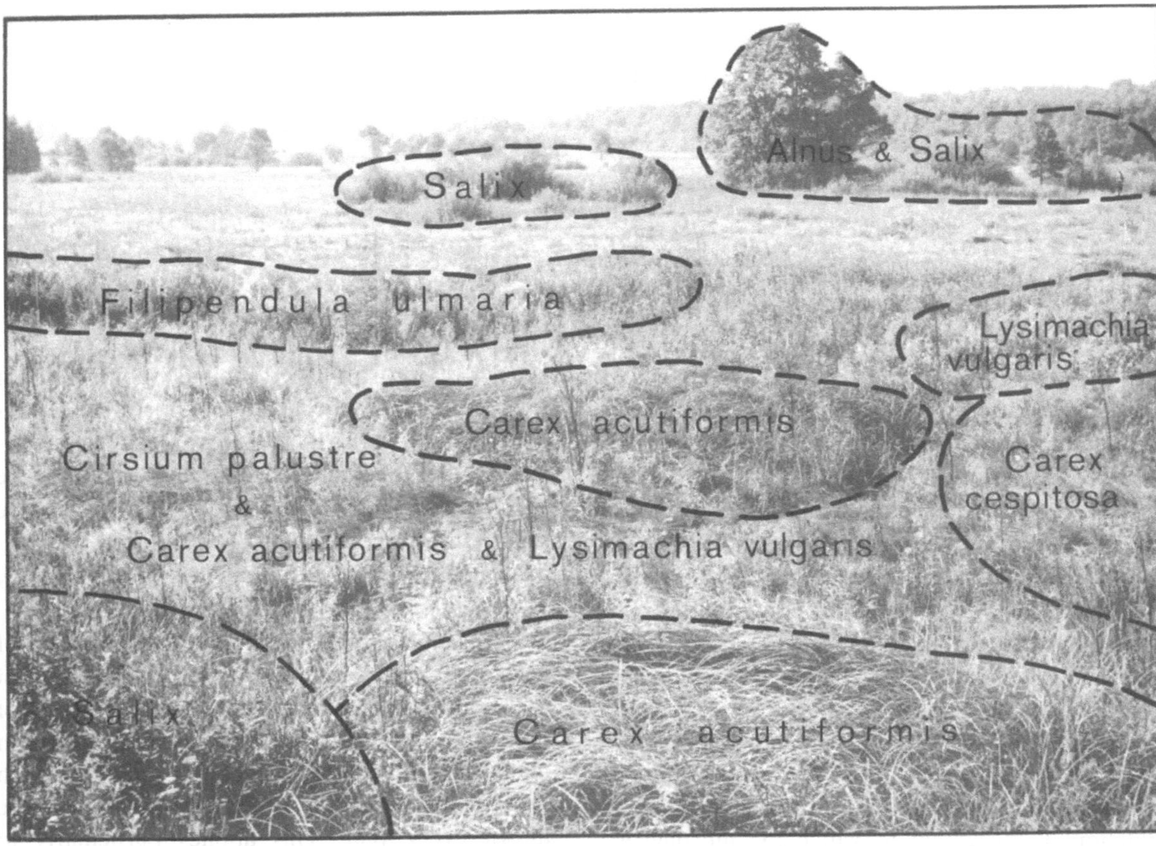

Photo 5.1. Mosaic of patches dominated by few species 10-years after abandonment of the meadow. Photo: J. J. Faliński

1986, Silvertown 1983). Hence, phytocoenoses that hold common species may also differ by the distinctness of their mosaic structure, e.g., meadows (*Cirsietum rivularis*) and macroforbs (*Lysimachio vulgaris-Filipenduletum*) (Figure 3.22). Species, which were in the first phytocoenosis, randomly distributed and formed small assemblages, in the second phytocoenosis, already formed large aggregations, e.g., *Filipendula ulmaria, Geum rivale, Potentilla palustris* and many others. This kind of community is termed polydominant. The two phytocoenoses widely differ in their spatial structure, both when they border on each other and, when in the process of succession, a meadow at a particular site is replaced by a macroforb community. A comparison of the structure of the phytocoenoses replacing one another in the process of succession, indicates that the mosaic pattern is most pronounced in the transition period, that is, the phase of the change of one species composition into another. It is the result of several species gaining dominance (Figure 3.22).

All these data indicate that the mosaic pattern within the phytocoenoses is largely the result of the colonization abilities of certain species, thus, of the ability to form mono- or poly-specific patches, and remain there for longer periods.

Generally, the mosaic pattern of the vegetation on a macroscale is determined above all by the heterogeneity of the habitat. Species with a given ecological amplitude colonize sites which satisfy their life requirements. The mosaic pattern on a microscale is to a greater degree dependent on the morphological and biological properties of the species. At the early stage of settlement, that is, the appearance of seedlings, the diversity of the niches plays an important role. For many species specific conditions must arise for the development of seedlings, which Harper (1977) describes as "a safe place for germination". The persistence of a particular species on a particular site depends to a large extent on the growth type of the plants. Species characterized by iterative growth, durable underground organs and vegetative reproduction have the greatest chances of not being eliminated and of attaining the status of dominant (cf. Part 3).

Diversity of meadow niches and rate of succession

Owing to the heterogeneity of the habitats, diverse niches occur in every environment, colonized by species with different life strategies (Drury and Nisbet 1973, Connell and Slatyer 1977, Grime 1979, Giller 1984).

The mosaic pattern of meadows mown every year is but little pronounced. When mowing is discontinued, however the mosaic pattern of meadow vegetation becomes more conspicuous from year to year. Its particular patches are dominated by different species. An analysis of the vegetation structure for the mosaic of dominants has demonstrated that a meadow seemingly homogeneous when mown every year is in fact heterogeneous. Cessation of mowing only reveals the heterogeneity of the habitats. Two types of niches have been distinguished: those that are easy to access, and those which are difficult to access by the colonizers.

Niches occupied by plants with a delicate structure and poorly developed mechanisms for survival at a site occupied by them are most rapidly taken possession of by one dominant species. Species like *Filipendula ulmaria* and *Carex acutiformis* most often take 5 years to colonize meadow patches occupied by *Caltha palustris, Myosotis scorpioides, Cirsium palustre, Lychnis flos-cuculi* and *Ranunculus acris*. In those habitats the turnover including the colonizers of species is the quickest, and the new colonizers are forest species. Ten years after meadow-abandonment, these sites are occupied by willows surrounded by perennial forest herbs, although there still is a large proportion of the earlier colonizers which have eliminated the floristic composition characteristic of meadows (Figure 2.2).

Niches occupied by grass and sedges are much less accessible to the colonizers. Their characteristic feature is a compact vegetation cover which makes difficult both the development of individuals from seeds and penetration of plants by stolons. the niches frequently remain with a little changed species compositon, even when willow-brushwood develops around them. They form "meadow islands" among the mosaic of macroforb communities and brushwood, greatly retarding, locally, the advance of succession (Figure 5.5).

Various types of mid-meadow mires, play a similar role. For instance, mires situated among sedge swamps in meadows mown yearly are filled with water and deprived of vegetation. Colonization of the mires becomes possible only after five years, owing to the accumulated necromass of the above ground parts of the adjacent sedge patches. Most frequently they are colonized by *Iris pseudacorus*. Seeds of the latter species must stay in water for several months and can germinate on the necromass during its decomposition. Therefore, *Iris pseudacorus* appears earliest in the mires, but the process of their filling continues for many years because the growth of the individuals is slow. After ten years, individuals of other species such as *Phragmites* and *Typha* may be found among them. As late as 15 years after the cessation of mowing the first individuals of *Alnus glutinosa* appear on these sites, that is when willow-alder brushwood has already developed. (Figures 5.6; 2.1).

Small mires distributed among the meadow vegetation are, as a rule, overgrown by way of vegetative propagation of *Carex acutiformis*, if this species is present (Figure 4.31). Sometimes mires of this type are overgrown with *Lythrum salicaria* because this species exhibits a high fecundity and high germination capacity under various conditions. Mires colonized by these species may locally

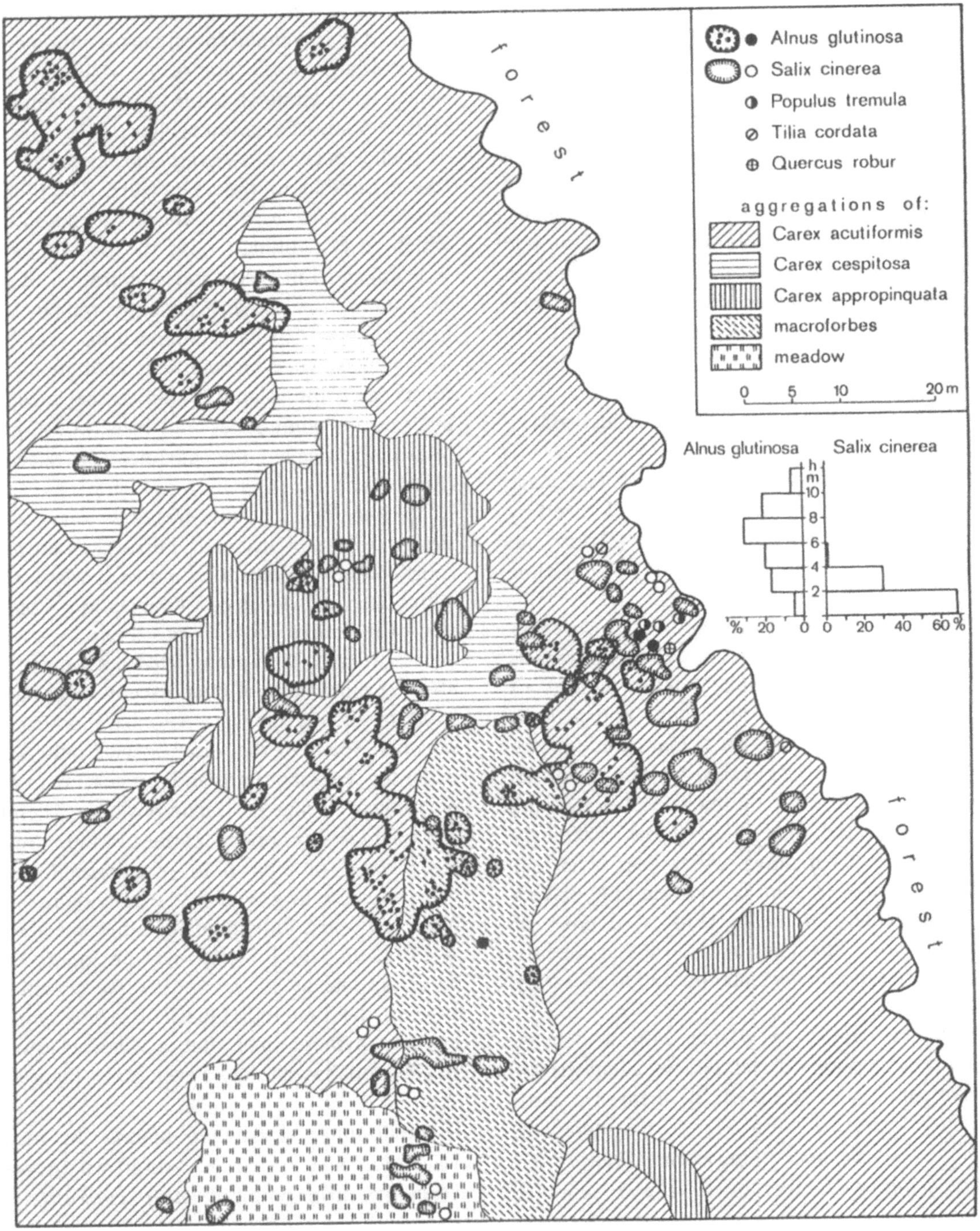

Fig. 5.6. Biogroups of trees in the patches dominated by macroforbes and by sedges on boundaries of forest 15 years after abandonment of the meadow.

impede species turnover, and thus, succession. They frequently still survive when willow brushwood surrounds them.

It should be stressed that the diversity of meadow niches determines the various possibilities of forest return to Reski Range. A retardation or acceleration of succession is, rather, of a local nature (Figure 2.1). On a macroscale the process of species turnover after cessation of mowing continues in the whole area, but at various rates.

The transition zone (ecotone) and its dynamic aspect

Spatial vegetation patterns usually present a mosaic of many types of phytocoenoses. Frequently the particular types of phytocoenoses are represented in various parts of a complex. The existence of diverse phytocoenoses class to each other is noticeable particularly in smaller areas. This phenomenon is illustrated by maps of the Reski Range vegetation, prepared during the 15th year of the research (1989), but showing the diversity of the vegetation as the effect of discontinued mowing of the meadows in the period 1960—1989. The diversity of the vegetation is, therefore, a reflection both of the heterogeneous habitat of the meadows and the vegetation dynamics, since phytocoenoses, representing various phases of succession, are neighbours here. The vegetation of the Reski Range is a mosaic of phytocoenoses expressing various dynamic tendencies. These phytocoenoses occupy biochores of different size, and their borders are sharply outlined or weakly delineated (Figure 2.1).

Among the investigators dealing with the spatial organization of vegetation, some support the continuum and others the discontinuum. In the present book, I do not intend to solve the controversy as to the kind of contacts — sharp borders or gradual transition of two phytocoenoses. This problem has been presented and critically discussed in the review paper of Goodall (1963). The present considerations based on the assumption that between two phytocoenoses *A* and *B* there is a certain transitional area in which neither of them is fully realized, but elements of both are present (Figure 5.7).

This phenomenon is known in the relevant literature under the name of ecotone, that is a transition zone between two or more biocoenoses (Odum 1963). According to the theory of ecological systems, one should rather say, between two ecosystems. For biotic elements, diversity is always accompanied by differentiation of abiotic elements. In investigations into the diversity of vegetation in the contact zone, a correlation has been repeatedly found between the changing properties of the soil, topography, microclimate and the changes in the spatial structure of the vegetation (Jakucs 1972). These changes have always been considered in relation to the environmental gradient (Traczyk 1960, Matuszkiewicz 1972). Phyto-sociological studies often call attention to the heterogeneity of the ecotone vegetation, manifested by a mosaic pattern and aggregateness of its spatial structure (Traczyk 1960, Polakowska 1966). A "motility" of boundaries has also been noted in the transition zone — a phenomenon resulting from the different seasonal dynamics of neighbouring phytocoenoses (Faliński 1962).

Ecotone types. With regard to the kind of contact between phytocoenoses in the mosaic vegetation complex of the Reski Range, various types of transition zone have been distinguished. On the basis of other investigations it was expected that the reaction of the population to different ecotone

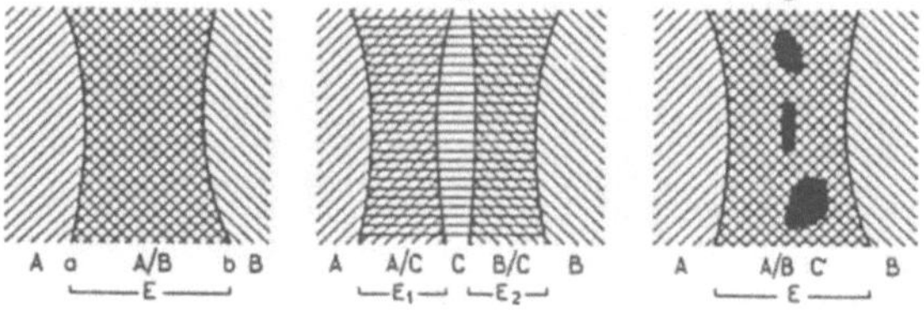

Fig. 5.7. Contact between two phytocoenoses of a different type (A, B) and formation of transition zone (ecotone — E). a, b — boundaries of phytocoenoses. The possibility of the development of an independent community (C) between A and B indicates that we are dealing with independent ecotones (2). Under specific conditions, in the transition zone, proper development of "islet" ecotone communities is possible (3C) (Faliński 1986).

types would vary (Jakucs 1972, Mahall and Park 1976, Falińska 1979, 1986c). The following types of ecotones have been distinguished:

(1) Ecotone between two phytocoenoses of similar species and spatial structure. This transition zone is of little pronounced diversity because the phytocoenoses gradually pass into one another. This type of contact is found between: *Cirsietum rivularis* and *Lysimachio vulgaris-Filipenduletum*, and *Cirsietum rivularis* and *Caricetum acutiformis.*

(2) Ecotone between phytocoenoses differing in species and spatial structure. The border is distinct as for instance between willow brushwood and meadow and reed-sedge swamps phytocoenoses as well as between forest phytocoenoses and the neighbouring ones (Photos 5.2; 5.3).

As regards to the length of the boundaries between the contacting phytocoenoses, macroecotones (forest boundary) and microecotones between phytocoenoses occupying small biochores (500—2000 m^2) are distinguished, and lengths of boundaries ranging from several score to several hundred metres.

Ecotone at the boundary between sedge swamp and forest (five years after the cessation of mowing)

Horizontal structure. It is at the boundary of the forest that closed *Carex acutiformis* patches end, while single plants or their small groups penetrate into the forest to a distance of 15 m from its border. A similar spatial distribution was noted at the forest boundary in *Scirpus sylvaticus* and *Phragmites australis* populations. The dispersed distribution of *Filipendula ulmaria* in the meadow changes significantly at the forest boundary (brushwood of *Salix cinerea* and *Alnus glutinosa*). In the forest it occurs sporadically. (Figure 5.8).

A reverse reaction was observed in *Carex cespitosa* populations. The number and size of the tussocks are distinctly higher in the meadow, and they decrease with diminishing distance from the forest. *Carex appropinquata* on the contrary, forms the largest and most numerous tussocks close to the forest edge. *Carex elongata* occurs, above all, at the forest boundary and in the forest, and only sporadically does it migrate to the meadow (Table 5.3).

In the nearly 15 m transition zone between the forest and sedge-swamp (Plots 11—13), a mosaic of the components of both neighbour communities

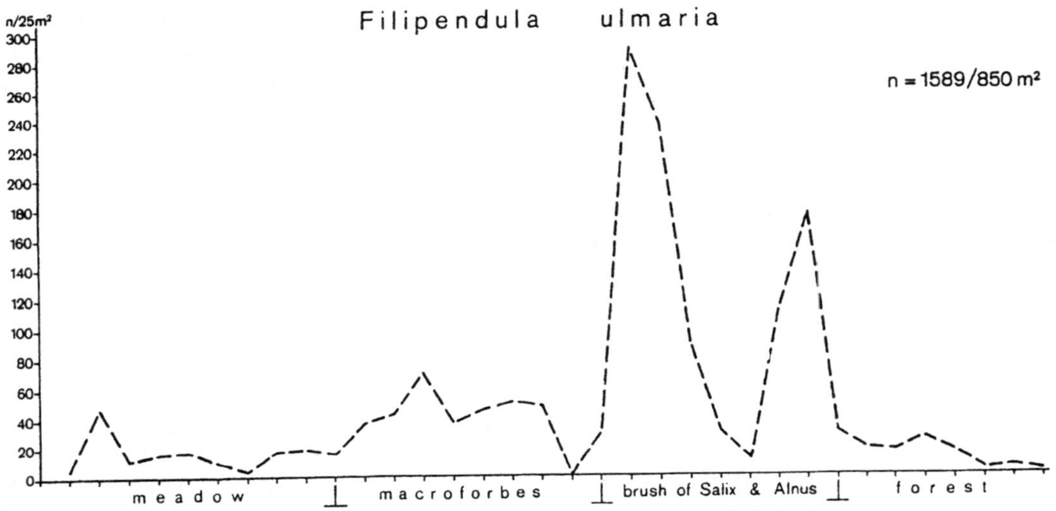

Fig. 5.8. Changes in the density of populations, observed on 34 plots (5 m × 5 m) of the transect (170 m long) through successive phytocoenoses.

Table 5.3. Changes in species composition on boundaries between two phytocoenoses 10 years after abandonment of the meadow, observed on 17 plots (5 m × 5 m) of the transect 85 m long

Species				macroforbes					ecotone		forest						
	1	2	3	4	5	6	7	8	9	10	11	12	13	14	15	16	17
Carex vesicaria	+	+															
Equisetum palustre	+	+	+	+	+												
Carex nigra	+		+	+	+	+											
Stellaria graminea	+	+			+												
Carex appropinquata	+	+		+				+	+	+							
Carex cespitosa	+	+	+	+	+	+	+	+	+		+						
Polygonum bistorta	+					+	+				+	+					
Carex acutiformis	+	+	+	+	+	+	+	+	+	+	+	+	+				
Scirpus sylvaticus	+	+	+	+	+	+	+	+	+	+	+	+	+				
Equisetum limosum	+	+	+	+	+	+	+	+	+	+	+	+	+				
Scutellaria galericulata	+	+	+	+	+	+	+	+	+	+	+		+				
Viola palustris	+		+	+		+				+	+	+	+				
Cirsium rivulare	+			+	+		+	+	+	+	+	+	+				
Lythrum salicaria	+	+	+	+	+	+	+	+	+	+	+	+	+		+	+	+
Lysimachia vulgaris	+	+	+	+	+	+	+	+	+			+	+	+		+	+
Filipendula ulmaria	+	+	+	+	+	+	+	+	+	+	+	+	+				+
Urtica dioica	+	+								+	+	+	+	+	+	+	+
Caltha palustris	+	+	+	+				+		+	+	+	+	+	+	+	+
Galium palustre	+	+								+	+	+	+	+	+	+	+
Stachys palustris		+								+	+	+	+	+	+	+	+
Poa trivialis		+			+					+	+	+	+	+	+	+	+
Lychnis flos-cuculi			+							+	+		+				+
Phragmites communis					+	+	+	+	+	+	+	+	+	+	+	+	+
Carex gracilis						+			+	+	+						
Carex elongata								+	+	+	+	+		+			
Thelypteris palustris									+	+	+	+		+			+
Solanum dulcamara									+	+	+	+	+	+	+	+	+
Impatiens noli-tangere									+	+	+	+	+	+	+	+	+
Cardamine amara										+							
Crepis paludosa										+	+						
Geum rivale										+	+						
Chrysosplenium alternifolium										+			+				
Sium latifolium										+	+	+	+				
Geranium robertianum										+				+	+	+	+
Alnus glutinosa										+	+	+	+		+		+
Myosotis scorpioides										+	+			+	+	+	+
Stellaria palustris											+						
Cirsium oleraceum											+				+	+	+
Lycopus europaeus											+	+	+	+		+	+
Ribes rubrum											+	+	+	+	+		+
Rubus idaeus											+		+	+	+	+	+
Ribes nigrum											+	+	+	+	+	+	+
Lysimachia thyrsiflora												+	+				
Humulus lupulus												+	+				
Stellaria nemorum												+	+	+	+	+	+
Oxalis acetosella												+	+	+	+	+	+
Galium aparine													+	+			
Maianthemum bifolium													+	+			
Calamagrostis canescens													+	+	+		
Glyceria fluitans													+	+		+	
Prunus padus													+	+			+
Sorbus aucuparia													+	+	+	+	+
Ranunculus repens													+	+	+	+	+

Table 5.3. Continued

Species			macroforbes						ecotone		forest						
	1	2	3	4	5	6	7	8	9	10	11	12	13	14	15	16	17
Angelica sylvestris													+	+	+	+	+
Dryopteris carthusiana													+	+	+	+	+
Milium effusum													+	+	+	+	+
Picea abies														+			
Corylus avellana														+			
Frangula alnus														+			
Stellaria media														+			
Quercus robur														+	+		
Deschampsia cespitosa														+		+	
Chaerophyllum hirsutum														+	+	+	
Athyrium filix-femina															+	+	+
Calla palustris																+	+
Veronica beccabunga																+	+
Epilobium palustre																	+
Fraxinus excelsior																	+
Circaea alpina																	+
Salix caprea																	+

and a more conspicuous aggregated plant distribution, can be seen. Tussocks of sedge and clumps of *Alnus glutinosa, Ribes nigrum* and *Oxalis acetosella* occur here. Moreover, *Caltha palustris, Urtica dioica, Thelypteris palustris, Galium palustre* and *Glyceria fluitans* grow here in larger or smaller groups. *Equisetum limosum* and *Lycopus europaeus* also grow in assemblages. Besides, there is a marked increase in density of *Sium latifolium* and *Impatiens noli-tangere* at the edge of the forest (Table 5.3). Species characterized in the meadow by an aggregated distribution, such as *Lysimachia vulgaris, Lythrum salicaria, Cirsium rivulare* and *Polygonum bistorta*, exhibit in the forest a random distribution.

Both in the meadow and in the forest, the presence of places poorly covered with vegetation is noteworthy: these are either depressions filled with water, or holes left by wind-fallen spruce trees.

Vertical structure. Analyses of the vertical structure of the vegetation at the boundary of forest and sedge swamp showed that in the contact zone significant changes occur in the number of layers and height of plants. The sedge swamp forms two distinct layers: the upper one (100—130 cm) is formed of *Carex acutiformis, Lysimachia vulgaris, Lythrum salicaria* and *Filipendula ulmaria*, the lower (50—80 cm) of *Cirsium rivulare, Urtica dioica, Equisetum palustre, Scirpus sylvaticus* (plots 1—6). At a distance of 20 m from the forest edge *Phragmites australis* reaches dominance, forming the highest vegetation layer (150—250 cm). At the forest edge four layers can be distinguished: (1) (120—200 cm): *Phragmites australis*; (2) (80—100cm): *Urtica dioica, Carex acutiformis, Poa trivialis, Filipendula ulmaria*; (3) (50—80 cm): *Impatiens noli-tangere, Sium latifolium, Angelica sylvestris*; (4) (10—30 cm): *Solanum dulcamara, Viola palustris, Lycopus europaeus, Geum rivale.*

In the forest (Plots 14—17; Table 5.3) the herb layer consists of three distinct layers. In the top one *Calamagrostis canescens, Milium effusum, Poa trivialis, Urtica dioica*, reaching a height of 100—150 cm, prevail. *Solanum dulcamara* and *Lycopus europaeus* form the second intermediate layer, 50—80 cm high, while on the forest edge they only reach 30—40 cm. They are accompanied, beside *Chaerophyllum hirsutum*, by *Cirsium oleraceum* and *Athyrium filix-femina*. Individuals of *Impatiens noli-tangere* which reached, at the forest-meadow boundary, 80—120 cm, attain in the forest barely 40—60 cm, and form, beside *Oxalis acetosella, Stellaria nemorum, Galium palustre, Myosotis scorpioides* and *Maianthemum bifolium*, the lowest herb layer.

The ecotone at the boundary between sedge swamp and forest (ten years after meadow abandonment)

In the case of a longer (10 years) disuse of the meadows, the horizontal structure of the vegetation at the forest boundary, is as follows: in the 40 m band adjacent to the forest a typical macroforb community develops. *Filipendula ulmaria* and *Urtica dioica* (abundant here) form large aggregations and at the very border of the forest-extensive patches. Among the macroforbes, in some places, willows (*Salix cinerea*) appear and clumps of *Frangula alnus* and *Prunus padus* appear even at a distance of 40 m from the forest. In the transition zone, *Lythrum salicaria* and *Lysimachia vulgaris* (Photo 5.2) as well as *Calamagrostis canescens*, are abundant. Immediately at the forest edge *Cardamine amara* forms patches.

A large patch of *Filipendula ulmaria* bordering on the forest is replaced in the forest by a mosaic of numerous species among which prevail: *Impatiens noli-tangere, Geum rivale, Aegopodium podagraria* and *Paris quadrifolia* There also is a high proportion of *Ribes nigrum, Rubus idaeus, Urtica dioica, Cardamine amara, Galium palustre, Lycopus europaeus* and *Oxalis acetosella. Filipendula ulmaria, Lysimachia vulgaris, Lythrum salicaria* and *Cirsium oleraceum* grow here sporadically.

Vertical structure. If macroforb communities develop at the forest boundary (*Lysimachio vulgaris-Filipenduletum*), the highest layer in the

Photo 5.2. Development of macroforbes community on the forest edge 10 years after abandonment of the meadow. Photo: J. B. Faliński

close vicinity of the forest is formed by *F. ulmaria* (120—140 cm) and the somewhat lower one by *Lysimachia vulgaris* and *Lythrum salicaria* (80—100 cm), and sometimes *Impatiens noli-tangere.*

The diverse height of individuals of the same species should be noted, which occur in the neighbouring phytocoenoses and in their ecotones. Some species attain the greatest height at the boundary of the forest as for instance *Filipendula ulmaria, Lysimachia vulgaris,* and *Impatiens noli-tangere* (Figure 5.9). On the other hand, *Carex acutiformis* becomes smaller as it approaches the forest, the height of the plants seldom exceeds 40 cm at the very border, whereas at a distance of 50 m it reaches a height of 80—100 cm. *Phragmites australis,* on the contrary, attains its maximal height (200—250 cm) at the forest boundary. In the forest and in the meadow, individuals of this species are by one half smaller.

The ecotone between sedge swamp and forest (fifteen years after the cessation of mowing)

The spatial structure of the vegetation in this case exhibits a mosaic of various phytocoenoses. Some of them occupy small biochores, repeatedly occurring at various places in the wide band of the contact zone between forest and sedge swamp (Photo 5.3). Most of the space is overgrown by well developed macroforb phytocoenoses and willow brushwood, and sedge swamps and the smallest parts are occupied by fragments of meadow. Between these phytocoenoses a specific mosaic is formed by populations of meadow and forest species. In the places where tree biogroups were present when the meadows were mown every year, 15 years after the ceasing of the utilization of the meadows, willow-alder brushwood appeared, that is, the initial phase of forest (Photo 5.4). The

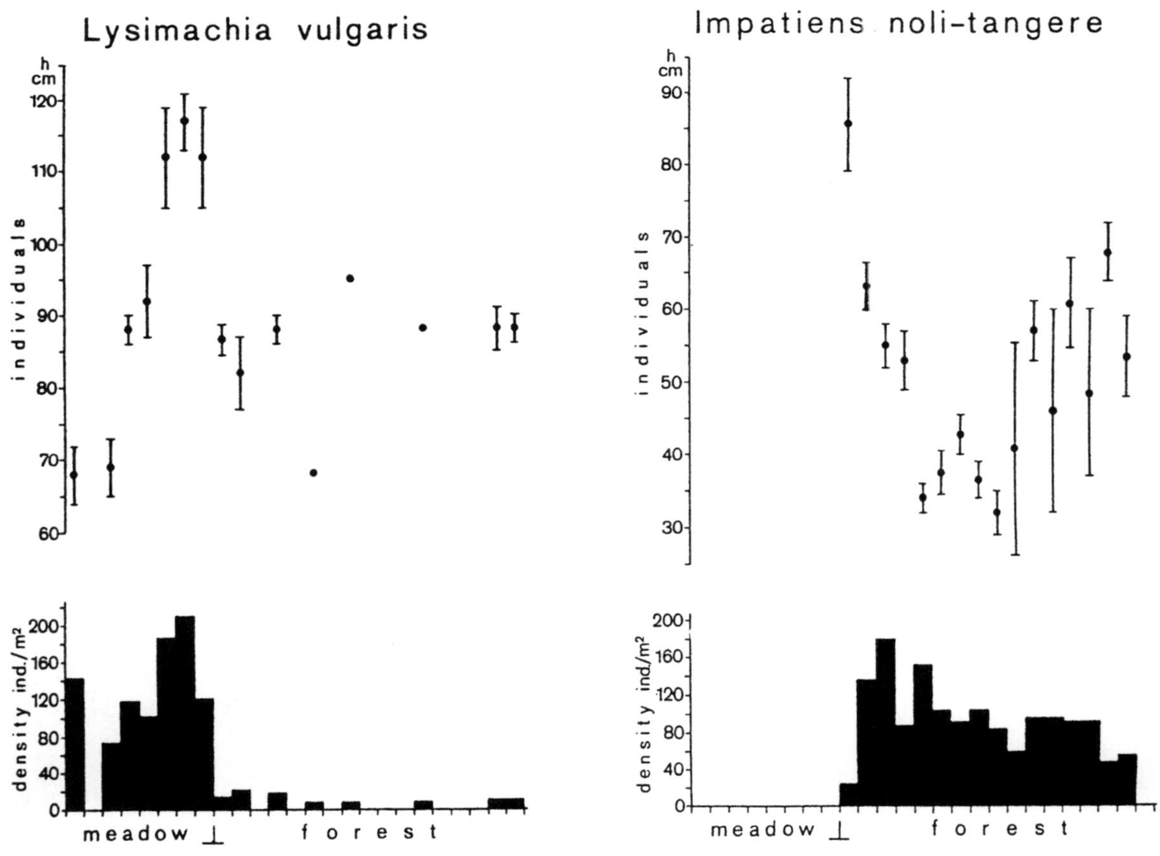

Fig. 5.9. Changes in the density and in the shoot size of two populations on boundaries of forest and meadow.

Photo 5.3. The sedge community on the forest edge 15 years after abandonment of the meadow.
Photo: J. B. Faliński

dynamic character of the transition zone between forest and the non-forest communities finds its reflection in the spatial structure of the phytocoenoses and populations as illustrated by the figures (Figures 2.1; 5.6).

Width of ecotone

The width of the transition zone at the boundary of the forest ranges from 10 to 150 m, depending on the time elapsed since the cessation of mowing. When this time was longer, the transition zone was broader and the boundary less distinct. If the meadow was abandoned at a relatively recent time (up to 5 years ago), the boundaries between the communities will be sharp (Table 5.3).

The contact zone between forest and meadow may present a mosaic of populations of certain forest and meadow species, or a mosaic of phytocoenoses representing various phases of succession.

The influence of both the neighbour communities on the transition zone may be similar or the influence of one of them may prevail. There are indications that in the first phases of succession the ecotone is under a strong influence of the meadow species, whereas, in the later phases, of forest species.

In the case of certain species, what is called the edge contact zone effect occurs, for instance in *Phragmites australis, Impatiens noli-tangere, Filipendula ulmaria, Lysimachia vulgaris* and *Lythrum salicaria*. Populations of these species

Photo 5.4. Densely growing woody species on the forest edge 15 years after abandonment of the meadow. Photo: J.B. Faliński

exhibit an increased density at the forest boundary, accompanied by an increased individual size and fecundity (Figures 5.9; 4.16; 4.17).

The succession rate at the forest boundary

According to Jakucs (1972), the forest edges form a separate edge community because at the boundary of forest and meadow the bioclimatic, soil and phytocoenotic conditions differ greatly from those on both sides of this zone, that is in the forest and on the meadow. This view found confirmation in investigations of long duration on the structure of the phytocoenoses and populations on abandoned meadows neighbouring the forest complex in the Białowieża National Park (Falińska 1979, 1986c).

Within several years after cessation of meadow

mowing the density of tall herbs increased and macroforbes communities rose in the ecotone almost simultaneously with the brushwood communities. The mosaic of macroforbes and brushwood phytocoenoses forms a characteristic "wrapping" at the forest boundary (Figure 2.1, Photos 5.2; 5.4).

The ecotone between the forest and the abandoned meadow is a dynamic system. The population and phytocoenotic processes occurring in the ecotone when mowing has ceased cause a gradual shift of the forest boundary. In the course of the 15 years when the meadow was not utilized, the forest boundary shifted in various places by 5 to 30 m. On the sites where the meadow had been abandoned 30 years ago the forest boundary shifted by 50—70 m (Figure 2.1). The advanced development of the willow-alder brushwood in various parts of the Reski Range seems to indicate that in the near future (5—10 years) this boundary will shift by at least 5—10 m. It is expected that the process of withdrawal of meadow phytocoenoses and their substitution by brushwood ones will take place in 25—30 years time. Over the whole area of the Reski Range in recent years, trees or biogroups of trees have appeared which, in future years, will initiate the development of willow or alder brushwood (Figures 2.1; 5.6).

In earlier considerations on the properties of herbaceous species populations in meadow and forest communities, it was suggested that at their boundary a distinct ecotone population arises, and the high density of the latter was attributed to the edge effect (Falińska 1979, 1986c), thus a phenomenon frequently noted in the contact zone of two ecosystems (Kershaw 1973, Odum 1969). Only through follow-up of the fate of labelled individuals of several species, was it possible to ascertain that it is the effect of the population attaining the growth phase. In the ecotone the peak of population size is much higher than in populations attaining this phase under different conditions, for instance on an open meadow (Figure 5.8). As a rule, in studies of these species the high density of the population was referred to as the fertility of the habitat and competition of other species, rather than the population developmental phase. The short duration of the macroforbes phytocoenoses (10—12 years) is due to the short phase of maximal population growth of many herbs, after which a drastic fall occurs, mostly after 8—10 years. Sometimes single species aggregations of *F. ulmaria* survived for a period of 10 years (Wolf 1979). Such a situation is possible when the macroforbes phytocoenosis develops in poorer habitats and not in close vicinity of the forest.

To better illustrate the observed differentiation of the vegetation in the transition zone a comparison is given below of the behaviour of populations and phytocoenoses in the forest-meadow ecotone.

Level of phytocoenosis. The mosaic pattern of vegetation is more pronounced. The influence of neighbouring phytocoenoses on the transition zone is not equal and variable, e.g. in the first period after cessation of mowing, meadow species play a more prominent role in the vegetation structure, and after a lapse of 10—15 years, forest species. A zone of strong influence of the phytocoenoses on both sides can be distinguished in the ecotone with many elements of both neighbouring phytocoenoses, that is a strict "transition zone".

Level of population. The plants in the ecotone exhibit tendencies to aggregation. The structure of the populations of species composing the neighbouring phytocoenoses differs from both or at least one of them. The individuals in the ecotone show a certain similarity to those of the neighbouring population (phytocoenosis A) and those of the neighbouring contralateral population (phytocoenosis B). In the strict transition zone significant differences are seen as compared with populations A and B, described as "Break-down of traits".

Type of reaction of plant populations

The response of plant populations to varying ecotone conditions is manifested in the first place by: (1) a change in population density, as compared with that of the neighbour phytocoenoses; (2) spatial structure of the population, especially by increasing aggregateness of the plants; (3) reproduction, that is, the fraction of individuals participating in reproduction, and by the number of vegetative and generative propagules produced.

The following types of plant reaction were seen. When the same species are components of several, or at lest two phytocoenoses (A and B), usually the following regularity is observed: the more distant are the individuals of the given species from A and B, the more they differ from the individuals of A and B populations. Sometimes a gradual change of the features of phytocoenosis A to those of B, from B to E, from E to D was noted, for instance changes in the density of individuals of the given species or fractions of fruiting individuals etc. (Figure 5.10).

In ecotones between communities that differ significantly frequently, a reaction of plants was observed described as the "break-down" of traits, manifested by a drastic decrease of density, fecundity and shapeliness (Figure 5.10), whereas in ecotones which connect certain elements of the neighbour communities the reaction is reversed, and the parameters of these traits become more pronounced. This is true especially of dynamic ecotones, that is those situated at the border of the forest.

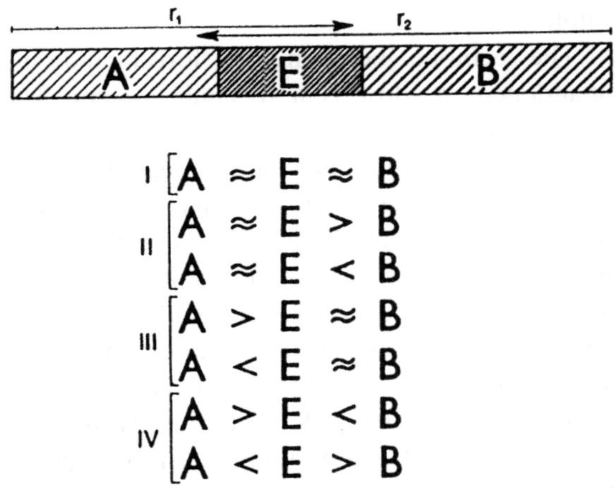

Fig. 5.10. Scheme of the reaction of a plant population to different conditions in forest ecosystems (A, B) and ecotones (E). (I) — no differentiation; (II), (III) type of undirectional reaction; (IV) type of bidirectional reaction to population E. R_1, r_2 — direction of changes in populations on the transect (Falińska 1986c).

CHAPTER 20

RETURN OF FOREST IN MEADOWS – CAUSES, MECHANISMS, PROCESSES

A hierarchical model of the causes of succession

Succession of abandoned meadows is a process of reconstruction of forest communities which starts as soon as human activity ceases. The return of forest is preceded by diverse floristic compositions of different durability, the earlier compositions usually influencing the subsequent specific compositions of the community, as documented in other studies (Shugart 1984). Interpretation of succession as a process of directional vegetation transformations, leading to the development of a climax community has been repeatedly criticized (Drury and Nisbet 1973, May 1973, Horn 1974).

In general, the differences in the opinions on succession are the result of adopting a holistic or reductionistic philosophy. It is also important in the interpretation of succession whether this process is analysed on the basis of micro or macroscale investigations and in studies of several or of many years duration.

Microscale, short-term researchers lead, as a rule, to oversimplifications, and do not take into account the whole variability and diversity of the succession processes. It results from numerous other studies in which there exists no universal model of succession (May 1973).

Short-lasting research also frequently leads to the conclusion that reconstruction of the original composition and its stabilization are, at the very least, probably (Horn 1974). The return of forest to abandoned meadows has, however, been described many times. More frequently the process has been observed up to the initial forest phase. Then its comparison with the dynamics and structure of stabilized forest communities made possible the forecasting of the further course of succession (cf. van der Maarel 1980, Shugart 1984, Faliński 1980, 1986). The diagram (Figure 5.11) presents the methodical requirements for research into vegetation dynamics in order to obtain correct conclusions. In the present study, at least two variants have been applied (II and VII), the fate of individuals of given species was

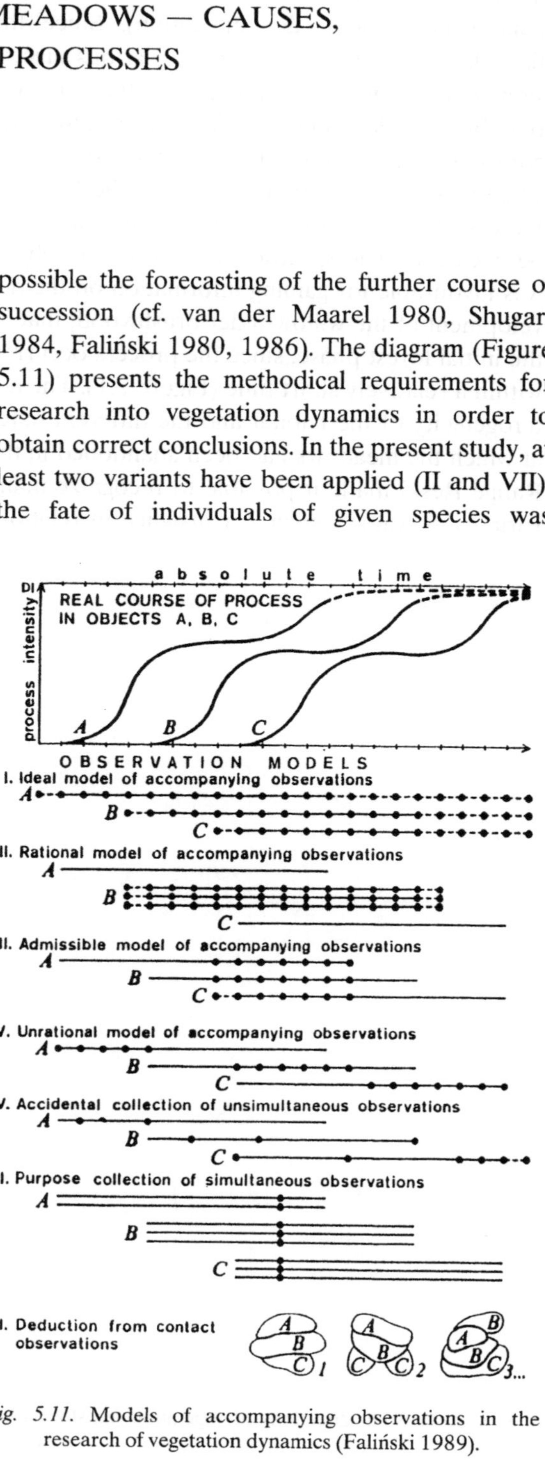

Fig. 5.11. Models of accompanying observations in the research of vegetation dynamics (Faliński 1989).

followed for 15 or 16 years in permanent plots set up immediately after the cessation of mowing and in meadows abandoned at different known times (Białowieża Station documentation). Since the floristic compositions representing various succession phases bordered on each other, it was possible to understand more fully the diversity and course of succession. The object of study was situated within the protected zone of the National Park, so any disturbances by human activity could be excluded. The specific situation of the object was favourable for gaining information on the development of the willow-alder brushwood, that is, the initial forest phase, since the process occurred within a relatively short time (Figure 2.1). The heterogenicity of the habitat and the different dates at which the meadows had been abandoned in the Range Reski made it possible to recognize many variants of forest return, for instance in relationship to the type of phytocoenosis, and life strategy of the dominant species in different conditions (Figure 5.12).

It was found in the present studies that in an even relatively small area (15 ha), succession may run differently and the following could be observed: (1) a characteristic floristic relay, that is, almost a model succession of meadow, macroforb and brushwood-phytocoenoses (Figures 2.2; 3.22); (2) forest return was preceded by a long-lasting dominance of one species (monopolist), e.g. *Carex acutiformis*, which within a few years occupied several hectares of a floristically rich meadow (Figure 5.6). In such cases trees appeared singly as late as after a dozen years or so, the return of fo: est was then delayed by about 15—20 years; (3) in some meadow patches forest species appeared immediately after the discontinuance of mowing (10 years later they formed biogroups

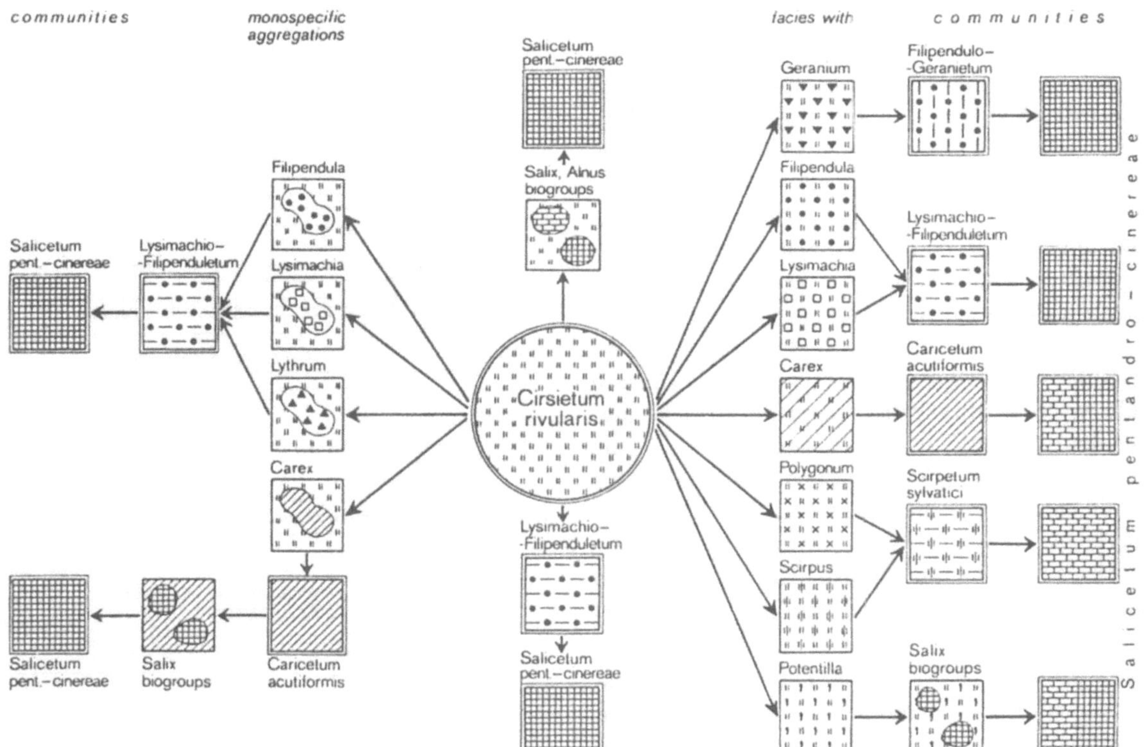

Fig. 5.12. Possible directions of vegetation transformations in the course of secondary succession in abandoned rich, moist meadows (Falińska 1989a).

composed of several tree species around which perennial forest herbs appeared). After 15 years in the parts of the meadow, willow-alder brushwood arose, that is the initial phase of forestation (Figure 2.1).

The authors of the hierarchic model of succession causes (driving forces) (Pickett *et al.* 1987) list the following as basic causes: (1) the availability of free space and of resources; (2) the different rate of species appearance (e.g. distribution accessibility of resources, seed bank); (3) different cycles and life strategies of species (e.g. life-span, rate of settlement, type of reproduction).

Succession in the abandoned meadows was analysed according to the model suggested by the above-mentioned authors.

1. What causes succession?

The immediate cause of the initiation of succession in meadows is the cessation of mowing. The meadow vegetation forms an almost 100 percent cover, which leaves no space for settling. Cessation of mowing is a factor disturbing the structure of the meadow communities, this becoming noticeable in the form of gaps only after several years.

2. What are the conditions and interactions participating in succession?

A loosening of the structure of meadow communities and the appearance of niches, occurs owing to the rapid rate of development of plants with iterative growth, which is fully realized only when mowing has ceased. Then individuals of certain species expand and reach maximal heights. A rapid or gradual change takes place in the quantitative and spatial relationships between meadow species, causing disturbances in the "perennial canopy".

There appear gaps in the meadow community which are occupied by species different from those which grew here when the meadows were utilized. These species, such as *Salix cinerea* and *Filipendula ulmaria*, cause disturbances.

They also play the role of "driving forces" of the internal disturbances in the vegetation structure, which lead to the appearance of free spaces, an essential condition for the initiation of succession. These gaps are at first occupied by species with wind-borne seeds such as *Cirsium palustre* and *Cirsium rivulare*, further by species having two ways of propagation, which soon colonize the free space, and then the neighbouring meadow patches, eliminating the neighbours. Since these species have durable rhizomes, their dominance may last over many years. Only the senescence of the underground organs may contribute to the next species turnover. New gaps arise which are taken over by forest species (Figures 5.13; 3.22).

3. What processes and factors are specific to the given biotope and determine the type of interaction between plants?

The meadows, in which forest return was followed, occupy fertile and moist habitats. Moreover, they lie in close vicinity of forest and include species with various life strategies, and species in common with forest communities. Seedlings of forest species are present in the meadow vegetation also in meadows mown every year, this considerably accelerating the replacement of meadow species by forest ones. Another factor

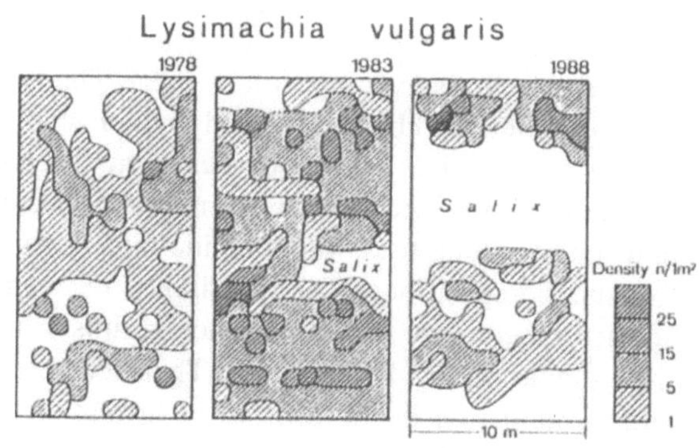

Fig. 5.13. Spatial dynamics of population during succession. Changes of coverage in a 200 m² plot 5, 10 and 15 years after abandonment of the meadow. Development of gaps within the aggregation of *Lysimachia vulgaris* permits colonization by willow.

responsible for the course of succession is the presence of species with iterative growth and two ways of propagation, characaterized by a varying distribution of assimilates into reproduction and underground organs of the individuals. In the first phases of succession individuals of these species allocate more biomass to the expansion of the above-ground parts, especially for flowering shoots (Figure 4.15). The number of shoots increases at that time in geometrical progression (Figure 3.2). In later phases of succession individuals of these species expand, above all, their underground organs, this allowing them to not only remain and keep the gained site, but to settle in new niches by way of expansion of the rhizomes and their migration. Twelve among the 142 species present in the meadow, exhibit high colonization abilities, owing to which they promote the transformation of the meadow communities towards forest ones. The presence of species of various growth types, reproduction and life history, determines their different role in the process of succession (Tables 5.1; 5.2).

Diversity and stability in the succession process

Elucidation of the processes of succession is closely connected with the search for the causes of the diversity and stability of floristic compositions. The return of forest into abandoned meadows is a process of long duration, during which variable and diverse floristic compositions arise. The rate and range of these changes depend both on abiotic factors and the biological properties of the species. All these jointly decide how many plant species and for how long they will combine into specific floristic compositions. The possibility of many species meeting at the same place results from their similar ecological amplitude. Their remaining and participation in the process of formation of one or many phytocoenoses in the process to succession is conditioned by their life strategy, thus the whole group of individual traits on which the individual competitive force depends. Thus, the development of a concrete composition of species at a given site is determined by the interaction between the individuals of all the component species occupying a common area. Rabotnov (1985) suggests to describe the results of these interactions as "phytocoenotic selections", while Whittaker (1972) calls them "community-integrating factors". The effect of the population of one species on that of another species is frequently stronger than the effect of abiotic factors, as demonstrated in numerous experiments (e.g. Zarzycki 1968, Harper 1977, Turkington *et al.* 1979, Faliński 1986). It has been found that an increase of the population size of some species usually limits the development of individuals of other species, sometimes eliminating the majority of the phytocoenosis components (Figure 3.15). The directional changes of the vegetation (succession) end by the formation of a climax community which can be realized under the given conditions. This means that as the final result of this process, species with specific habitat conditions and requirements, capable of coexisting, will remain at the particular site. After the end of the process of selection of the species in a definite area the phytocoenosis attains a phase which is described as a relative stabilization or dynamic equilibrium (Figure 2.5). This does not mean at all that this system is stagnant. Stabilization is understood as a relatively stable species and spatial structure of the phytocoenosis, expressed by relatively constant quantitative relations between the populations. This structure, however, is of a dynamic nature, since phytocoenoses are subject to seasonal and long-term variations. These changes do not, in general, lead to a modification of the structure of the phytocoenosis, but ensure its stabilization and give it an individual character. Presented in Figure 5.14 are the diversity and nature of numerous processes determining the vegetation dynamics, one of which is succession. The individuality of a phytocoenosis is the effect not only of a definite species composition, but also — to a high degree — of the number and distribution of the individuals of all its components. Even phytocoenoses with a similar species composition preserve a distinct nature, owing to the variability of the traits of the individuals and properties of the populations. Individuals of the same species are charac-

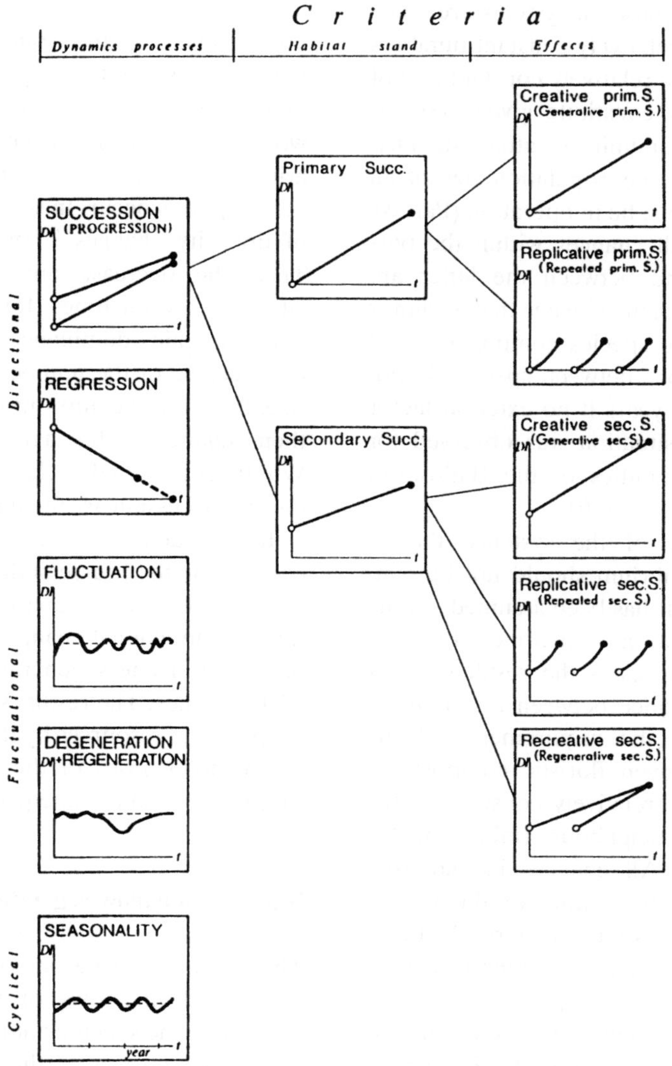

Fig. 5.14. Basic dynamic processes in vegetation. DI — dynamics index (Faliński 1986).

terized by a high plasticity of size, habitus, fecundity and course of life-history. Owing to this variability, populations of the same species exhibit wide differences in respect of many features, and this in turn leads to the development of a specific nature of these phytocoenoses (Figure 2.2).

An example of this are phytocoenoses arising in the course of succession, which have many common species in their composition (Figure 2.2). The differences in size and spatial structure of these populations result in their varying proportions and role in the structure of the successively forming phytocoenoses. It is obvious that the greater the differences in the species composition in the succession process, the more pronounced is their diversity.

The term "stability" in reference to climax communities is criticized by some authors. They raise doubts as to the reported facts that the communities, after the return to the original species composition, reach, after temporary disturbances, a stability of structure and processes

(Horn 1974). Stable ecological systems are characterized not only by relatively constant numbers of species, but also by a relatively constant size of the populations which constitute given systems. Therefore, for the community to attain stability, mechanisms regulating the population size of all the species included must be in operation (Part 4). On the other hand, all changes within the particular populations and between the latter are fluctuations. There is much evidence that in climax forest communities the species composition and the quantitative relations between species do not change over scores of years if no external factor affects them. Such a situation was observed for instance in the communities of the Białowieża Forest Complex (Faliński 1986).

Of course stability of the systems, that is relative dynamic equilibrium should not be confused with stagnation. It has been assumed, on the basis of the investigations performed that the stability of phytocoenoses is the result of their attaining a certain balance between resources of the environment and the requirements of the species forming the given floristic composition, which is expressed by relatively constant spatial relations between the neighbours (mild, not dislodging competition). All changes in stabilized phytocoenoses are of the nature of fluctuation. Phytocoenoses of various types may reach such a state in various times. A macroforb phytocoenosis, for instance, may persist in the process of succession in meadows for about ten years, whereas willow-alder brushwood for about 25—30 years.

In general, in the early phases of succession the rate of species turnover is quicker and the durability of the floristic compositions shorter. In later succession phases new species compositions arise in a long-lasting process, and the phytocoenoses formed last for several scores of years. The rate of these changes frequently depends on the age of the trees which compose it. The trend of the changes in the floristic diversity during succession has been found many a time (May and MacArthur 1972, Bazzaz 1975, Huston 1979).

It is worth noting that diversity changes in the process of succession have different, and not always directional, courses. It has been found, for instance, that during a succession the most diverse floristically, are the transition phases, that is, phytocoenoses of the stage of floristic decomposition of the meadows and initiation of the brushwood phytocoenoses. Then a floristic composition arises in which meadow species, macroforb and forest species occur. In some situations the course of diversity changes is of undulating character. Then the decrease and increase of diversity alternate several times, for instance when *Carex cespitosa* gains dominance in the meadows. With the increase of the number of tussocks and their size, the species composition on them and around them changes. This agrees with the views of Whittaker (1975) that changes in floristic diversity during succession are fluctuations.

Independent of the temporary fluctuations observed in the course of the vegetation changes in the succession period analysed, marked tendencies towards directional changes were found: a decline of species richness, general diversity and increase of dominance. The results suggest that the sharing of space and resources by the populations building the phytocoenoses of the consecutive phases of succession becomes unequal (Figures 2.2; 2.3).

Nature of meadow-vegetation changes

The fast and varying rate of forest return into the abandoned meadows of the Reski Range depended on the specific nature of this entity (Figure 5.15). The history of the meadows is relatively short because the Narewka River valley was deforested in relatively recent times, i.e., 200 years ago (Faliński 1966). In the past, the meadows located in this part of the Białowieża Glade were not intensively used, that is, they were only mown once per growing season, and not every year, because in some years the water level was above the ground. The proximity of the Białowieża National Park, on which the Reski Range borders, ensures a steady input of forest species seed, as indicated by the presence of seedlings and juvenile individuals of tree species in the meadows at the time of mowing. Moreover, in the period of their use there were trees randomly distributed in the

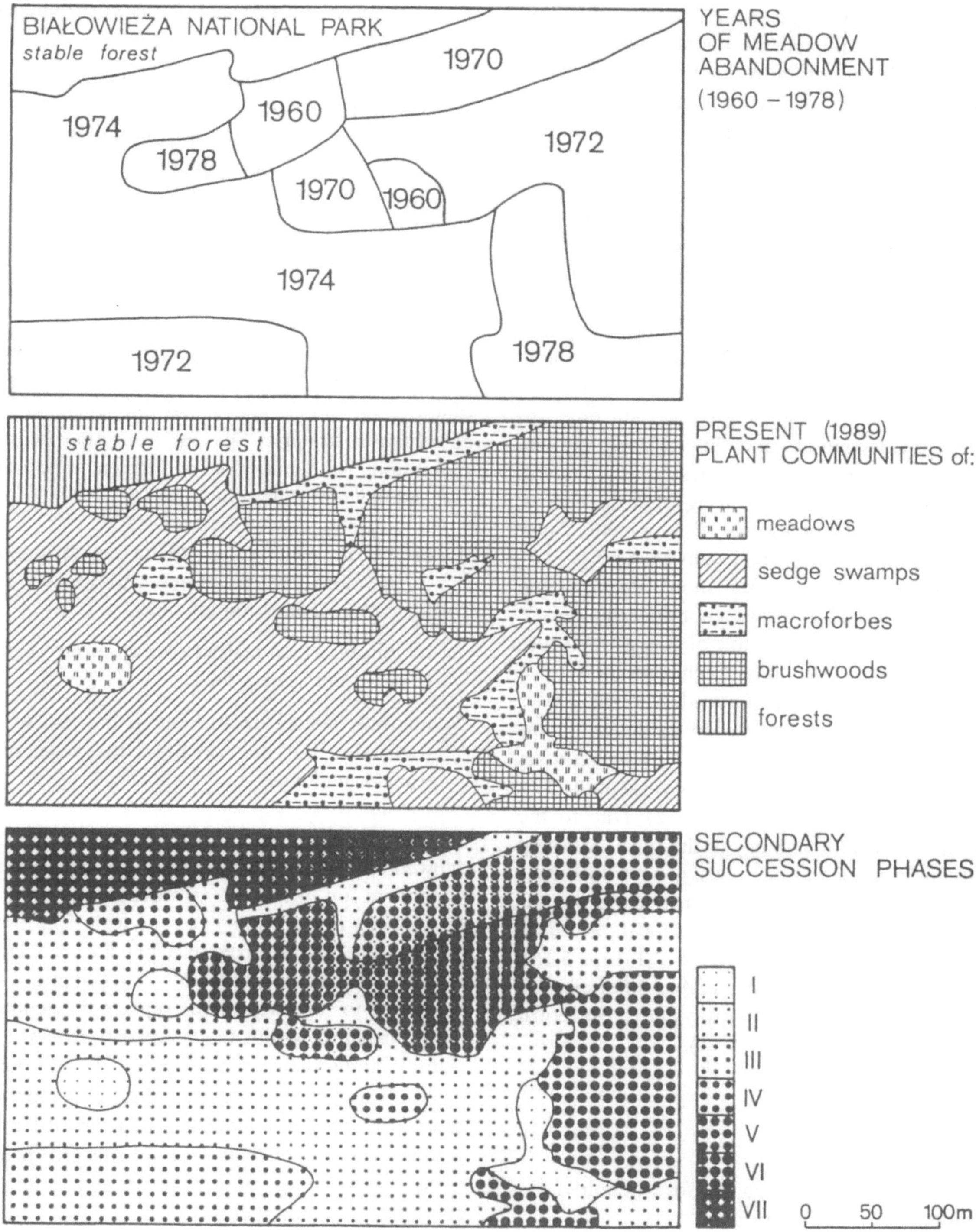

Fig. 5.15. Relationship between period of meadow abandonment and succession phases and type of communities.

meadows, with many woodland species clustered around them. Following the abandonment of the meadows they initiated the development of bio-groups made up of individuals of several tree species which became the forest outposts. The factor that has significantly speeded up the changes in the meadow communities after the cessation of mowing was, however, the presence of species which contribute to the decomposition of meadow phytocoenoses and initiate the formation of herbaceous and scrub communities, i.e., those species which show considerable colonizing capabilities after the cessation of mowing. They are species that attain the maximum level of numbers within several years following meadow abandonment, and can maintain it for a certain time. In forest phytocoenoses and in regularly mown meadows their proportion, expressed as a percent cover, rarely exceeds 5%. During the formation of macroforbes phytocoenoses the percent cover comes up to 50—70%, and locally even 90%.

In abandoned fertile and moist meadows the course of succession shows a considerable degree of complexity. The initiation of forest phytocoenoses in them may have been preceded by a gradual reconstruction of meadow communities towards herbaceous and willow-scrub communities, or by the formation of biogroups or a mosaic of monospecific aggregations of several of the 124 meadow components (Figure 5.12). Due to the varied colonization strategy of the species present in still utilised meadows, as well as to the environmental heterogeneity that becomes manifest after the cessation of mowing, there appears a clearly marked vegetation mosaic pattern (Figure 5.15). This may explain the considerable differences in the succession progress and ways of forest return to the Reski Range (Figure 5.12).

Summary

The rates modes in which the forest entered the abandoned meadows varied:

1. Meadow phytocoenoses gradually passed into macroforbe and then into shrub communities. It lasted ca. 15 years.
2. The mosaic of monospecific aggregations of some of 124 meadow components was formed. The aggregations may last from 5 to 15 years. They initiate the development of macroforbes and shrubs later on, and lead to the mosaic pattern.
3. Biogroups of trees directly developed in the meadow and formed the outposts of the forest.

The diverse course of succession in meadows result from the heterogeneity of their habitats (microrelief, water level) and the richness of colonization strategies of species present before succession. The significance of different species in succession varies. They play roles of:

1) promotors, which stimulate the transformation of meadow phytocoenoses and initiate the formation of later communities;
2) inhibitors, which restrict species turnover due to persistent aggregations;
3) neutrals, which are present in the course of succession but do not affect the vegetation transformations.

The species present in the course of succession showed either the directional changes in numbers dependent on the stage of succession, or the fluctuations not promptly related to it. The promotors were characterized by a long life span, high fertility, strong and persistent root systems, and the ability to support or even replace the generative reproduction with the vegetative one. These features allow them to rapidly colonize the new places. However, the changes in the architecture of promotors in the course of their growth and gradual decaying of their root systems give place to other species, e.g. to forest plants with different microhabitat requirements.

BIBLIOGRAPHY

Andrzejewska L and Falińska K (1983) Struktura wielkości osobników w populacji [Size structure of individuals in a population], Wiad. ekol. 29.1, 3—31.

Antonovics J and Levin DA (1980) The ecological and genetic consequences of density-dependent regulation in plants, Ann. Rev. Ecol. Syst. 11, 411—452.

Barkham JP (1980a) Population dynamics of the wild daffodil (Narcissus pseudonarcissus). I. Clonal growth, seed reproduction, mortality and the effects of density, J. Ecol. 68, 607—633.

Barkham JP (1980b) Population dynamics of the wild daffodil (Narcissus pseudonarcissus). II. Changes in number of shoots and flowers, and the effect of bulb on growth and reproduction, J. Ecol. 68, 635—664.

Bazzaz FA (1975) Plant species diversity in old field successional systems in Southern Illinois, Ecology 56, 485—488.

Begon M and Mortimer M (1981) Population ecology. A unified study of animals and plants, Blackwell Scien. Publ., Oxford, London, Edinburgh, Boston, Melbourne.

Bell G (1984) Evolutionary and nonevolutionary theories of senescence, Am. Nat. 124, 600—603.

Belsky AJ (1986) Revegetation of artificial disturbances in grasslands of the Serengeti National Park, Tanzania, I. Colonization of grazed and ungrazed plots, J. Ecol. 74, 419—437.

Bishop GF, Davy AJ and Jefferies RL (1978) Demography of Hieracium pilosella in a Breck grassland, J. Ecol. 66, 615—629.

Buss LW (1979) Habitat selection, directional growth and spatial refuges: Why clonal animals have more hiding places. In Larwood G and Rosen BR (eds.), Biology and systematics of clonal organisms, Academic Press, London, pp. 459—497.

Cahn M and Harper JL (1976) The biology of the leaf mark polymorphism Trifolium repens L. I. Distribution of phenotypes at a local scale, Heredity 37, 309—325.

Callaghan TV (1980) Age-related patterns of nutrient allocation in Lycopodium annotinum from Swedish Lapland, Oikos 35, 373—386.

Callaghan TV and Emanuelsson U (1985) Population structure and processes of tundra plants an vegetation. In White J (ed.), The population structure of vegetation, Dr W. Junk Publishers, Dordrecht, pp. 399—439.

Callaghan TV, Headley AD, Svensson BM, Li Lixian, Lee JA and Lindley DK (1986) Modular growth and function in the vascular cryptogram Lycopodium annotinum, Proc. Roy. Soc. Ser. B. 228, 195—206.

Caswell H (1986) The evolutionary demography of vegetative reproduction. In Jackson J, Buss L, and Cook RE (eds.), Population biology and evolution of clonal organisms, Yale University Press, New Haven, CT.

Caswell H and Werner P (1978) Transient behaviour and life history analysis of teasel (Dipsacus sylvestris Huds.), Ecology 59, 53—66.

Charlesworth B (1980) Evolution in age-structured populations, Cambridge University Press.

Chessel D (1977) La description nom paramétrique de la dispersion spatiale des individuals d'une espece, Journée francaise de biométrie, 28 avril 1977.

Clements FE (1916) Plant succession, Carnegie Inst., Washington Publ.

Clements FE (1936) Nature and structure of the climax, J. Ecol. 24, 252—284.

Colinvaux PA (1973) Introduction to ecology, Wiley J, New York.

Connel JH and Slatyer RO (1977) Mechanisms of succession in natural communities and their role in community stability and organization, Am. Nat. 111, 1119—1144.

Cook RE (1983) Clonal plant populations, Am. Sci. 71, 244—253.

Czarnecka B (1986) Biological properties of Majanthemum bifolium (L.) F. W. Schm. polycormones under various ecological conditions. Acta Soc. Bot. Pol. 55.4: 659—678.

DeSteven D (1989) Genet and ramet demography and Oenocarpus mapora ssp. mapora, a clonal palm of Panamanian tropical moist forest, J. Ecol. 77, 579—596.

Drury WH and Nisbet ICT (1973) Succession, J. Arnold Arbor. Harv. Univ. 54, 331—368.

Egler FE (1954) Vegetation science concepts. I. Initial floristic composition — a factor in old-field vegetation development, Vegetatio 4, 412—417.

Ehrendorfer F (1973) Liste der Gefässpflanzen Mitteleuropas, Gustav Fischer Verlag, Stuttgart.

Falencka M (1983) The effect of the size of cotyledons on the development of individuals in Impatiens noli-tangere L. population, Ekol. pol. 31.1, 123—143.

Falińska K (1978) Behaviour of Caltha palustris L. populations in forest and meadow ecosystems of the Białowieża National Park, Ekol. pol. 26, 85—109.

Falińska K (1979) Experimental studies of the reproductive

strategy of Caltha palustris L. populations, Ekol. poll 27, 527—543.
Falińska K (1982) The biology of Mercurialis perennis L. polycormones, Acta Soc. Bot. Pol. 50, 443—513.
Falińska K (1985) Demography of coenopopulations of forest herbs. In White J (ed.), The population structure of vegetation, Dr W. Junk Publishers, Dordrecht, Boston, London, pp. 241—264.
Falińska K (1986a) Pojęcie osobnika w demografii roślin [The notion of an individual in plant demography], Wiad. ekol. 32, 361—380.
Falińska K (1986b) Demography of Iris pseudoacorus L. populations in abandoned meadows, Ekol. pol. 34, 583—613.
Falińska K (1986c) Dynamics and structure of plant populations in forest ecosystems. In Faliński JB, Vegetation dynamics in temperate lowland primeval forests, Dr W. Junk Publishers, Dordrecht, Boston, London, pp. 307—346.
Falińska K (1989a) Plant population processes in the course of forest succession in abandoned meadows. I. Variability and diversity of floristic compositions, and biological mechanisms of species turnover, Acta Soc. Bot. Pol. 58, 439—465.
Falińska K (1989b) Plant population processes in the course of forest succession in abandoned meadows. II. Demography of succession promoters, Acta Soc. Bot. Pol. 58, 467—491.
Falińska K (1990) Osobnik, populacja, fitocenoza [Individual, population, phytocoenoses], PWN, Warszawa.
Falińska K and Priożnikow E (1983) Ecological structure of Geranium robertianum L. populations under natural conditions and in the garden, Ekol. pol. 31, 93—121.
Faliński JB (1962) Variabilité saisonniere des frontiers de phytocénoses, Acta Soc. Bot. Pol. 31, 239—263.
Faliński JB (1966) Antropogeniczna roślinność Puszczy Białowieskiej jako wynik synantropizacji naturalnego kompleksu leśnego [Végétation anthropogene de la Grande Forét de Białowieża comme un résultat de la synanthropisation du territoire silverstre naturel]. Dissertationes Universitatis Varsoviensis 13, 1—256.
Faliński JB (1980) Vegetation dynamics and sex structure of the populations of pioneer dioecious woody plants, Vegetatio 43, 23—38.
Faliński JB (1986) Vegetation dynamics in temperate lowland primeval forests, Ecological studies in Białowieża forest, Geobotany 8, 1—537. Dr W. Junk Publishers, Dordrecht, Boston, London.
Faliński JB (1989) Le temp et l'espace dans les recherches ecologiques sur la dynamique de la vegetation, Giorn. Bot. Ital. 123, 81—107.
Fekete G and Melkó E (1981) Reproductive allocation in the stages of sandy succession, Acta Bot. Acad. Scien. Hung. 27, 351—364.
Fekete G, Tuba Z and Melkó E (1988) Background processes at the population level during succession in grasslands on sand, Vegetatio 77, 33—41.
Fenner M (1987) Seed characteristics in relation to succession. In Gray AJ, Crawley MJ and Edwards PJ (eds.), Colonization, succession and stability, Blackwell, Oxford, pp. 103—114.
Fiala K (1978) Underground organs of Typha angustifolia and Typha latifolia, their growth, propagation and production, Acta Sc. Nat. Brno. 12, 1—43.
Gadgil PM and Solbrig OT (1972) The concept of r- and K-selection: evidence from wild flowers and some theoretical considerations, Am. Nat. 106, 14—31.
Giller PS (1984) Community structure and the Niche. Chapman and Hall, London, New York.
Goodall DW (1963) The continuum and the individualistic association, Vegetatio 9, 297—316.
Gray AJ, Crawley MJ and Edwards PJ (eds.) (1987) Colonization, succession and stability, Blackwell Scien. Publ., Oxford, London, Edinburgh.
Grime JP (1977) Evidence for the existence of three primary strategies in plants and its relevance to ecological and evolutionary theory, Am. Nat. 111, 1169—1194.
Grime JP (1979) Plant strategies and vegetation processes, J. Wiley Sons, Chichester, New York, Brisbane, Toronto.
Gross KL (1980) Colonization by Verbascum thapsus (Mullein) of an old-field in Michigan: experiments on the effects vegetation, J. Ecol. 68, 919—927.
Gross KL and Werner PA (1982) Colonizing abilities of "biennial" plant species in relation to ground cover: implications for their distributions in a successional sere, Ecology 63, 921—931.
Grubb PJ (1977) The maintenance of species-richness in plant communities: the importance of the regeneration niche, Biol. Rev. 52, 107—145.
Grubb PJ (1982) Control of relative abundance in roadside Arrhenatheretum: results of a long-term garden experiment, J. Ecol. 70, 845—861.
Hamilton WD (1966) The moulding of senescence by natural selection, J. Theoret. Biol. 12, 12—45.
Harberd DJ (1961) Observations on population structure and longevity of Festuca rubra L, New Phytol. 60, 184—206.
Harper JL (1977) Population biology of plants, Academic Press, London, New York, San Francisco.
Harper JL (1980) Plant demography and ecological theory, Oikos 35, 244—254.
Harper JL (1981) The concept of population in modular organisms. In May RM (ed.), Theoretical ecology principles and applications, 2nd edn, Blackwell Scien. Publ., Oxford. pp. 53—77.
Harper JL (1986) Modules, branches and the capture of resources. In Jackson JBC, Buss LW and Cook RE (eds.), Population biology and evolution of clonal organism, Yale University Press. New Haven, London.
Harper JL and Bell AD (1979) The population dynamics of growth form in organisms with modular construction. In Anderson RM, Turner BD and Taylor RL (eds.), Population dynamics, Blackwell Scien. Publ. Oxford. pp. 29—52.
Harper JL and Ogden J (1970) The reproductive strategy of higher plants. I. The concept of strategy with special reference to Senecio vulgaris, J. Ecol. 58, 681—698.
Harper JL and White J (1971) The dynamics of plant populations. In den Boer PJ and Gradwell GR (eds.), Dynamic of populations, Wageningen: Centre for Agricultural Publishing and Documentation, pp. 41—63.
Harper JL and White J (1974) The demography of plants, Ann. Rev. Ecol. System. 5, 419—463.
Hartnett DC and Bazzaz FA (1985a) The genet and ramet

population dynamics of Solidago canadensis in an abandoned field, J. Ecol. 73, 407—413.
Hartnett DC and Bazzaz FA (1985b) The integration of neighbourhood effects by clonal genets in Solidago canadensis, J. Ecol. 73, 415—427.
Hartnett DC and Bazzaz FA (1985c) The regulation of leaf, ramet and genet densities in experimental populations of the rhizomatous perennial Solidago canadensis, J. Ecol. 73, 429—443.
Holt BR (1972) Effect of arrival time on recruitment mortality, and reproduction in successional plant populations, Ecology 53, 668—673.
Horn HS (1974) The ecology of secondary succession, Ann. Rev. Ecol. Syst. 5, 25—37.
Horn HS (1976) Succession. In May R (ed.), Theoretical ecology: Principles and applications, Blackwell Scientific Publishers, London, pp. 187—204.
Houssard C, Escarré J and Romane F (1980) Development of species diversity in some mediterranean plant communities, Vegetatio 43, 59—72.
Huston M (1979) A general hypothesis of species diversity, Am. Nat. 113, 81—101.
Huston M and Smith T (1987) Plant succession: life history and competition, Am. Nat. 130, 168—198.
Hutchings MJ and Barkham JP (1976) An investigation of shoot interactions in Mercurialis perennis L., a rhizomatous perennial, Herb. J. Ecol. 64, 723—743.
Inghe O and Tamm CO (1985) Survival and flowering of perennial herbs. IV. The behaviour of Hepatica nobilis and Sanicula europaea on permanent plots during 1943—1981, Oikos 45, 400—420.
Jakucs P (1972) Dynamische Verbindung der Wälder und Rasen, Akadémiai Kiadó, Budapeszt.
Kershaw KA (1962) Quantitative ecological studies from the Landmannahellir, Iceland. II. The rhizome behaviour of Carex bigelowii and Calamagrostis neglecta, J. Ecol. 50, 171—179.
Kershaw KA (1973) Quantitative and dynamic plant ecology. 2nd edn., Edward Arnold, London.
Knapp R (ed.) (1974) Vegetation dynamics, Dr W. Junk Publishers, The Hague.
Kotańska M (1986/1987a) Współwystepowanie populacji Crisium rivulare (Jacq.) All. i C. oleraceum (L.) Scop. w dolinie Wierzbanówki na Pogórzu Wielickim. I. Warunki występowania i struktura przestrzenna [Co-existence of populations of Cirsium rivulare (Jacq.) All. and C. oleraceum (L.) Scop. in the Wierzbanówka valley in the Wieliczka Foothills. I. Conditions of occurrence and spatial structure], Fragm. Flor. et Geobot. 31—32.3—4, 355—421.
Kotański M (1986/1987b) Współwystępowanie populacji Cirsium rivulare (Jacq.) All. i C. oleraceum (L.) Scop. w dolinie Wierzbanówki na Pogórzu Wielickim. II. Cykl życiowy i struktury wieku [Co-existence of populations of Cirsium rivulare (Jacq.) All. and C. oleraceum (L.) Scop. in the Wierzbanówka valley in the Wieliczka Foothills. II. Life cycle and age structure], Fragm. Flor. et Geobot. 31—32.3—4, 423—442.
Law R (1979) The cost of reproduction in annual meadow grass, Am. Nat. 113, 3—16.
Leck MA, Parker VT and Simpson RL (1989) Ecology of soil seed banks, Academic Press, Inc., Harcourt Brace Jovanovich Publishers, San Diego, New York, Berkeley, Boston, London, Sydney, Tokyo, Toronto.
Lee TD and Bazzaz F (1980) Effects of defoliation and competition on growth and reproduction in the annual plant Abutilion theophrasti, J. Ecol. 68, 813—821.
Levin SA and Paine RT (1974) Disturbance, patch formation, and community structure, Proceedings of the National Academy of Sciences, USA, 71, 2744—2744.
van der Maarel E (ed.) (1980) Succession. Symposion on advances in vegetation sciences, Nijmegen, The Netherlands, May 1979, Dr W. Junk Publishers, The Hague, Boston, London.
MacArthur RH (1972) Geographical ecology, Harper and Row, New York.
MacArthur RH and Wilson EO (1967) The theory of island biogeography. Princeton University Press, Princeton, N.J.
Mack RN and Harper JL (1977) Interference in dune annuals: spatial pattern and neighbourhood effects, J. Ecol. 65, 345—364.
MacMahon J (1980) Ecosystems over time: succession and other types of change. In Waring R (ed.), Forests: Fresh perspective from ecosystem analyses, Oregon State University Press, Corvallis, Oregon. pp. 27—58.
Mahall EB and Park RB (1976) The ecotone between Spartina folios Trin. and Salicornia virginical L. in salt marshes of Northern San Francisco bay. I. Biomass and production, J. Ecol. 64, 421—433.
Margalef R (1968) Perspectives in ecological theory, Univ. Chicago Press, Chicago, London.
Matuszkiewicz J (1972) Analiza zmienności przestrzennej runa w strefie kontaktowej dwu fitocenoz [Analysis of the spatial variation of the field layer in the contact zone of two phytocoenose], Phytocoenosis 1, 121—150.
Matuszkiewicz W (1981) Przewodnik do oznaczania zbiorowisk roślinnych Polski [Guide-book for the identification of the plant communities of Poland], PWN Warszawa.
Maun MA (1984) Colonizing ability of Ammophila breviligulata trough vegetative regeneration, J. Ecol. 72, 565—574.
May RM (1973) Stability and complexity in model ecosystems, Princeton University Press, Princeton.
May RM and MacArthur RH (1972) Niche overlap as a function of environmental variability, Proc. Nat. Acad. Sci. USA 69, 1109—1113.
McMillan C (1964) Ecotypic differentiation within four North American prairie grasses. I. Morphological within transplanted community fractions, Amer. J. Bot. 51, 1119—1128.
Mithen R, Harper JL and Weiner J (1984) Growth and mortality of individual plants as a function of "available area", Oecologia 62, 57—60.
Newell SJ and Tramer EJ (1978) Reproductive strategies in herbaceous plant communities during succession, Ecology 59, 228—234.
Noble JC, Bell AD and Harper JL (1979) The population biology of plants with clonal growth. I. The morphology and structural demography of Carex arenaria, J. Ecol. 67, 983—1008.
Noble JC and Marshall C (1983) The population biology of plants with clonal growth. II. The nutrient strategy and

modular physiology of Carex arenaria, J. Ecol. 71, 865—877.
Noble JR and Slatyer RE (1980) The use of vital attributes to predict successional changes in plant communities subject to recurrent disturbances, Vegetatio 43, 5—21.
Noodeén LD (1980) Senescence in the whole plant. In Thimann KV (ed.), Senescences in plants, RCK Press, pp. 219—258.
Odum EP (1963) Podstawy ekologii [Fundamentals of ecology], PWN, Warszawa.
Odum EP (1969) The strategy of ecosystem development, Science 164, 262—270.
Osbornová J, Kovárová M, Lepś J and Prach K (eds.) (1990) Succession in abandoned fields studies in Central Bohemia, Czechoslovakia, Geobotany 15, 1—168, Kluwer Acad. Publ., Dordrecht, Boston, London.
Peart DR (1989a) Species interactions in a successional grassland. I. Seed rain and seedling recruitment, J. Ecol. 77, 236—251.
Peart DR (1989b) Species interactions in a successional grassland. II. Colonization of vegetated sites, J. Ecol. 77, 253—265.
Peart DR (1989c) Species interactions in a successional grassland III. Effects of canopy gaps, gopher mounds and grazing of colonization, J. Ecol. 77, 267—289.
Peet RK and Christensen NL (1980) Succession: a population process, Vegetatio 43, 131—140.
Penzes A (1960) Über die Morphologie, Dynamik und zönologische Rolle des Sprosskolonienbilden — den Planzen (Polycormone). Fragm. flor. geobot, 6, 505—515.
Pianka ER (1970) On r- and K-selection, Am. Nat. 104, 592—597.
Pickett STA (1976) Succession: an evolutionary interpretation, Am. Nat. 110, 107—119.
Pickett STA, Collins SL and Armesto JJ (1987) Models, mechanisms and pathways of succession, Bot. Rev. 53, 335—371.
Pirożnikow E (1983) Seed bank in the soil of stablilized ecosystem of deciduous forest (Tilio-Carpinetum) in the Białowieża National Park, Ekol. pol. 31, 145—172.
Platt WJ (1975) The colonization and formation of equilibrium plant species associations on badger disturbances in a tall-grass prairie, Ecol. Monogr. 45, 285—305.
Polakowska M (1966) Analyse der Übergangszone zwischen Waldgesellschaften, Ekol. pol. 14, 1—24.
Prentice IC (1986) Some concepts and objectives of forest dynamics research. In Fanta J (ed.), Forest dynamics research in Western and Central Europe, Pudoc, Wageningen, pp. 32—41.
Rabotnov TA (1985) Dynamics of plant coenotic populations. In White J (ed.), The population structure of vegetation, Dr W. Junk Publishers, Dordrecht, Boston, London, pp. 121—142.
Raynal DJ and Bazzaz FA (1973) Establishment of early successional plant populations on forest and prairie soil, Ecology 54, 1335—1341.
Sarukhan J (1980) Demographic problems in tropical systems. In Solbrig OT (ed.), Demography and evolution in plant populations, Blackwell Scien. Publ., Oxford, pp. 161—188.
Sarukhan J and Harper JL (1973) Studies on plant demography: Ranunculus repens L., R. bulbosus L. and R. acris L. I. Population flux and survivorship, J. Ecol. 61, 675—715.
Schenkeveld AJM and Verkaar HJPA (1984) On the ecology of short-lived forbs in chalk grasslands, Utrecht.
Sharitz RR and McCormick JF (1975) Population dynamics of two competing annual plant species, Ecology 54, 723—740.
Shugart HH (1984) A theory of forest dynamics, Springer Verlag, New York, Berlin, Heidelberg.
Silander JA and Antonovics J (1982) Analysis of interspecific interactions a coastal plant community — a perturbation approach, Nature 298, 577—600.
Silvertown J (1983) Why are biennials sometimes not so few? Am. Nat. 121, 448—453.
Silvertown J (1987) Introduction to plant population ecology, Longman Sci. Technic, London, New York.
Smirnova OV, Zaugol'nova LB and Ermakova IM (1976) Cenopopulacii rastenij (osnovnyje ponjatija i struktura) [Plant coenopopulations: basic concepts and structure], Nauka, Moscow.
Smith BH (1983a) Demography of Floerkea proserpinacoides a forest-floor annual. I. Density-dependent growth and mortality. J. Ecol. 71: 391—404.
Smith BH (1983b) Demography of Floerkea proserpinacoides a forest-annual. II. Density-dependent reproduction. J. Ecol. 71: 405—412.
Soćava VB (1972) Ućenie o geosistemach — sovremennyj etap kompleksnoj fizićeskoj geografii. — "Izv. AN SSSR", serija geogr. 3.
Solbrig OT (ed.) (1980) Demography and evolution in plant populations, Blackwell Scien. Publ., Oxford.
Stenseth NC (1980) Spatial heterogeneity and population stability: some evolutionary consequences, Oikos 35, 165—184.
Symonides E (1983/1984) Population size regulation as a result of intra-population interactions. I. Effect of density on the survival and development of individuals of Erophila verna (L.) C.A.M., Ekol. pol. 31, 839—881.
Symonides E (1985) Population structure of psammophyte vegetation. In White J (ed.), The population structure of vegetation, Dr W. Junk Publishers, Dordrecht, Boston, London, pp. 265—291.
Szmeja J (1987a) The structure of a population of Lobelia dortmanna L. along gradient of increasing depth in an oligotrophic lake, Aquatic Botany 28, 1—13.
Szmeja J (1987b) The seasonal development of Lobelia dortmanna L. and annual balance of its population size an oligotrophic lake, Aquatic Botany 28, 15—24.
Tamm CO (1972a) Survival and flowering of some perennial herbs. II. The behaviour of some orchids on permanent plots, Oikos 23, 23—28.
Tamm CO (1972b) Survival and flowering of some perennial herbs. III. The behaviour of Primula veris on permanent plots, Oikos 23, 159—166.
Titov JV (1978) Efekt grupy u rastenii, Nauka, Leningrad.
Traczyk T (1960) Badania nad strefą przejścia zbiorowisk leśnych. [Über die Übergangszonen zwischen Waldgesellschaften], Ekol. pol. 8, 88—125.
Trathan P (1983) Clonal interactions of Trifolium repans and Lolium perenne, Ph.D. diss., University of Wales.

Turkington R, Cahn HA, Vardy A and Harper JL (1979) The growth, distribution and neighbour relationships of Trifolium repens in a permanent pasture. III. The establishment and growth of Trifolium repens natural and perturbed sites, J. Ecol. 67, 231—243.

Turkington R and Harper JL (1979) The growth, distribution and neighbour relationships of Trifolium repens in a permanent pasture. IV. Fine-scale biotic differentiation, J. Ecol. 67, 245—254.

Turkington R, Harper JL, de Jong P and Aarssen LW (1985) A reanalysis of interspecific association in an old pasture, J. Ecol. 73, 123—131.

Turner CI (1950) The reproductive potential of a single clone of Pelmatohydra oligactis, Biol. Bull. 99, 285—299.

Vorontzova LI and Zaugolnova LB (1985) Population biology of steppe plants. In White J (ed.), The population structure of vegetation, Dr W. Junk Publishers, Dordrecht, pp. 131—135.

Walker LR and Chapin JL (1987) Interactions among processes controlling successional change, Oikos 50, 131—135.

Watkinson AR and Harper JL (1978) The demography of a sand annual Vulpia fasciculata. I. The natural regulation of populations, J. Ecol. 66, 15—33.

Watkinson AR and White J (1985) Some life-history consequences of modular construction in plant, Phil. Trans. R. Soc. Lond. 313, 31—51.

Werner PA (1977) Colonization success of a "biennial" plant, teasel (Dipsacus sylvestris Huds.), Can. J. Plant. Sci. 55, 783—794.

Werner PA (1979) Competition and coexistence of similar species. In Solbrig OT, Jain S, Johnson GB and Raven PH (eds.), Topics in plant population biology, McMillan, London, New York, Ch. 12, pp. 287—310.

White J (1979) The plant as a metapopulation, Ann. Rev. Ecol. Syst. 10, 109—145.

White J (1980) Demographic factors populations of plants. In Solbrig OT (ed.). Demography and evolution in plant populations, Blackwell Scien. Publ., Oxford, pp. 21—48.

White J (ed.) (1985) The population structure of vegetation, Dr W. Junk Publishers, Dordrecht, Boston, London.

White J and Harper JL (1970) Correlated changes in plant size and number in plant populations, J. Ecol. 58, 467—485.

Whittaker RH (1972) Evolution and measurement of species diversity, Taxon 21, 213—251.

Whittaker RH (1975) Communities and ecosystems, McMillan Publ. Co., New York.

Whittaker RH and Levin SA (1977) The role of mosaic phenomena in natural communities, Theor. Pop. Biol. 12, 117—139.

Wilkoń-Michalska J (1976) Struktura i dynamika populacji Salicornia patula Duval-Jouve [Structure and dynamics of the populations of Salicornia patula Duval-Jouve], Rozprawy UMK, 1—156.

Wilkoń-Michalska J (1985) Structure and dynamics of the inland populations of Salicornia patula, Vegetatio 61, 145—154.

Wolf G (1979) Veränderung der Vegetation und Abbau der organischen Substanz in aufgegebenen Wiesen des Westerwaldes, Schr. Reihe Vegetationskde 13, 1—115.

Woolhouse HW (1972) Ageing processes in higher plants. In Head JJ and Lowenstein OE (eds.), Oxford Biology Readers: 1—30.

Yoda K, Kira T, Ogawa H and Hozumi K (1963) Self thinning in overcrowded pure stands under cultivated and natural conditions, J. Biol., Osaka City Univ. 14, 107—129.

Zarzycki K (1964) Biological and ecological studies in Carpathian beechwood, Bull. Acad. Pol. Sci. Cl. II, 12, 15—21.

Zarzycki K (1968) Eksperymentalne badania zdolności konkurencyjnej roślin [Experimental investigation of competition between forest herbs], Acta Soc. Bot. Pol. 37, 393—411.

Zhukova LA and Ermakova IM (1985) Structure and dynamics of coenopopulations of some temperate grasses. In White J (ed.), The population structure of vegetation, Dr W. Junk Publishers, Dordrecht, pp. 179—205.

SUBJECT INDEX

(*page numbers in italics refer to figures, photos, and tables*)

SYSTEMATIC INDEX
(Plant species and communities)

(*page numbers in italic refer to figures, photos, and tables*)

MIX
Papier aus verantwortungsvollen Quellen
Paper from responsible sources
FSC® C105338

If you have any concerns about our products,
you can contact us on
ProductSafety@springernature.com

In case Publisher is established outside the EU,
the EU authorized representative is:
Springer Nature Customer Service Center GmbH
Europaplatz 3, 69115 Heidelberg, Germany

Printed by Libri Plureos GmbH
in Hamburg, Germany